Indian Forester, Scottish Laird

H.J. Noltie

Indian Forester, Scottish Laird

The Botanical Lives of
Hugh Cleghorn of Stravithie

Royal Botanic Garden Edinburgh

MMXVI

First published 2016 by the Royal Botanic Garden Edinburgh

ISBN 978-1-910877-10-4

Designed and typeset in Garamond and Optima by Caroline Muir

Printed by Meigle Colour Printers Limited

© H.J. Noltie/RBGE 2016. All images, other than those individually credited, are reproduced by permission of the Trustees of the Royal Botanic Garden Edinburgh

All rights reserved. No part of this publication may be reproduced, stored in a retrieval system or transmitted, in any form, or by any means, electronic, mechanical, photocopy, recording or otherwise, without the prior written consent of the copyright holders and publishers.

Frontispiece. Design by John Duarte (a pupil then teacher at the Madras School of Art) for the title-page of the 1863 *Illustrated Madras Almanac*. Sprot Family Collection.

Front cover: Cleghorn by T. Blake Wirgman (detail of Fig. 66).

Back cover: The Forest Mountain of Huttoo by Mrs W.L.L. Scott (detail of Fig. 37).

Contents

Foreword   vii

Introduction   ix

Chapter 1.   Family, Childhood & Education, 1820–1841   1

Chapter 2.   East India Company Surgeon, 1842–1848   17

Chapter 3.   First Furlough & the British Association Report, 1848–1851   29

Chapter 4.   Madras Polymath, 1851–1856   53

Chapter 5.   Forests of South India, 1856–1860   81

Chapter 6.   Gardens of South India   105

Chapter 7.   Second Furlough & Marriage, 1860–1861   117

Chapter 8.   Forests of North India, 1861–1864   141

Chapter 9.   Third Furlough, Officiating Inspector-General & Farewell to India   175

Chapter 10.   Conjugal Reunion & a Grand Tour   195

Chapter 11.   Retirement to Britain   205

Chapter 12.   Laird of Stravithie   215

Chapter 13.   Scottish Societies   225

Chapter 14.   Grand Old Forester   237

Chapter 15.   Cleghorn's Legacy   263

Post Script   284

Acknowledgements   285

References   287

Endnotes   293

Appendix 1. Bibliography of Cleghorn's publications (with references to obituaries and book reviews; membership of societies)   305

Appendix 2. Cleghorn eponymy   313

Index   314

For my sister, Dr Kirsty Noltie, whose pre-clinical medical training, like Cleghorn's, was undertaken at the University of St Andrews.

Published with this is a companion volume, *The Cleghorn Collection: South Indian Botanical Drawings, 1845 to 1860*, which includes a selection of some 200 drawings reproduced in colour from Cleghorn's illustrations collection. Many of these are related to botanical, historical or biographical information in the present work: cross-references to individual illustrations are cited in the form '*Cleghorn Collection* (no. 35)'.

Foreword

Biography is to history what taxonomy is to botany, namely a means to order, to impart a sense of system regarding the connectedness of living things in their families and classes. Hugh Francis Clarke Cleghorn (1820–1895) or, to use as did he the simpler binomial Hugh Cleghorn, was born to a prosperous Edinburgh family and spent his life within and among those social classes who made Britain imperial. Cleghorn had one life but many lives: lecturer in medical education and in botany, plant collector, forester in India, economic botanist, leading member of the Botanical Society of Edinburgh, confrère to men and women who shaped Britain's empire in nineteenth-century India, and, not least, Scottish estate owner. In *Indian Forester, Scottish Laird*, Henry Noltie wonderfully captures a world in which both plants and improved knowledge about them were, at once, objects of scientific study and political economy and the subject of spiritual reverence and cultural authority.

As was the case for many of his contemporaries and several members of his immediate family, Hugh Cleghorn was shaped profoundly by the exigencies of the British presence in nineteenth-century India. India in its turn profoundly shaped him. In several studies of this period, and not just of India, scholars have identified in Britain's civil servants, soldiers, engineers, military doctors and, of course, forest conservators such as Cleghorn what has been termed a culture of 'imperial careering.' By this is meant that conjoint sense of useful service, moral duty and obligation unto others that sustained one's daily routine even when, as was the case with Cleghorn, his responsibilities in managing Indian forests were interrupted by the requirements of geographical mobility and long-distance travel. This is not to say that Cleghorn was a careerist: he achieved his status by virtue of hard work, sound common sense and very considerable ability. It is to observe that while we are treated here to the detailed documentation of one man's botanical life, we are also given telling glimpses into more general circumstances, into the correspondence and postings and lives of countless others who have not yet found themselves to be, and may never find themselves being, the subject of a considerate biographer.

Scotland and India had particularly close connections, intellectually, militarily and, as was the case for the Cleghorns, in established family links in which botanical interests had a central place. Born in Madras, Cleghorn encountered India and Indian botany in Scotland, at home, and at the universities of St Andrews and Edinburgh before he first worked abroad, from 1842, as a surgeon in the employ of the East India Company. Cleghorn travelled a route laid down by others: his first position was the same as that occupied by William Roxburgh (1751–1815), the Ayrshire-born 'father of Indian Botany', and it reflects that of the botanist and medical Scot, Robert Wight (1796–1872), who spent over thirty years in India. It is easy to forget now just how important botanical knowledge was then as an essential element in a medical career and in medical treatments. And if his first sojourn made clear his ability as a medical educator, it also laid down Cleghorn's interests in economic botany, in cultivated plants, and, crucially, in forests and forestry.

Cleghorn was a 'Forest Conservator' from the later 1850s – not at all the same thing as a 'Conservationist' in modern terms – a benevolent custodian armed both with a sense of what, with reference to his intellectual training in late Enlightenment Scotland, we might call 'statistical', even 'arboreal accountancy', and a sensitivity to the need for negotiation over what a forest represents: economic resource, source of botanical diversity, locus of customary rights for indigenous peoples, to name only a few. His Forest Reports of the late 1850s are rooted in extensive travel and expressed in a rhetoric that admonished wastefulness. They and his other publications laid the basis for the establishment of a whole new department in Indian botanical science and in the British administration of India. Cleghorn used numerous exhibitions, in India and in Britain, to display the importance of forest products to different audiences.

In recounting the life and work of Hugh Cleghorn, the 'father of Indian Forestry', Henry Noltie has done much more than recover for posterity someone whose botanical works and world have been too long overlooked, or, where looked at at all, rather misunderstood. A window into British social and imperial history, Cleghorn's biography reveals a life shaped by its geography: as St Andrews' scholar, Edinburgh medical student, Indian forester, Madras gardener, botanical lecturer, and Fife-based man of letters long after retirement. *Indian Forester, Scottish Laird* is the product of years of patient archival study, of piecing together fragmentary traces in order to establish connections personal and scientific and, almost literally, of following in Cleghorn's botanical footsteps in order to see, where it is possible to see, something of what Cleghorn himself saw and understood of India's botanical riches. This is a book about why and how the historical study of botany has significance beyond the plants themselves.

Professor Charles W.J. Withers FBA FRSE
Ogilvie Chair of Geography, University of Edinburgh, and Geographer Royal for Scotland

Introduction

This book became an inevitability as long ago as 1998: it came out of my daily use of the Indian library and herbarium collections of the Royal Botanic Garden Edinburgh (RBGE), while undertaking research for the *Flora of Bhutan*. In pursuing this work I became familiar with a series of South Indian specimens mounted on paper of a strong blue hue and labels written neatly in black ink, with the riches of the Indian botanical drawings in the Illustrations Collection, and with an ink stamp reading 'Cleghorn Memorial Library' on the title-page of some of the most interesting books in the library. These all proved to originate from a common source: Hugh Francis Clarke Cleghorn. Here, clearly, was a major RBGE benefactor and yet his name was almost entirely unknown within the organisation; he was not even mentioned in the Garden's official history.[1] This, at least in part, has emerged to be due to the fact that Cleghorn's benefaction was indirect – in old age he seems to have become vague about what should happen to his valuable collections of books, drawings and specimens *post mortem*. The RBGE was not mentioned in his will and (being childless) it fell to his nephew-heir to distribute the collections – the herbarium went to RBGE, and the books and drawings were split between Edinburgh University and the Museum of Science and Art, but the latter's share was transferred to RBGE in 1940.

The primary motivation for this work has been as a contribution to the history of the RBGE collections, a preliminary outcome being an exhibition at Inverleith House of a selection of the botanical drawings and a small book about them in 1997/8.[2] As part of the research for this I visited Hugh and Elizabeth Sprot at Stravithie, who, having sold the 'big house' in 1979, were by then living nearby in what Cleghorn knew as Stravithy Mill Farm, but by then renamed Nether Stravithie. Gerard Hugh Cleghorn Sprot, in his late seventies, was very much the retired major, whose regiment, like that of his grandfather Sir Alexander Sprot (Cleghorn's nephew) and his father (*né* Hereward Sadler), was the Carabiniers. In World War II he had seen active service in Burma, which his great-uncle Hugh Cleghorn had visited a century earlier to study its teak forests. Like his male Cleghorn forebears Major Sprot was a member of the Royal & Ancient Golf Club. Elizabeth was a keen Friend of the St Andrews Botanic Garden and in 1985 had been on one of the early botanical trips to Yunnan after the opening-up of China. The Sprots proved excellent custodians of the Cleghorn heritage, presenting a rich archive of papers to St Andrews University library and maintaining the parts of the Stravithie library that had not been sent to Edinburgh in 1896. In Aylwin Clark the papers found an outstanding cataloguer.[3] A history mistress in a famous St Andrews girls' school, in which Cleghorn had shares, Aylwin used the papers to excellent effect in her little-known biography of Cleghorn's eponymous grandfather. One of the sadnesses of the long gestation of this biography of the grandson is that neither the Sprots nor Aylwin are alive to see it. I was unaware of the Sprots' deaths and the subsequent sale and dispersal, in 2007, of the parts of the Cleghorn library that they had kept at Nether Stravithie. A second corollary of the 1997 exhibition (part of the semi-centenary celebrations of Indian independence) was that it led to my meeting Richard Grove and his wife Vinita Damodaran. Richard was the first environmental historian of recent times to realise Cleghorn's significance – as part of his thought-provoking work on the role of Scottish East India Company (EIC) surgeons in the field of forest conservancy and, more controversially, in the origins of the modern environmental movement, on which topic he lectured at RBGE as part of a series of exhibition-related talks.

The second reason for this biography was to investigate more deeply Cleghorn's role as a pioneering forest conservator – the few to whom his name will mean anything at all are likely to approach it with this interest foremost in mind. In the two decades since Grove's inspirational *Green Imperialism*,[4] other historians – including Gregory Barton,[5] David Lowenthal,[6] and Pallavi Das[7] – have rightly challenged Grove's over-statement of climatic concerns as being the primary motivation of a group of Scottish EIC surgeons that led to forest protection in India. Grove's view underplayed the extent to which such issues were already well-known and acted upon on the European continent. Grove's view of Forest Conservancy as a precursor of modern 'Conservation' is also a debatable one in which he underestimated the role of George Perkins Marsh, who did become a friend of Cleghorn, but only on the latter's return from India for the last time in 1867. In my book on Alexander Gibson,[8] the first in a trilogy devoted to Edinburgh-trained EIC surgeons, I showed Grove's view to be untenable, most notably his underestimation of Gibson's straightforward economic and resource-shortage motivation and an anachronistic concept of surgeons 'lobbying' their paymasters to take an interest in conservation least of all for climatic reasons. A much closer reading of original sources was called for, with regard to precise dating and provenance of documents, which has led to the realisation of other long-standing misunderstandings on the origins of Indian forest conservancy in the literature, such as the myth of the 'Dalhousie Minute' (see p. 83). Care is essential: it really does make a difference whether a letter was issued in the name of 'His Honour in Council' or 'His Lordship in Council'. My only regret in this regard is that a severe injury in a road accident means that it is no longer possible to discuss these matters with Richard Grove. He was one of those figures from whom knowledge and ideas streamed, from a dauntingly wide range of sources, to come up with imaginative syntheses and big pictures. In his company I always felt like an ant grubbing away at small, pedantic details; but I have no doubt that he would have relished the new information and most probably put it to better use.

The fact that Cleghorn is not a household name raises the question of why to devote years to the study of someone who might (other than in RBGE institutional history and the history of Indian forestry) be considered a 'minor' figure? I have to admit to being rather bored with an unhealthily exclusive emphasis on the 'major' figure (a symptom of our celebrity-driven culture?), for which it is easier to find funders and publishers – the Darwin 'industry' is a case in point. I am therefore particularly grateful to the RBGE for enabling and allowing me to pursue such byways. Not only is the 'winner takes all' school of historiography unbalanced, but the major figures can only be fully understood by also considering the groundswell of what was going on alongside (*not* beneath) them. There is also a satisfaction in rediscovering and resurrecting such foot-soldiers of European applied science, so this volume succeeds ones on Alexander Gibson and Robert Wight.[9] Those works too had for starting points the botanical drawings commissioned as offshoots of their official Company activities, and led (in so far as was possible) to a documentation of their fuller lives – in the first case primarily as another early forest conservator (Cleghorn's *precursor*, in the Western Presidency of Bombay), and in the second as a major taxonomist and economic botanist.

In my view the devil is in the detail; it is surely the biographer's duty to study and record a life in its fullest possible richness, and chronologically, in order to build up an accurate picture. This is the way I proceeded – by piecing together, slowly, over a period of 15 years a detailed chronology of Cleghorn's life and publications from as many sources as possible, looking at the full range of the man, his religious interests, his wife and her remarkable family, then fleshing out this skeleton. As an Appendix to 'the life' is a detailed bibliography of his publications (far more numerous than previously imagined) and a catalogue of his outstanding library is available on the RBGE website. Originally the biography was going to stop in 1868, the date Cleghorn returned to Scotland from India, but in the course of writing I came to realise that one could not ignore 25 years of a man's life, which led to further delay. Inevitably it emerged that his links with India did not stop in 1868 and, moreover, that Cleghorn's views, as expressed in his later writings, continued to develop. In the past some of these later statements have been taken out of context – for example, later thoughts on the importance of the climate question used as evidence for his earlier beliefs. The three works together can be seen as a 'prosopography' of less well-known figures that, taken together, provide a detailed picture of botany and its application to economic ends in colonial India in the period 1820 to 1870.

The sources for all three of these Company Surgeons are limited largely to their reports and publications. For none of them, regrettably, are there extant diaries, though there are some letters. In the case of Gibson and Wight those of major importance, though scarcely intimate, were addressed to William Hooker, Director of Kew. With Cleghorn the most revealing are ones written to John Hutton Balfour, Regius Keeper of RBGE and a fellow evangelical Christian (if of a different denomination). These, however, do not amount to enough to be able to presume to say very much about Cleghorn's character or psychology. That he was unforthcoming is suggested by the fact that he merits not the briefest mention in the memoirs of any of his contemporaries so far examined, though his omission from the first edition of the *Dictionary of National Biography* was probably because he lived too long. Obituaries are almost bound to be eulogistic, but he does seem to have been lovable and held in considerable esteem and affection by his fellow Europeans, and possibly more so than were others of his contemporaries by the Indians he encountered. A passionate sincerity comes over in the letters to Balfour, but sadly none of those to his family have survived – only the replies to some from his first Indian period by his father Peter. His father considered Hugh almost morbidly pious and this is likely to be true; it is perhaps significant that in old age, after the death of his wife, he stayed when in Edinburgh in a club and not with his in-laws, the Cowans who though pious were outgoing.

What emerges is Cleghorn not as a proto-conservationist, but, perhaps primarily as a result of his early upbringing by his grandfather, the first Hugh Cleghorn of Stravithie, as a latter-day figure of the Scottish Enlightenment, a figure from the Age of Reason, who, infected with a Protestant evangelical fervour, lived into that of High Imperialism. Cleghorn, through his paternal grandmother, was a cousin thrice-removed of Adam Smith and his philosophy and actions were built on the tradition of useful knowledge, the development of natural resources, seen as justified by, if not imposed as, a religious duty. This also included the commissioning of botanical drawings, following the tradition of William Roxburgh and Robert Wight, in which light his interest in horticulture and economic botany can also be understood. For him one of the major justifications of preserving forests was the usefulness of the 'non-timber' products such as gums, resins and dyes. From the conclusions of his pioneering, and notably modern-sounding, British Association report of 1851 on the 'Probable Effects in an Œconomical and Physical Point of View of the Destruction of Tropical Forests' his priorities are perfectly clear – that as much forest should be cleared as possible for the good of man (and the development of European capital), with only the minimum left necessary for the maintenance of rainfall, the protection of springs and the prevention of erosion. It should be remembered that forest conservancy in the nineteenth century was far from being the same as what 'conservation' became in the twentieth or twenty-first; rather, the management of a natural resource for the benefit of man.

1
Family, Childhood & Education, 1820–1841

Hugh Francis Clarke Cleghorn was born in Madras on 9 August 1820 and baptised the following month, on 19 September, in St George's Church.[1] When the boy was only 11 years old his grandfather, another Hugh, while teasing him about his multiple forenames, outlined his destiny:

> When you write to me you need not sign with all your Christian Names of which you have as many as a German Prince, while you can only be a small Fife Laird, and that only if by your future Industry & good fortune you shall be able to keep possession of this pretty place.– Sign yourself Hugh which is sufficient.[2]

The poor boy could hardly be blamed for his polynomial, for which his binomially named father Peter, following normal Scottish practice, was entirely responsible. Of these forenames, the inspiration behind the first was his grandfather, the first Laird of Stravithie, that for 'Francis' is unknown, and the third was for Richard Clarke (c. 1785–1868), a Madras contemporary of his father.[3] For Cleghorn belonged to the third generation of his family whose life and career would be intimately connected either with the Subcontinent or its major offshore island.

Grandfather, Hugh Cleghorn of Stravithie (1752–1837)

The Cleghorns were a distinguished Edinburgh family who, both in their own right, and through marriage with the Hamilton family, had, since the late seventeenth century, been significant players in the academic life of the town. The family's prosperity lay in brewing, an activity undertaken in premises called 'Society' on the site of the present Museum of Scotland, opposite Greyfriars Kirk.[4] In 1773, at the tender age of 21, Cleghorn's eponymous

Fig. 1. Three generations of the Cleghorn family.
Hugh Cleghorn senior, photogravure after a pastel by Archibald Skirving (from *The Cleghorn Papers*, ed. the Rev. William Neil, 1927) (left).
Peter Cleghorn, oil sketch by David Wilkie (Sprot family collection) (centre).
Hugh Cleghorn junior, chalk drawing by Alexander Blaikley, c 1838 (Sprot family collection) (right).

CHAPTER 1

grandfather (Fig. 1), straight from undergraduate studies at Edinburgh University, was appointed to the chair of Civil History at the University of St Andrews, where he lectured conscientiously for 15 years. In 1788, aged 36, the Professor was struck with an acute attack of wanderlust and ambition. He abandoned his wife, Rachel McGill, and their family of seven children for twelve years, transforming himself, in the words of Aylwin Clark his biographer, into a 'traveller, government intelligence agent, empire-builder and administrator'.[5] During this period he returned to St Andrews only once – in 1798 – for three days. Cleghorn's initial employment was as tutor to the young Earl of Home on a Grand Tour through France, Switzerland, and Italy as far as Sicily. His observations of the political situation in these disturbed times made him useful to the British Government and on returning to London he put himself at the disposal of Henry Dundas with little guarantee of any adequate financial recompense.

His masterstroke was to make use of a contact fortuitously made in Switzerland with the Comte de Meuron, whose family regiment was in the pay of the Dutch Government in the garrisoning of Ceylon. With the French invasion of the Netherlands, and the Stadholder William V's flight to Kew Palace, the possibility of a British take-over of Ceylon (important for the security of EIC Indian territories) became a priority, and if this could be done on a permanent basis (rather than one which might revert when a treaty was signed with France), and with minimal bloodshed (through getting the de Meuron regiment to change sides), then so much the better. An extraordinary journey followed: de Meuron and Cleghorn travelled to India via Egypt and the Red Sea – negotiations in Southern India through the Madras Government – and a three-day, hare-brained escapade by Cleghorn to Ceylon in which he attempted to smuggle a letter to Colonel de Meuron (the Comte's brother, commander of the regiment) in a Dutch cheese. Cleghorn here met up with his elder son John of the Madras Engineers, now aged 17, whom he had not seen for seven years; he also encountered a large number of Scots, many with St Andrews connections. These included two former pupils, the Rev. Andrew Bell later noted for his educational system and philanthropy, and (by correspondence) with Alexander Walker, later 'of Bowland', a notable Indian administrator. In January 1796 Cleghorn returned to Ceylon on a short, one-man, statistical survey, collecting information on administration and revenue systems while also taking an interest in its natural resources. With him, as Tamil translator and plant collector, he took Dr Johann Peter Rottler, one of the Tranquebar Missionaries, whose collections representing 500 species were sent to Joseph Banks via Andrew Bell in Madras. Cleghorn was not himself present at the taking of Colombo, though his son John, and John's commanding officer, the great surveyor Colin Mackenzie, were. Mission accomplished, Cleghorn returned to Britain, though even *en route* he had further adventures, cooped up for 37 days in the Lazaretto on Malta (having passed through a plague-ridden Egypt) and being used by Sir William Hamilton in Naples to take letters to Sir Gilbert Elliot (later first Earl of Minto), Governor of Corsica, whose secretary was Frederick North (son of Lord North, erstwhile Prime Minister and loser of the American Colonies). Back in London Cleghorn submitted a report on the constitution and history of Malta to the government and became, in Clark's words, one of those 'persons who hang about [the Dundas] office'.

Cleghorn's reward came with the promise of £5000 (to be paid from the revenues of Ceylon) for services already rendered, the promise of a pension for his wife and daughters, and his appointment as Secretary of State to the new Government of Ceylon that carried a £3000 salary. At this stage the governance of Ceylon was a joint venture between the EIC Madras Government and the British Crown. The Governor was to be the intelligent, but brattish, young aristocrat Frederick North – whose courtesy style as 'the Honourable' was to prove, at least so far as Cleghorn was concerned, purely nominal. So in 1798 Cleghorn returned to Ceylon – he managed to extract his £5000, which was sent home through Alexander Walker, by now in Bombay. The secretaryship, however, ended in tears over the 1799 annual pearl fishery, to which Cleghorn had been appointed as one of three commissioners. The proceedings were of immense complexity, with numerous opportunities for fraud on the part of all the participants. Traditionally many had taken advantage of such temptation, though it would have been quite against Cleghorn's nature to have done so, and he refused many offers of bribes. However, the fishery went badly wrong and should never have been embarked on in the first place, having been over-exploited in the two previous years. When it failed to produce as much revenue as the government had hoped, accusations started to fly. North, rather than believe Cleghorn, chose to take the word of a ne'er do well Indian called Narrain, himself heavily implicated in the corruption. Over and above negligence of the management of what was a complex process, Cleghorn was accused by Narrain of stealing pearls and having them bored to make into two necklaces. Cleghorn had fully intended to pay for these, but had unfortunately not yet done so, though he returned them to the Treasury immediately the accusation was made. In fact, one of the sets of pearls was a commission from Dundas. The vitriol aimed at Cleghorn – 'an owl of a secretary' and 'beloved Hugs' North facetiously called him[6] – is such as to make it almost certain that there was more to this incident than can ever be known, and probably more than a mere difference in outlook and temperament (brilliant, aesthetic, probably not entirely heterosexual English aristocrat v. a conscientious, pedantic, canny Scot). Although Dundas saw through North there was little he could do other than to write challenges and reproofs. It was not until 1802 that Cleghorn's name was cleared by a counsel acting for the EIC. From this the only slur that stuck was that he should have kept his eye more closely on the ball in terms of financial accounting, which had never been, nor ever would be, one of Cleghorn's strengths.[7]

The storm weathered, in 1800 Cleghorn returned to his long-suffering wife and family in Fife, where he purchased St Leonard's Lodge, one of the grandest houses in St Andrews (in which Dr Johnson had once stayed). The aspiration, however, was to purchase an estate, not only for the status of gentility and the right to vote (qualification for a superiority required a rental income of over £400), but to try his hand in the role of an improving laird. He had initially bought land to the west of St Andrews at Dewarsmiln and Denbrae but wanted something larger and in 1806 purchased the 1000-acre estate of Stravithie in the parish of Dunino. Its previous proprietor had not been resident and the old castle had fallen down, so Cleghorn enlarged the large farm of the southern part of the estate, called Wakefield, into a country seat, into which he moved his family. Despite ever-precarious finances Cleghorn threw himself into his role as improver: draining, enclosing, tree-planting, and rebuilding farms and mills. In view of his grandson's future activities in India it is worth quoting some statistics relating to his planting activities at Stravithie. In 1809 he made a major plantation (the 'North Plantings', later known as Stravithie Wood) for which, in February 1809, he purchased from the nurseryman John Galloway 12,000 larch, 10,000 oak, 20,000 birch, 30,000 Scotch fir, 1000 [Norway] spruce and 500 Balm of Gilead (*Populus*), supplemented in April with more spruce and firs, and 2000 mountain ash trees, with 1000 blackthorn bushes for hedging.[8]

CHAPTER 1

As Laird of Stravithie, Cleghorn was one of two principal heritors of the parish of Dunino (having failed in a ploy to have the estate transferred into the parish of St Andrews). This led to a feud with the minister of Dunino, an unscrupulous character called the Rev. James Roger, a cousin of James Playfair, Principal of the United Colleges of St Salvator and St Leonard, in whose gift the living lay. This incident, characteristic of the abuses of patronage that would later lead to the Disruption of the Church of Scotland, led to the family, at least sometimes, attending the Episcopal chapel in St Andrews (though the grandfather kept his options open by also renting a pew in the Church of Scotland parish church of St Andrews, Holy Trinity).

Father & Uncle

Hugh's eldest son, John Ross Cleghorn (1778–1825), as already noted, had embarked on an Indian career with the Madras Engineers. Following Ceylon he was in 1799 present at a much more violent affray, the storming of Seringapatam; in 1814 he was Superintendent of Tanks (artificial water bodies for irrigation) for the Madras Presidency; and in 1824, by now a Major, he married a woman with a reputation as a gold-digger who had arrived in Madras in 1819 as Mrs Selina Day (1797–1889). John's younger brother Peter (1783–1863, confusingly also known as Patrick), was more academic – after studies at the universities of St Andrews and Glasgow (with Professor James Mylne), he followed the footsteps of Adam Smith (his mother's cousin at one remove) by winning a Snell Exhibition to Balliol College, Oxford, which he held for three years from 1804. Peter Cleghorn (Fig. 1) qualified as a barrister at Lincoln's Inn in 1810, and in 1816, after a spell in London and with the help of his father's oldest friend, Lord Commissioner William Adam, he followed the example of his father and elder brother and headed east to the 'Cornchest of Scotland'. In Madras Peter was appointed Coroner and in 1820 Registrar and Prothonotary of the Supreme Court. It was here that he met and, in 1819, married Isabella (b. 1796), daughter of Thomas Allan a Leith merchant, whose two sisters had preceded her to India in search of husbands. Both elder sisters chose EIC surgeons – the eldest, Margaret (1793–1841), married Dr William Mackenzie of Culbo, a minor laird from Dingwall and the middle sister, Helen (1795–1873), married Thomas Wyllie (who died in Ceylon three years later). The union of Peter and Isabella would also prove a short one: having borne two sons and two daughters she died of cholera on 1 June 1824 aged only 28. Less than five months before their mother's death, the 3½ year old Hugh and his younger brother Allan Mackenzie, had been sent home on the ship *Britannia* in the care of their *ayah* Fatima, and Susan Foulis (*née* Low), wife of a Fife neighbour and Madras soldier (eventually Major General Sir David Foulis of Cairney Lodge near Cupar).[9] For three years after his wife's death Peter Cleghorn stayed on in Madras, in order to acquire some much needed capital to plough into Wakefield and, perhaps, to be close to his two infant daughters Isabella ('Isa') and Rachel, who were initially brought up in the family of their uncle John and aunt Selina. John, by now a Lieutenant Colonel, died in 1825 on a voyage to St Helena while trying to recover his health, and his widow Selina returned to Scotland with the two girls the following year. Isa would spend her adult life as an invalid-spinster, including much time at spas including Clifton, Germany (Baden, Heidelberg, Wiesbaden, Dresden) and from 1861 largely in Italy in Naples, and in Rome (where she died), and it was left to her younger sister Rachel to propagate half of the family genes (if not the Cleghorn name), when, in 1850, she married Alexander Sprot of Garnkirk.

1. FAMILY & EDUCATION, 1820–41

Childhood & Youth

The two young brothers, Hugh and Allan, were initially brought up at Wakefield (later renamed Stravithie) by four maiden aunts, Rachel, Anne, 'Hugh' (Hughina), and Jessie, with Jane Dinwoodie (daughter of William Dinwoodie, Peter's man of business in London) as nanny, and a doting grandfather (who may have over-compensated for the neglect of his own children thirty years earlier).[10] The human family was extended by a small menagerie – cats, bull terriers, ponies and rabbits. In due course James Cruickshank was brought into the household as tutor for the brothers, whose small class also included Robert Campbell, a stepson of their aunt Jane, and – brightest of the party – David Thomson, son of their grandfather's personal servant. Cruickshank had studied Arts at St Andrews followed by Divinity in Edinburgh.[11] He taught the young Hugh Latin to university-entrance level and also made a start with Greek. In the meantime, in May 1827, the boys' father had returned to Britain, with £2000 in his pocket,[12] but was defrauded of a further £2561 by Dinwoodie. The latter sum was said to be but one fortieth of the fortune Peter had amassed in Madras, which would imply that he had about £100,000. This may have been an exaggeration but the sum was certainly vastly more than Hugh (senior) had ever managed to extract from the British government for his diplomatic services. Despite paternal remonstrances, Peter returned to Fife apparently sapped of ambition, determined not to continue practising law in London, and, after three months of travelling in Europe in 1828, ended up what his go-ahead father described as 'that most deplorable of characters, a Country Laird'.[13] Perhaps unhealed grief over the death of his young wife contributed to this self-imposed rustication, or perhaps his father's peripatetic lifestyle had been an off-putting example. Peter's capital allowed Stravithie to be kept together; in November 1825 he effectively paid off some of his father's debts by purchasing from him the largest of the estate's farm, Mains of Stravithie, for the enormous sum of £18,000 (the estate at this point was valued at £40,000, and was made over entirely to Peter in 1831).[14]

Of the towns closest to Stravithie, St Andrews, 3½ miles to the north, was notable for its ancient university. Five miles to the south, the small East Neuk port of Anstruther had proved to be a remarkable nursery for intellectual talent – in the late eighteenth century it had produced the theologian Thomas Chalmers, the poet and linguist William Tennant, and the medical Goodsir family. The Goodsirs were close friends and it was to Anstruther – to the care of the town physician Dr John Goodsir – that the juvenile Cleghorn, aged five, was sent to escape from an 'epidemical fever' at Stravithie that carried off his aunt Anne.[15] Goodsir's own father had been a physician at nearby Largo, and no fewer than three of his sons received MDs, the third generation of their family to be medically qualified. Of these John (1814–1867) would achieve fame as Professor of Anatomy at Edinburgh; Harry (b. 1819), only a year older than Cleghorn and so a particular friend, became assistant surgeon and naturalist on *Erebus* on the fatal 1845 Franklin expedition in search of the North-West Passage; and it fell to his younger brother Robert Anstruther (1823–1895) to take part in two of the numerous expeditions sent in search of them.[16] It was not until 1854, when John Rae found relics and remains, and heard Inuit reports, that the expedition's fate (and chilling hints of cannibalism) was finally known, the melancholy news reaching Cleghorn in Madras the following year.

In September 1831 the two Cleghorn brothers were sent south across the Firth of Forth to Edinburgh to attend the city's High School (Fig. 2), an ancient foundation that two years previously had moved into magnificent new premises at the foot of Calton Hill – the Athens of the North's *acropolis* – designed in suitably severe Grecian style by Thomas Hamilton.

CHAPTER 1

The boys entered the fourth-year class of the Classical Master, William Pyper.[17] Cruickshank accompanied them and they lodged with their paternal aunt Jane, who in 1820 had married (as his second wife) William Campbell, a Writer to the Signet (solicitor), and they lived in the family property at 'Society'. Aunt Jane's two younger stepsons Robert and William Hunter (b. 1814) were still at home. William, and his eldest brother James (b. 1806) by now in the Madras Army, shared their younger step-cousin's botanical interests and William would later become Secretary of the Botanical Society of Edinburgh. In September 1832, the twelve-year old Hugh entered the class of Dr Aglionby Ross Carson, the school's Rector (headmaster).[18] Family connections in Edinburgh were strong, and the brothers socialised with old family friends

Fig. 2. Cleghorn's schools.
Royal High School, Edinburgh by Thomas Hamilton. Recto of alumnus card for the year 1832–3, etching by W.H. Lizars (above).
Madras College, St Andrews by William Burn. Etching by W.H. Lizars after a drawing by W. Banks reproduced in Roger, 1849 (below).

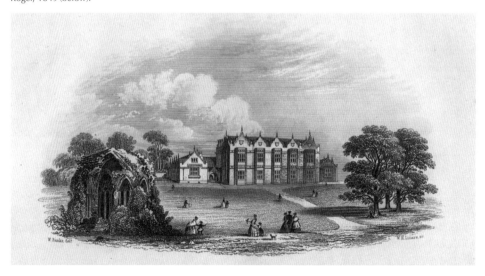

including Sir George Ballingall (Professor of Military Surgery at Edinburgh University who had overlapped for two years with Peter Cleghorn as an Arts undergraduate in St Andrews),[19] with Mrs Foulis (the boys' guardian on their childhood voyage) and her colonel husband, and also with their maternal aunt Helen Wyllie, who, as a widow, lived at 14 Carlton Terrace with her sister Margaret and brother-in-law Dr William Mackenzie.

After two years in Edinburgh, Cruickshank left the boys to become the parish minister of Manor, Peebles-shire, and after succumbing to measles Hugh and Allan were summoned home to Fife. Perhaps the close-knit family would have missed them too much to let them return to Edinburgh, but a new educational establishment had just opened in St Andrews: Madras College. Hugh Cleghorn (senior) gave careful consideration to his grandsons' education, particularly that of his heir. The old man was conscious of his property and the matter of its succession, but knew that the position of 'a small Fife laird' could only be maintained if Hugh could bring in additional financial resources, to be gained only through a profession (the estate brought in only £800 a year in 1845). Before making any change, Grandfather carefully considered the relative teaching methods and syllabus of the Edinburgh High School and the new Madras College (Fig. 2), and whether the new establishment was likely to provide a good enough education.[20] A local school did, of course, have the advantage of being more economical (fees for each of the major subjects were 7s 6d per quarter), and family finances were tight.[21]

The new school had been founded with a bequest of £50,000 to the town of his birth by the Rev. Andrew Bell, who had in India, as a (plural) chaplain, and lecturer on natural philosophy, managed to amass a substantial fortune (for which his salary must have been augmented by entrepreneurial activities). In Madras, at the Male Orphan Asylum, Bell had developed a monitorial system, where older pupils taught the younger ones. The bequest may owe something to Bell's friendship with his old teacher, the elder Hugh Cleghorn, as after his return to St Andrews in 1800 Cleghorn was on the committee of the English School whose teacher, a Mr Smith, used Bell's monitorial system and Cleghorn had asked Bell for the money that allowed them to build a new school behind the Town Church (Holy Trinity). After Bell's death his bequest enabled the erection of a handsome new building, at a cost of £17,000, designed in the Jacobean style by William Burn, in generous grounds, set back from South Street, with a ruined transept of the Blackfriars monastery as an eye-catching gothic garden feature. The young Cleghorn may just have been able to take advantage of the grand new premises though it was not completed until the year he left school.

Unlike the Edinburgh High School, Madras took both boys and girls, from the age of 5 to 15, of whom some were taught free, though ones from Cleghorn's background paid fees. Subjects taught included the basics of maths, drawing, writing and arithmetic, and for older pupils Latin, Greek and modern languages. Standards were high, the masters having been recruited with care, and they presided over their own classes according to subject, so Bell's monitorial system, ironically, was *not* a feature of the school. The teaching of French and German had been made possible because the school had, in 1832, appointed as a master the Swiss Samuel Messieux (previously in Dundee where he had been a friend of the poet Thomas Hood). The young Hugh took advantage of this, studying both modern languages, winning a prize in French in 1833.[22] This training in living foreign languages would stand him in good stead in India. As by this stage Cleghorn was 14, he probably had no need of the classes in writing and arithmetic, and probably devoted most of his efforts to Latin and Greek. In addition to formal education the younger generation also had to be instructed in the traditional rural pursuits of the gentry – riding, shooting and fishing. The ponies served for the first, and the estate provided an arable habitat for partridge and hare, and the North Plantings for roebuck.

CHAPTER 1

Undergraduate Years
St Andrews

The transition from Madras College to the University of St Andrews was an obvious one for the Cleghorn brothers: Hugh would spend the next three years at the United College of St Salvator and St Leonard (Fig. 3), and Allan would follow him in 1836. At this time the university was tiny with only about 130 students between the United College and the divinity school of St Mary's. At the age of 14, in February 1834, Hugh matriculated in the faculty of *Litterarum Graecarum et Humaniorum*, taking the second half of the junior classes in Latin and Greek in the session 1833/4 and advanced Latin and Greek in 1834/5.[23] The Rev. Thomas Gillespie, Professor of Humanity, taught Latin and Professor Andrew Alexander, Greek. Fees of three guineas were due to each of the professors and classes were held in a handsome new building to the north of the eastern apse of the magnificent mediaeval chapel of St Salvator. This new building, completed only in 1831 and designed by Robert Reid,[24] last of the King's Architects, was, like the Madras College in the Jacobean style – buckle quoins, Dutch gables and gargoyles – doubtless with a nod to Oxbridge colleges, though in striking contrast to the prevailing Neo-Classicism of Edinburgh academic architecture.

A summer holiday in 1835 took the Cleghorn family to Loch Lomond and Glasgow, then to visit the boys' old tutor, the Rev. James Cruickshank, at Manor; on the way home they stopped in Clackmannanshire to see another old friend of the family.[25] William Tennant had until recently been master of classical and oriental languages at the Dollar Institution (later Dollar Academy), a charitable foundation, which, like the Madras Academy, had been handsomely endowed by a local-boy-made-good, John McNabb. It catered both for the local poor and for prosperous outsiders who paid substantial fees of up to 50 guineas a year for boarding. Like Madras it was mixed-sex and housed in a handsome new building, though this one was in Neo-Classical style, the work of W.H. Playfair, who in 1821 is also believed to have designed Devon Grove, a villa on the outskirts of the village, for Tennant. The syllabus was adventurous and there can't have been many schools in the country that offered 'Persic Languages', even if at an 'additional charge'.[26] 'The Professor', as the Cleghorns knew Tennant, is best known for his monumental, mock-heroic poem *Anster Fair*, in which he revived the form of *ottavo rima*, first published anonymously in 1812. From 1813 to 1816 Tennant had been the parish schoolmaster at Dunino and something of a soul-mate for the elder Hugh Cleghorn. Earlier in 1835 he had been appointed by Lord Jeffrey to the chair of Hebrew and Oriental Languages at St Mary's College, St Andrews, but he retained his house at Dollar until his death in it in 1848, and must have been there for the summer vacation at the time of the Cleghorn visit. Another of the enlightened aspects of the Dollar school was its teaching of horticulture: in the grounds was a 'botanical and oeconomical' garden established by the Institution's first principal, the Rev. Dr Andrew Mylne, where some of the local boys could serve apprenticeships. Two pupils who benefitted from this horticultural element were James McNab and William McIvor. McNab would later succeed his father William as Curator of RBGE and one day redesign the Stravithie garden, and had actually boarded with Tennant in 1825/6 and 1827/8. McIvor attended the school from 1831 to 1839, so was there at the time of the Cleghorns' visit. This was a curious prefiguring of the dealings over the cinchona experiment in the Nilgiris that Cleghorn and McIvor would have in the early 1860s.

At St Andrews Hugh continued with advanced Latin and Greek in session 1835/6, and was made an honorary member of the University's Literary Society.[27] It was at this point that he may have started to consider medicine as a future career, as in this academic year he took Thomas

1. FAMILY & EDUCATION, 1820–41

Fig. 3. Cleghorn's universities.
United College, St Andrews. Etching by W.H. Lizars after a drawing by W. Banks, from Roger, 1849 (above).
University of Edinburgh, interior of Old College in Cleghorn's day. In the centre is the Museum, in the quadrant to its right the Anatomy Department, in the quadrant to its left the Natural History and Chemistry Departments; on the left the University Library. Etching by J. and J. Johnstone, from J. Stark's *Picture of Edinburgh*, 5th edition, 1831 (below).

Duncan's mathematics class, and chemistry with Robert Briggs, Professor of Medicine. Though medicine, at least in a rustic form, was also in his blood: his grandfather was in the habit of purchasing stocks of medicines to dispense to his tenantry and to the poor of Dunino. During his final year at St Andrews (1836/7), Cleghorn continued with science, attending the physics lectures of Thomas Jackson, Professor of Natural Philosophy. A fellow student on the course was Robert Adamson (1821–1848), and it may have been through him that Cleghorn met Robert's elder brother Dr John Adamson (1809–1870) who was to become a good friend. The handsome, bright, young Adamson brothers came from a farm in the nearby village of Boarhills and were interested in chemistry; an interest that would lead to their significant contributions to the early development of photography.

CHAPTER 1

In this final St Andrews year are the first suggestions that Cleghorn's interests had a humanitarian slant. In addition to science he attended the ethics lectures given by the Rev. George Cook, another family friend, best known as an historian of the Church of Scotland, who held the Chair of Moral Philosophy. Cook's influence on the serious, and perhaps already liberal-minded, young man would appear, if anything, to have been negative. Cook, at this point, was leader of the conservative, Moderate Party of the Church of Scotland, and supported the principle of patronage for appointments to parish livings. This inevitably led Cook (and conservatives like Peter Cleghorn) to oppose the Rev. Thomas Chalmers (leader of the Evangelicals), and Cook later became leader of the parts of the church that remained in the establishment after the Disruption of 1843. Cleghorn (in contrast to his father) would become a committed supporter of the Free Church.

For Cook, Cleghorn wrote an essay 'On the study of mind, and the causes which have given an erroneous direction to it', which advocated painstaking Baconian induction over the easier route of 'visionary theory', which he denoted the 'erecting a fabrication of his own fancy and imagination'.[28] In order 'to investigate truth – the love of truth must be our governing principle', but errors could be made from imperfections of language, from using inappropriate analogy (for example, that because a planet such as Mars orbited the sun it might support living and intellectual beings), by the influence of prejudice from ideas outwith the realm of reason, but above all by using non-inductive methods. Cook marked the essay of the 16-year old 'excellent', and showed 'great soundness of judgement' in a style that, with very few exceptions, 'would do credit to a veteran composer'. The student was urged to persevere, in order to 'attain the excellence to which he has so strong a hereditary claim'. The last comment was a reference to the boy's beloved grandfather, whose death at the advanced age of 85 took place at Stravithie on 19 February 1837 and which no doubt cast a shadow over Cleghorn's final year at St Andrews.

Edinburgh

It was common for boys in the north and east of Scotland to attend their local university, St Andrews or Aberdeen, and then to progress to Edinburgh for further studies, especially in the field of medicine. And so for Cleghorn. Edinburgh in 1837 had for a century been a pre-eminent centre for medical education and Cleghorn matriculated in the Medical Faculty in October of this year (Fig. 3);[29] he also became apprenticed for a period of five years to Professor James Syme, for which Peter Cleghorn must have paid a hefty fee (Fig. 4). Since 1833 Syme had held the university's Regius Chair of Clinical Surgery, which gave him responsibility for 30 surgical beds in the Royal Infirmary. These beds were in the former High School building (next to the Infirmary), which had been turned into a surgical hospital – but Syme was also still involved with a private hospital that he had started in Minto House in 1829. His consulting rooms for private patients were at 1 Shandwick Place at the west end of Princes Street (the building was replaced in the late nineteenth century, of which the upper floors are now a backpackers' hostel). Syme was a somewhat controversial figure, known for inspirational teaching, but

Fig. 4. Professor James Syme. Mezzotint by James Faed, from R. Paterson's *Memorials of the Life of James Syme*, 1874. (Courtesy of Royal College of Physicians of Edinburgh).

also for his old-fashioned methods and bitter and very public spats (in print and verbally) with colleagues.[30] Despite this Syme attracted deep loyalty and affection from his pupils and associates, as recounted by Dr John Brown in *Horae Subsecivae*.[31] Although no evidence survives to indicate the relationship between master and pupil in this case, botany was a shared interest, and Syme was known for the garden and hothouses at his substantial suburban villa, Millbank, which lay between Morningside and the Grange on a site now occupied by the Astley Ainslie hospital.

Bright Edinburgh medical students intent on a professional career usually joined the undergraduate Royal Medical Society (RMS, founded 1737), which had its own premises in a handsome, purpose-built hall in Surgeon's Square, with a museum and library, where debating skills could be honed and knowledge exchanged, attended both by undergraduates and former members, including professors.[32] Cleghorn signed his 'obligation' to the RMS on 10 November 1838, when he probably paid the 'composition fee' of £8. The following year he submitted to it a dissertation entitled 'What are the processes by which the reparation of wounds is effected, and how far are they to be regarded as the results of inflammation' – probably an early version of his lost doctoral thesis on the same topic. The subject was much discussed at the time, since, prior to the discovery of antisepsis, the success of surgery, as performed at virtuosic speed by practitioners such as Syme, was severely compromised as a result of wound infection. In 1840 Cleghorn and Thomas Anderson (later Professor of Chemistry at Glasgow, not to be confused with the eponymous botanist) were appointed as the first two honorary (as opposed to salaried) Secretaries of the RMS. It is a tribute to Cleghorn's standing with his fellow undergraduates (if not necessarily his intellect), that with Anderson and two others, he was elected as one of the four Annual Presidents of the Society for its 104th session of 1840/1. His association with the RMS continued after his undergraduate days and in 1851,[33] while on his first home leave from India, he was involved in the Society's move from Surgeon's Square to a new, larger, hall, designed by David Bryce, in Melbourne Place on George IV Bridge (the G&V Hotel now occupies the site). On Cleghorn's second furlough, in 1862, came another minor interaction when (with old teachers such as Handyside and Christison, old friends including J.H. Balfour and John Goodsir, and a new generation including Alexander Dickson who would succeed Balfour as Regius Keeper of RBGE), he signed the diploma for Charles Darwin's honorary membership.[34]

From records kept by Edinburgh University it is possible to chart Cleghorn's undergraduate career in considerable detail – even down to his examination results. At this time there were also several extra-mural medical schools (in both Brown Square and Surgeon's Square, on either side of the College buildings), which provided often more radical teaching than the largely conservative university professors (Fig. 5), and Cleghorn availed himself of the tuition on offer in both systems.[35] In his first session (1837/8), in the Surgeon's Square extra-mural school, he continued the chemical studies begun in St Andrews with an inspirational young extra-mural lecturer called Kenneth Treasurer Kemp, and the anatomy lectures of Dr Peter Handyside with whom he would continue in the following two years. Handyside was probably a kindred spirit who pioneered medical-missionary work in the slums of the Cowgate and Grassmarket in the 1840s. At the University (then usually referred to as the 'College'), for this and the next three years Cleghorn attended his own master Syme's lectures in Clinical Surgery, with practical studies in the Royal Infirmary that still occupied its original William Adam building in Infirmary Street, across Nicolson Street from the College.

CHAPTER 1

Fig. 5. The Edinburgh Medical Faculty, c 1850. In the front row are four of Cleghorn's lecturers – Left to right: James Young Simpson, Robert Jameson, William Pulteney Alison, Thomas Stewart Traill; back row centre John Hutton Balfour. Lithograph by L. Ghémar, published by Schenk & McFarlane, Edinburgh. (RBGE).

Origin of botanical interests

Cleghorn's interest in botany is likely to have started in childhood, in the dens and hedgerows around Stravithie, and on the nearby links of St Andrews where his grandfather played golf (he was Captain of the Company of Golfers in 1802/3). But an Indian slant to this interest could also date from this early period of his life – for Indian botany was 'in the air' in landed and intellectual circles in this remote corner of Fife. Grandpapa Hugh may himself have encouraged this interest, having attended John Hope's botanical lectures in 1770 and employed J.P. Rottler to collect botanical specimens in Ceylon in 1796. While in India Cleghorn had in 1795 corresponded with his former St Andrews pupil Alexander Walker (1764–1831), son of the minister of Collessie in Fife. In western India in the 1790s Walker took a serious interest in botany resulting in a series of botanical drawings by an Indian artist that will be described later (p. 269). Also friends of grandfather Hugh were the Patton family, one of whom, Colonel Robert Patton had (with funds obtained in Bengal) purchased the estate of Kinaldie, adjacent to Stravithie. Robert's brother, Captain Charles Patton, who lived in St Andrews, had in the 1790s been asked to keep an eye on Rachel Cleghorn's affairs while her husband was on his travels in Europe, London and India. By the 1830s the Patton family had let Kinaldie and relocated to Hampshire, but the Cleghorns must have kept in touch with the family, and one of Colonel Robert's daughters Anna Maria and her husband Colonel George Warren Walker, were in the 1820s and 1830s making extensive plant collections in Ceylon,[36] and in the 1840s and 1850s grandson Hugh knew two of Anna Maria's children in Madras. The Laird of Balmungo, another nearby estate, was Alexander Kyd Lindesay, who collected plants in Lohaghat in Kumaon in the early 1830s (the specimens were given to George Walker-Arnott and are now at RBGE), with whom Cleghorn was still corresponding from India in 1858.

1. FAMILY & EDUCATION, 1820–41

Even for Company employees who didn't manage to save enough money to become landed proprietors, St Andrews in the early nineteenth century was a popular place of retirement for old India military, civilian or medical hands, for its bracing climate and its golf links. One of this number had strong Indian botanical credentials: George Govan (1787–1865) had been the founding superintendent of the Saharunpur Garden in northern India, and had retired to Pilmour Cottage beside the Links – and one of his sons, also George, was in the Indian Medical Service at the same time as Cleghorn. There was also the Playfair family, rather similar to the Cleghorns, with an academic father (John Playfair had been Principal of the United College) and military/medical sons – George had ended up as head of the Bengal medical establishment, his brother Hugh Lyon ('The Major') had been in the military in India then became a mover and shaker as Provost of St Andrews (from 1842 until his death in 1861). Cleghorn knew at least three of George's sons: Lyon (the chemist, later Baron Playfair), Robert Lambert (with whom he later corresponded over botany in Aden – see *Cleghorn Collection* Fig. 18), and George (a Bengal surgeon). St Andrews, with its small intellectual community (the total population in 1838 was about 5000), was therefore a stimulating place for the young Cleghorn to be raised, though its major blossoming began just as he was leaving it in the late 1830s.

Botanical Society of Edinburgh (Part 1)

Cleghorn's formal botanical studies began in the summer of 1838 when he took Professor Robert Graham's class at the Botanic Garden in Inverleith Row. Two years previously, on 17 March 1836, Graham, with a group of like-minded botanists, pre-eminent among whom were John Hutton Balfour and Robert Kaye Greville, had founded the Botanical Society of Edinburgh (BSE), with Cleghorn's young step-cousin, the young lawyer William Campbell as its secretary. Hugh was balloted and elected a member of the Society on 14 June 1848, for which he would have paid a one-guinea admission fee, thereafter a guinea a year, or, more likely, a ten-guinea 'composition fee' for life membership.[37] As this body was to be one of the cornerstones of Cleghorn's botanical life, not least in his long retirement, it is necessary to say more about it. Its aims were wide-ranging, based around the subjects of practical, physiological and geographical botany, but also its application to 'the Arts and Agriculture'. This was to be achieved by means of monthly meetings; in 1838 the winter ones, from November to April, were held on the second Thursday of the month in the room of the Society of Antiquaries in the Royal Institution, and the summer ones, from May to July, in Graham's classroom at RBGE. At these meetings papers were read and within a few years the initially planned publication of annual reports and proceedings had been expanded to a full-blown *Transactions*. Another aim was to run field excursions. Various membership categories existed – Honorary, Resident, Non-Resident and Foreign; Associate status was available to artisans who could not afford the subscription, and on the payment of two guineas, a woman could become a 'Lady Member' for life.

Links between the Society, the University and the Botanic Garden were strong, and have remained so ever since. In these early days the Society used rooms in the University to house its collections – its library, herbarium and museum that, from the outset, were another of its chief *raisons d'être*. In 1838 the Society's herbarium was amalgamated with that of the University, later being transferred to the Government and relocated to RBGE. The library would later (1872) follow the herbarium, the two forming major elements of the present RBGE collections. At this stage the primary aim of the Society was not only to assemble its own herbarium, but also to act as a distribution network for the exchange of specimens. A series of Local Secretaries was designated stretching from Devon to Inverness-shire via Dublin and Belfast, with outposts at the Cape of Good Hope and in Switzerland. There were strict rules over the contribution of specimens by the various categories of membership, in terms of how many were expected each year and the entitlements that these earned the donor; the specimens were contributed to a central pool, run by a Curator. The Society was rapidly successful in terms of membership and specimen-exchange, so that in 1838, the year Cleghorn joined, there were 185 members (of all categories) which had increased to 351 by the time he set sail for India three years later. In the first year 30,000 British and 30,000 foreign specimens were contributed, and Christian, Lady Dalhousie, had donated her herbarium of Indian plants (collected mainly around Simla and in the Calcutta Botanic Garden). To ensure uniform nomenclature the Society published a *Catalogue of British Plants*, and, in 1840, a set of *Directions for Collecting and Preserving Botanical Specimens* was written by Greville (which Cleghorn would later reprint in Madras[38]).

Cleghorn threw himself into the activities of the Society, donating British herbarium specimens in October 1839 and December 1840. More unusually for an undergraduate, but an early indication of his bibliographical interests, he gave books for the library in 1839, 1840 and 1841. For the first year he was appointed Assistant-Secretary to Campbell, and for the year 1841/2 he held the onerous post of Curator, which meant dealing with a vast influx and parcelling out of specimens.

Continuing Medical Studies

In his second academic year (1838/9), in addition to taking botany again, Cleghorn attended the following university classes: Materia Medica with Robert Christison, Surgery with Sir Charles Bell, Natural History with Robert Jameson, and the Theory of Medicine (i.e., physiology, pathology and therapeutics) with Dr William Pulteney Alison. These teachers reflect the changes affecting Edinburgh medical education at the time – the 64-year old Bell was an anatomist of the old school, who in 1833 had written a Bridgewater Treatise on *The Hand: its Mechanism and Vital Endowments as Evincing Design* (illustrated by David Wilkie a friend of Peter Cleghorn when both were young men in London, see Fig. 1). The same age as Bell, Jameson was also old-fashioned (as famously indicted by Charles Darwin), but had broad interests in zoology and geology, and, more importantly, had the curation of, and control over access to, the University's outstanding museum. Of the new breed – men in their forties – were Christison and Alison. Robert Christison, who had previously held the chair of Medical Jurisprudence, had interests in toxicology and therefore the forensic aspects of medicine. Alison's main interest was in social health, and the improvement of living conditions in the slums of Edinburgh, which led to disputes with the Rev. Thomas Chalmers. The latter held idealistic ideas on poor-relief provided by parish churches, whereas Alison saw an urgent need for intervention by the state or civic authorities to provide financial aid to alleviate the desperate poverty in the newly industrialised cities.

In October 1839 Cleghorn matriculated for a third session at Edinburgh, continuing studies with Handyside and Syme, and practical work in the Infirmary. University classes taken for the first time, and compulsory for graduation, were in the Practice of Physic (what would now simply be called 'Medicine' – clinical work undertaken in the Infirmary examining patients and their symptoms, and working out appropriate treatments) with James Home, Medical Jurisprudence with Thomas Stewart Traill, Pathology with Allen Thomson and Anatomy with the 66-year old Alexander Monro *tertius* who reputedly lectured from the antique notes of his grandfather. During his Christmas holiday in Fife, however, Cleghorn attended in St Andrews twelve altogether more stimulating lectures – a course on systematic zoology, given by the brilliant young naturalist Edward Forbes (1815–1854). Forbes had also trained as an artist and his lectures were illustrated with chalk drawings and doubtless also with living specimens of marine organisms – at the time he was preparing his engagingly illustrated *History of British Starfishes*. While in St Andrews, walking on the sand with the Cleghorns' friend the anatomist John Goodsir, Forbes found two species of comb jelly, which he studied with the help of local microscopical skills and instruments – those of Sir David Brewster and Major Playfair ('old eyes at looking through glasses'[39]).

In Edinburgh, in May 1840, Cleghorn took his first examinations in medicine.[40] His examiners were Sir George Ballingall, Monro, Alison, Jameson, and the venerable Professor of Chemistry – the 74-year old Thomas Charles Hope. Cleghorn passed 'cum nota', being judged 'S.B.' (*satis bona*) in most subjects, but, surprisingly, for botany Robert Graham had to record:

> a more perfect specimen of forgetfulness I never saw.– From familiar intercourse, I know that the subjects of my examination are perfectly known to him.

Despite his early interest in the subject it was Chemistry, however, that let Cleghorn down most seriously: he had to be re-examined by Hope in June, and again in July, when he was eventually judged to have 'made a better appearance, yet barely S.B.' As he had not attended Hope's lectures, but the extra-mural ones of Kemp, it is possible that this might say more about professional rivalry (or differing scientific views) between the teachers than about the student's knowledge.

Cleghorn's fourth and final academic session at Edinburgh, for which he was one of the Presidents of the RMS, was that of 1840/1. His address was 21 Brown Square, a new address for his Aunt Jane's old house in 'Society'. New subjects this year were Military Surgery (with Ballingall) and Midwifery with the extrovert James Young Simpson, in what was his first year as holder of the chair. The first of these was a necessity for someone considering employment with the East India Company, and the second would have been useful as Company surgeons often also had civilian medical duties. This was the outset of Simpson's distinguished career, seven years before his discovery of the anaesthetic power of chloroform, but many years later Cleghorn would encounter him again over his wife Mabel's persistent gynaecological problems. In June came the second of Cleghorn's examinations,[41] and the submission of his thesis, which was on the subject of wound-healing (this must have been printed, but no copy can be found). For his *viva* he was examined by four professors – Traill, Simpson, Sir Charles Bell and Christison. It would appear that Cleghorn did not shine in oral examination: only Bell gave him a 'B' (*bona* = good) and he was passed for the degree of MD in the 'second division'. In the meantime, from May to August 1841, he was one of the House Surgeons at the Infirmary. Some of his case notes have survived:[42] six-year-old Isabella Polston happily recovered from scarlatina, but Jane Davies, aged 29 from Musselburgh, suffering from diarrhoea and swollen feet, did not. The notes are revealing in showing how Edinburgh students were taught to make careful observations of symptoms (with great attention to stool and urine). Such training would prove useful when comparing taxonomic characters of plants, or of any other of the many objects – both natural and artificial – that came under the purview of the more curious surgeons when they reached India.

Fig. 6. The Falls of Gersoppah, Karnataka. (Author's photograph, 2007).

2
East India Company Surgeon, 1842–1848

Cleghorn's activities over the winter of 1841/2 are unknown but by the summer of 1842 he was preparing for entry into the EIC's medical service, obtaining a certificate that he was 'competent to perform the operation of Cupping' from H.C. Betts of University College Hospital, and a licentiate diploma from the Edinburgh College of Surgeons, for which he was examined by John Robert Hume.[1] The requisite tropical clothing was purchased from Christian & Rathbone of Wigmore Street, London and the necessary recommendation for an appointment to the Madras Presidency was provided by his father Peter. So, on 10 August 1842, Cleghorn was formally nominated by John Loch, a Company Director with strong Scottish and familial connections. Loch's mother was Mary Adam, sister of the elder Hugh Cleghorn's friend, Lord Chief Commissioner William Adam, who, with his brother James Loch (the notorious clearer of the Sutherland estates), had been brought up at Blair Adam in Kinross-shire. Cleghorn had to undergo a physical examination and two doctors deemed that:

> his eye-sight and hearing are perfect; that he is without deformity ... that he has no appearance of any constitutional disposition or tendency to disease; and that he does not appear to have any mental or bodily defect whatever to disqualify him from Military Service.

Cleghorn's constitution must, indeed, have been robust as, apart from two episodes of 'jungle fever' (leading to periods of sick-leave of three and 12 months respectively), his health stood up remarkably well to 25 years of active life in India, to which, aged 22, he set sail from Spithead on the ship *Wellington* on 15 August 1842. The ship put in at Madeira where he made notes on the scenery and vegetation (being struck by the large number of introduced plants and the difficulty of distinguishing these from natives), which he sent as a letter to the Botanical Society of Edinburgh. This communication was read at a meeting chaired by Douglas Maclagan on 12 January 1843, and, as reported in the *Edinburgh Evening Post*, became Cleghorn's first publication; it was also summarised in the Society's *Transactions* published the following year.[2]

At the Cape of Good Hope, rather surprisingly, Hugh met his younger brother Allan, whom, after leaving St Andrews University had in 1841, aged 19, joined the 4th Native Infantry of the Madras Army and was returning home wounded.[3] This was probably the last time they met; poor Allan turned out to be something of a disaster – in the view of his (far from unkind) father he was a drinker, a spendthrift and a smoker; an episode of 'debauchery' in Edinburgh before his return to the East suggests additional sins of a carnal nature. From Madras, where he rejoined his regiment and caused a further scandal, Allan went to Singapore and then to Hong Kong where he died on 4 November 1844, just short of his twenty-third birthday, leaving big brother to pay off his debts.

On arrival at Madras on 6 December Cleghorn was assigned to duty under the Surgeon of the General Hospital,[4] thereby re-enacting the launch of an Indian botanical career that had taken place 66 years earlier, as this was the same position occupied in 1776 by William Roxburgh, the 'Father of Indian Botany'. In the intervening generation Robert Wight had, in many ways, inherited Roxburgh's mantle both as India's pre-eminent botanical taxonomist and as a commissioner of botanical drawings from Indian artists. Although Wight had at this point recently been appointed to a post supervising cotton cultivation in Coimbatore, where he was largely to be based until retiring 11 years later, he cannot have yet left Madras as he was there to welcome young Hugh. A tangible souvenir of their meeting survives in the form of a book now in the RBGE library – a copy of the first (and only) volume of Wight & Arnott's *Prodromus Florae Peninsulae Indiae Orientalis* (1834), which Cleghorn enthusiastically inscribed:

> *Donum mei Wightii, amici carissimi, Botanici meretissime,*
> *liberalissimi, ardentissimi. Receptum Dec. 1842.*

Wight must have known Peter Cleghorn in Madras in the early 1820s, and when Wight was on home leave in September 1831 he had made a detour to St Andrews to visit him, though as at this time Hugh was away at school in Edinburgh, this Madras meeting may have been their first one in person.

The first half of 1843 was spent in Madras, where the young surgeon lived with Colonel Duncan Sim of the Madras Engineers, a one-time colleague of his late uncle John. Cleghorn was already moving in evangelical Christian circles and would later express gratitude for the benefits received at prayer meetings in Madras (and Bangalore). However, it can't have been all work and piety. Cleghorn had taken a gun with him, and his brother Allan recommended the snipe shooting at Poonamallee; he also asked his father to send out a frock coat and shoes, which hints at some sort of social life.

On 28 July Cleghorn was appointed to the post of Superintending Surgeon in the Mysore Division and he would be based largely in Mysore for the next four years: the first half in the army, the second in a civil position. Since 1831, after an unsuccessful restoration of the Hindu Woodyer dynasty following the defeat of Tipu Sultan in 1799, which had ended in a rebellion, the kingdom of Mysore had been taken over by the Company and administered by a 'Mysore Commission' reporting directly to the Supreme Government in Calcutta. The administrative capital was moved to Bangalore, and the deposed Raja pensioned off to Mysore. The state, which covered an area the size of Ireland, with a population of around four million, was divided into four administrative Divisions, and since 1834 the Commissioner had been the widely admired Mark Cubbon. Cleghorn was initially (4 August 1843) assigned to the medical charge of a British regiment, Her Majesty's 25th Regiment of Foot, but may not have taken this up as later in the same month he was appointed to a Company regiment, the 2nd Madras (European) Light Infantry stationed at Bangalore. It was doubtless a relief to move from hot and sticky Madras to the pleasanter climate of the Deccan, though Cleghorn's (over-)fastidious character comes out in the replies from home to his (not-surviving) letters – his father chided him for aloofness over complaints about the 'dissipation of the mess'. In the autumn of 1844 Cleghorn was seconded to another Madras regiment (the 35th Native Infantry) based at Harihar 100 miles north-west of Bangalore, but by April 1845 he was back with the Madras Light Infantry, stationed at Trichinopoly far to the south in the Tamil country. At Trichy he was in 'medical charge of artillery' and had the additional duties of a 'large garrison', where at one point 'pestilence ... [raged] in the Cantonment' – the heat was intense, reaching over 100 degrees, and Cleghorn bought a web of China grass cloth (imported from Singapore) to make a cool summer jacket, a subject on which he would read a short paper to the British Association in 1850. The fibre came from the nettle *Boehmeria nivea*, and was known as 'chu ma'. There can have been no time for plant collecting in this peripatetic period of his life, but there was time for some botanical reading and one of the books he enjoyed on the many long marches was J.S. Henslow's *Descriptive and Physiological Botany*, of which he kept a copy in the top of his 'bullock trunk'.

2. COMPANY SURGEON, 1842–8

The Shimoga Period

Given Cleghorn's dislike of the rough and tumble of military life, he longed for a civil appointment. In July 1845 his hopes were answered with an appointment in the Mysore Commission – to the medical charge of the rural Nugger Division,[5] under its Superintendent, Captain Charles Francis Le Hardy. Given his short period of service, and the desirability of the job, this can only have been the result of patronage in Madras. Shortly after the medical appointment, the duties of which included the supervision of Indian vaccinators and of a hospital, Cleghorn was also made a magistrate, which added the responsibility for overseeing a jail in Shimoga. The magisterial role would have involved trials with five-man juries of local men, who would have spoken 'Canarese' (i.e., Kannada).[6] Cleghorn therefore applied himself to learning the language, which he mastered at least enough to be able to write a letter as a 'puzzler' to Professor Tennant in St Andrews.[7] Since the British takeover of Mysore, and the appointment of the four divisional Superintendents in 1834, great efforts had been made at the restoration of law and order, and bringing to justice members of what were believed to be criminal castes, especially one called the Koramers.[8] Each month in the 1840s there were, on average, 170 men in the Shimoga jail, of whom 18 were sick, attended by native doctors or 'hakeems'.

The Nugger Division of Mysore

One of the four administrative divisions of Mysore, this region covered some 6500 square miles (108 miles north to south, 116 east to west). Within it were 14 talooks including more than 5000 villages, and a population in 1849 of 610,105, which had increased by about 25% over the previous ten years – a period of political stability after a rebellion in the early 1830s. Cleghorn noted proudly that 'the cessation from war has enabled our resources to be devoted more assiduously to *the triumphs of peace*' among which he counted:

> the bridging of rivers and nullahs; the formation of Ghauts, by which the inland traffic reaches the coast; the abolition of transit duties; the extension of made roads; the increased number and better construction of labour-carts ... and the completion of other public works, as Moosaffir-Kanahs, Choultries, Travellers' Bungalows, &c.[9]

The road from Kadur to Shimoga with seven bridges had been completed shortly before Cleghorn's arrival, and among other 'improvements' were the destruction of shocking numbers of tiger and leopard (which predated on livestock and humans), and the repairing of irrigation tanks.[10] The area was a prosperous one, with a rich diversity of exportable products (both natural and manufactured), a good climate, and relatively fertile soil. Even in the year 1844, when the rains failed, the *ryots* had enough stored grain,[11] or enough money to buy it, so there had been no famine though large numbers of cattle had perished. The most prevalent illnesses faced by Cleghorn were fever and stomach/bowel complaints, to which he himself was not immune and years later he would still suffer from bouts of dysentery. Other prevalent illnesses that he would have had to treat were skin diseases and syphilis; and of minor complaints one that he noted was the removal of thorns from travellers' feet.

Part of the Division represented the remains of the higher parts of the ancient Hindu kingdom of Keladi, once a tributary state of the Vijayanagar Empire with Nayaka rulers; it also included areas formerly ruled by the Tarikere Polligars and two districts (Chikmagalur and Kadur) that had traditionally been part of the kingdom of Mysore. The Keladi kingdom had originally been based at Ikkeri and thereafter at the fort of Bidanur on the eastern

CHAPTER 2

Fig. 7. Nagar Fort, Karnataka. (Author's photograph, 2007).

edge of the Ghats (which was renamed Hyder Nagar when taken by Tipu's father, from which the British name 'Nugger' was derived). The fort of Bidanur was large, with stone walls (on which the star-fruit, *Averrhoea carambola*, was found growing by Cleghorn); its great circular bastions remain magnificent structures to this day (Fig. 7), though the adjoining and once great town is now a mere village. The final headquarters of the Keladi rulers was at Shimoga (or Shiva Moga) on the banks of the large, eastward-flowing Tunga River. In the 1840s Shimoga was a small town of about 7000 inhabitants, with a riparian fort, whose mud walls were already in decay in Cleghorn's time (the composite *Pluchea ovalis* grew on them, as recorded by Cleghorn's artist – see *Cleghorn Collection* no. 14). The only part of the Shimoga fort that survives today is a palace with attractive woodwork, said to date from the period of Shivaka Nayaka in the seventeenth century. Today this is Shimoga's only ancient secular building, though a few mid-nineteenth century houses still stand, some of which may have been familiar to Cleghorn.

The Shimoga region is still one of considerable natural beauty and interest to the botanist, and many of the plants drawn for Cleghorn can still be seen in the localities where he collected them. Rising from a narrow coastal strip are the still thickly forested Western Ghats, with eastern extensions including the peaks of Kudremukh, and Muliyangiri (in the Baba Booden Hills) rising to over 1890 metres. This seasonally very wet (annual rainfall at Agumbe 8300 mm), forested area is known as the 'malnad', through which only a few routes (Cleghorn's 'ghauts') cross the mountains, plunging down steep slopes to the coastal plain (then known as Southern Canara, now Dakshina Kannada) the major route being at Agumbe down which a 'carriage road' had been constructed only shortly before Cleghorn's arrival. In the north of the district one of the very few westward-flowing rivers (the Sharavati) takes a more precipitous course over a thousand-foot cliff, forming the most spectacular natural feature of the area the Falls of Gersoppah (Fig. 6). Since 1956, less romantically known as the Jog Falls, the cataract has been sadly reduced due to the impounding of the river by a hydro-electric dam upstream. This wet, forested area is still, however, the source of the three chief natural products of the district – cultivated areca nuts and cardamom, and native sandalwood. In the nineteenth century it also yielded lesser products such as wild pepper and gum from trees such as *Vateria indica*. Four species of bamboo provided famine food – as and when, with different periodicities, they synchronously flowered and fruited. In the *malnad*, rice, grown without irrigation, formed the staple diet of the people.

From the *malnad*, the land slopes gently eastwards, forming a plateau at about 800 metres, known as the 'maidan' (or 'byla shimi'); this region is drier (annual rainfall 610 mm at Honnali, three-quarters of it falling during the summer south-west monsoon) with a natural vegetation of scrub.[12] The prevailing red soils ('kengalu' or 'kempu'), through hard labour, yield a wide range of crops – the staple grain is *ragi* (finger millet – *Eleusine coracana*), traditionally intercropped with legumes (including pigeon pea, chick pea, cow gram) and the oil-crop 'til' (*Guizotia abyssynica*). Other crops known to, and illustrated for Cleghorn, and still grown around Shimoga, are *cholum* (*Sorghum*), millet (*Panicum miliaceum*), cotton, and the dye-plant safflower (*Carthamus tinctorius* – Cleghorn Collection no. 24). These crops are grown largely unirrigated, but tanks (with sophisticated sluice mechanisms) allow for local watering, especially for the cultivation of sugar cane. Livestock raised in the mid-nineteenth century included sheep, cattle (including a breeding establishment for the Amrit Mahal breed for the haulage of artillery) and camels; wool was used to weave 'cumblies' (black woollen blankets) that, with cotton textiles, iron/steel, and glass bangles (made from soda naturally efflorescing from the soil), were among the manufactured items exported from the area. The *ryots* lived in villages surrounded by hedges of thorny shrubs, on which Cleghorn would later write a paper[13] – for this purpose Hyder Ali had favoured a legume called Mysore thorn (now *Caeaslpinia decapetala*), but the hedge around Shikaripur was of *Capparis sepiaria*.[14] Isolated, scrub-covered, granitic hills rise above the plain, the most spectacular of which is the massively fortified Chitaldroog, then headquarters of the adjacent Division under the charge of Richard Dobbs of the 9th Madras Native Infantry, an evangelical friend of Cleghorn.[15]

Botanical exploration of Shimoga

Following the defeat of Tipu, as part of a rapid, one-man survey of the territories acquired thereby, commissioned by the Governor-General, Francis Buchanan had traversed the Shimoga district in April 1801. The 'statistical' information he obtained from local Brahmins (on agriculture, forestry, minerals and many other subjects) was written up as an extensive report, which the Company began to publish in London, in undigested form and without Buchanan's knowledge. This appeared in 1807, in three-decker form, as *A Journey from Madras, through the Countries of Mysore, Canara and Malabar*. That this contains very little botany is because Buchanan had intended not only to condense and revise his reports, but to publish as an appendix a Flora illustrated with at least some of the drawings of 131 species (made by an unknown artist who accompanied him, now at the Linnean Society of London). When on leave in Britain in 1806 Buchanan realised that he would probably never write up this Flora himself, so he gave his drawings and plant descriptions to J.E. Smith. However, Smith did nothing with them and so they were unknown to Cleghorn. It was at this time that Buchanan first became interested in the seventeeth-century *Hortus Malabaricus* of Hendrik van Rheede, which using a rich combination of vernacular sources (plant collectors and ayurvedic medical practitioners) and Dutch artists, documented a significant proportion of the rich flora of the Malabar coast, including many species that occur in the forested parts of Shimoga.[16] Cleghorn must also have known about this pioneering tropical Flora at this stage, but would have to wait several years before having his own copy of the first six volumes. At the same time as Buchanan's survey a much more detailed survey of Mysore was being undertaken (1799–1808), under Colin Mackenzie; the botanist for its first two years being Benjamin Heyne (who originally went to India as doctor to the Moravian Mission at Tranquebar), who studied and published on similar subjects to Buchanan though his particular interest was in minerals.[17] Cleghorn knew Buchanan's *Journey* and while in Shimoga was well aware that he was following in his footsteps. In the early 1830s another Edinburgh botanical surgeon, Alexander Turnbull Christie (1801–1832), had botanised in the Southern Maratha country, immediately to the north of the Shimoga district – the information was published in the *Edinburgh New Philosophical Journal*,[18] and his specimens are in the RBGE herbarium. But despite this earlier work, the flora of the Mysore Plateau and the Western Ghats was still rather poorly known in the 1840s.

Excursions from Shimoga

Almost nothing would be known of this period of Cleghorn's life (other than from a few brief scientific papers), but for the discovery and identification of nearly 400 anonymous drawings scattered through the RBGE Illustrations Collection. When reassembled, and put into chronological order, these were found to represent a day-by-day, visual, botanical diary that sprung from a suggestion made by Sir William Hooker who had encouraged Cleghorn to study 'one plant a day', which he did 'for a quarter of an hour … after the morning's duty in the jail and hospital was over'.[19] The drawings were made for Cleghorn by a 'Mahratta' draughtsman, and from the precise annotations of dates and localities it is evident that, rather than sending collectors out from Shimoga to bring plants for him to study, Cleghorn himself, between 1845 and 1847, made a series of journeys including both *malnad* and *maidan* areas. Some of these journeys may have been undertaken to supervise vaccination (Alexander Gibson had similarly travelled in the Deccan of the Bombay Presidency twenty years earlier), though in the regional headquarters of Kadur and Chikmagalur, Cleghorn may also have had magistrate's duties to perform; he also had to accompany his Superintendent (Le Hardy, followed by William Campbell Onslow) on an annual tour.

From the information on the drawings, five major excursions from Shimoga can be reconstructed for the period July 1845 to July 1847 (Fig. 8). Each lasted between one and four months, and they covered the whole district from Harihar in the north-east, to the Munzerabad Ghat in the south, with westerly side-trips to three of the ghats that led down to the coastal region – at Agumbe, Kollur and Gersoppah. The first excursion, of about two months, took him south to the Baba Booden Hills, Vastara, Chikmagalur, and the Munzerabad Ghat then eastwards to Kadur, Yagati, Tarikere and Ajampur. The second was longer, of about four months, starting at Sagar (a centre for sandalwood carving), then southwards along the eastern edge of the *malnad* through Nagar, Tirthahalli, Koppa and back via Chikmagalur and Tarikere. The third excursion was also extensive, of around three months, and started by heading east through the *maidan* to Channagiri, then south (Ajampur, Kadur) and west (Chikmagalur) reaching as far as Agumbe then back to Shimoga. The fourth excursion, of about a month, probably accompanying Onslow, was to the north of the district, to Shikarpur, Sorab and Sagar, from which Cleghorn made two side trips into the

2. COMPANY SURGEON, 1842–8

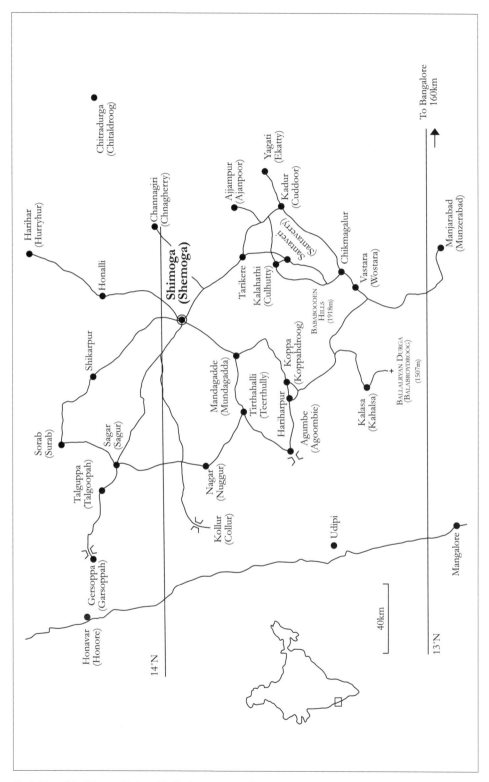

Fig. 8. Map of the Shimoga district, with Cleghorn's journeys between August 1845 and April 1847.

CHAPTER 2

hilly forests: to Gersoppah and to the newly constructed Kollur Ghat – where Onslow was shocked by the huge areas burnt for *coomri* cultivation on the Canara side of the mountains.[20] The final excursion was to Kadur again, where he spent almost a month. One further journey of Cleghorn is known – from his earliest days in Shimoga – when he accompanied Captain Le Hardy on his autumnal tour of 1845 to the north-west of the district. It was on this occasion that Cleghorn found wild gamboge between Gersoppah and Nagar.

Field botanist

Cleghorn's study of the botany of the Shimoga district doubtless took his mind away from uncongenial duties, unsympathetic colleagues (his father, a realist, expected that some would 'live in open sin', that is, with an Indian *bibi*), with no church closer than Bangalore.[21] Cleghorn's evangelical piety was church- and congregationally based, and it was his father who had to remind him that 'there is a temple made not with hands – the upright heart and pure – where God may at all times be worshipped in sincerity and truth'. In addition to the drawings made by his artist, Cleghorn also made herbarium specimens (said to represent 2000 species,[22] though this seems very unlikely and certainly only a tiny fraction of this number survive). It was a great satisfaction to the present author, having identified the plants in the drawings, to look for specimens of the same species in the RBGE herbarium and to find Cleghorn specimens cross-referenced with annotations such as 'D 157' – a reference to drawing number 157. In 1848 these specimens were taken home to Britain, where most were put into the Edinburgh University herbarium, though some duplicates were available for distribution through the Botanical Society of Edinburgh.

Fig. 9. *Gloriosa superba*, between Sagar and Keladi, Karnataka. (Author's photograph, 2007).

2. COMPANY SURGEON, 1842–8

Much can be deduced from a listing of the species drawn and the localities from which they were collected, of which only the briefest summary can be given here. The major provenances in the plains were Shimoga itself (some 156 drawings), Kadur, Ajampur and Santaveri – the collections represent a fairly comprehensive survey of what is not a rich flora. The many cultivated plants include fruit trees, cereals, pulses, cucurbits, oil-seed plants, ornamentals, even some medicinal plants, and the *tulsi* 'held sacred to Vishnoo' found in 'enclosures around Hindoo temples'. In the much more species-rich *malnad* areas Cleghorn barely scratched the surface – his major collecting areas in this habitat were the Baba Booden Hills (27 drawings), Vastara, Sagar, Koppa and Tirthahalli. Among the species recorded here were several of the region's useful trees, including *Vateria indica* and sandalwood. It was also in this region that he found a wild species of gamboge (now *Garcinia morella*), on which he wrote to his old professor of Materia Medica, Robert Christison, who analysed samples of it chemically and published his findings.[23] The most significant discovery, though it would not be realised for another seventy years, was also from this region, at Ballarayan Droog on 13 April 1846: a wild relative of the caper, which Cleghorn identified as *Capparis roxburghii*, but which in 1915 was described as a new species *Capparis cleghornii* (*Cleghorn Collection* no. 12).[24]

Development of useful resources: coffee, cotton and teak

A major concern of the EIC was to develop local natural resources, and introduce exotic ones, with a view to export. Onslow, who succeeded Le Hardy as head of the Nugger Division was, as an enlightened administrator, keen to assist in this project. Two such commodities were cultivated in the Division – coffee had been grown in the Baba Booden Hills on a small scale since its introduction from Yemen around 1670, but cultivation had recently been expanded. J.H. Jolly of Parry & Co. is said to have started a plantation at Shimoga in 1823–5, but the first major British plantation was set up by Thomas Cannon at the Mylimane Estate, Chikmagalur in 1832, and in 1843 Frederick Daniel Meppen had opened an estate near Yemmedoddi in the same area[25] – it was probably from one of these recent plantations, at Vastara (still a coffee-growing area today), that Cleghorn had a drawing of a coffee plant made in 1846 'cultivated under much shade' (*Cleghorn Collection* no. 25). Indigenous, diploid species of cotton were grown in the *maidan* (in the talooks of Honalli, Hiriyur, Tarikere and Channagiri), and it was one of these (*Gossypium arboreum*) that Cleghorn had drawn for his collection in Shimoga (*Cleghorn Collection* no. 26). But it was at this time that the EIC was trying to persuade Indian *ryots* to grow long-staple New World cottons suitable for spinning and weaving in Manchester – to be sold back to India in the form of woven cloth. At Kadur Onslow himself undertook an experiment with seed of American cotton supplied by Robert Wight from his experimental farms at Coimbatore, and Cleghorn would report on this work to the Botanical Society when back in Edinburgh in 1850.[26]

The edge of the Ghats to the south-west of Shimoga had traditionally been a source of supply of teak for the Mysore Ordnance Department, and Francis Buchanan had commented on a thick teak forest on the left bank of the Tunga river between Tuduru (now Tirthahalli) and Baikshavani Mata (now Mantagadde) when he had passed through the area on 1 April 1801.[27] Forty-six years later Cleghorn noted a serious decline, which he blamed on the practice of shifting, slash and burn, cultivation (known variously as 'kumari', 'koomri' or 'coomri'), and he brought the matter to the attention of his superior. As a result, on 5 May 1847, Onslow wrote a letter to the Mysore Commission, which shows that he had also been making observations of his own and was not entirely dependent on advice from the young Cleghorn in such matters.[28] The problem of the teak forests was not *coomri*, but the fact that they had been let to the highest bidder, with no subsequent control over what was felled (though teak, ebony and blackwood were supposed *not* to be cut). In Onslow's mind (though not apparently Cleghorn's) the serious consequences of unchecked *coomri* applied to other areas; it had become a major problem in the *malnad* forests, as seen when looking down from the new Kollur ghat, with tens of thousands of acres of forest destroyed. The problem came from increased pressure of population, and the taking of crops and burning in two successive years (rather than the traditional one, followed by a 12- to 20-year fallow) – this killed the tree stumps preventing forest regeneration and leading to soil exhaustion. Onslow's recommendations led to the banning of *coomri* in Mysore and Coorg (except in forests that had already been spoiled, and then only under supervision) – more than a decade before a similar ban was to be achieved (under Cleghorn's own recommendation) in the Madras Presidency. The other recommendation was that especially the Ghat forests should be kept under government control. Onslow had also sought advice from H.V. Conolly of Malabar on the (re)planting of teak forests and had himself collected teak seed which Cleghorn was asked to grow and from which Onslow made new plantations on the banks of the Tunga on the site of the former forests. Onslow similarly made plantations of *sissoo* (*Dalbergia sissoo*), the hard wood of which was used for making bullock carts and gun carriages. This was the dawning of Cleghorn's interest in forestry, so the outlines of his subsequent Indian career, including his interest in useful and cultivated plants, can be seen to have been established in his Shimoga period.

Farewell to Shimoga & Adieu to India

The first of Cleghorn's professional periods in India ended in sickness – Mysore fever. Like the similarly riverine Seringapatam that had led to Robert Wight's first sickness in 1825, Shimoga's was an unhealthy situation. The reason was then unknown, but while the riverbed in the dry season provided a habitat for some interesting plants (including *Crinum defixum*, Fig. 10) and for the seasonal cultivation of water melons ('a staple article of food during the hot months'), pools of stagnant water lurked between the rocky reefs. These formed an ideal habitat for the breeding of the *Anopheles* mosquito though, at the time, one contemporary theory (credited to Humboldt) was that it was the granite rocks themselves that were source of unhealthy *miasma*. Cleghorn returned to Bangalore to be examined by a doctor and it was there, on 14 September 1847, that the Mysore Commissioner, Major-General Mark Cubbon, signed a sick certificate allowing him three months in Madras.[29] As Cubbon was known for keeping open-house in Bangalore, Cleghorn may well have met him personally, and they would have dealings ten years later over the Lal Bagh garden. It must also have been in Bangalore that Cleghorn met Alphonso Bertie (who died there the following year). Bertie, an apothecary, had been one of Robert Wight's correspondents, and annotated three of Cleghorn's Shimoga drawings with native plant names.

Cleghorn reached Madras on 1 October and after five years in India was granted leave on 'medical certificate'. At the same time he resigned from the Mysore Commission, which meant that when he returned he would be back on the Madras establishment. Around this time he must have had some of his Mysore drawings copied and sent to Robert Wight in Coimbatore, as while Cleghorn was in Britain, and greatly to his delight, Wight published two of these (*Osbeckia hispidissima* and *Mitreola paniculata* – now *M. petiolata*) in one of his long-running botanical publications, the *Icones Plantarum Indiae Orientalis*.[30]

Fig. 10. *Crinum defixum*, Tungabhadra River, Honalli, Karnataka. (Author's photograph, 2007).

Fig. 11. The *Sutlej*, East Indiaman.
'In a Hurricane off the Cape on the Morning of April 1st 1848 at ½ past 3'. Hand-coloured lithograph by Thomas Goldsworth Dutton, printed by Day & Sons. (Sprot family collection) (above).
In Table Bay – right, under jury rigging as it arrived after the hurricanes; left, ready to sail following restoration. Watercolour by B. Jandrell, 'Marine painter of Cape Town'. (Sprot family collection) (below).

3
First Furlough & the British Association Report, 1848–1851

Wreck of the *Sutlej* and the Cape of Good Hope

With memories of his outward voyage from London in 1842, and its pleasant stop at Madeira, Cleghorn was probably looking forward to the journey home when he boarded the ship *Sutlej* in Madras on 18 February 1848. The vessel was a splendid wooden sailing clipper, one of the Blackwall Frigates, built only the previous year by Richard & Henry Green at Blackwall, opposite Greenwich, for the firm's London to Calcutta route.[1] The ship was three-masted, 1150 feet in length and weighed 1159 tons; it had set sail for its return voyage from Calcutta under Captain A. Parish two weeks earlier, laden with a cargo of 'sundries'. The passengers included 23 named individuals (i.e., members of the middle-class, some of them military, others civil servants) with 28 of their children and 11 servants, together with a detachment of 83 men of Her Majesty's 50th Regiment (with seven attendant women and eight children).[2]

All went well until 1 April. The ship was off the coast of Mozambique (36° S, 24° E), when, at midnight, a north-westerly gale struck, which in two hours had reached hurricane force. The description of the events, taken from the ship's log as reported in the South African newspapers, makes for startling reading and one can only imagine the terror faced by the men, women and children as the ship was battered by the gale. The bowsprit was broken, as were all three masts – main, fore and mizzen – which, with their rigging, had to be cut free and cast adrift into the seething ocean. Water poured in through one of the stern ports, through the starboard quarter gallery doors, and through the poop skylights; only by dint of the soldiers' manful attendance of the pumps was the ship saved. The wind did not slacken until 10 o'clock the following morning, and to lighten the burden some of the cargo was ditched, but the danger was not over for the stricken ship. On midnight of 6 April the *Sutlej* survived a second hurricane and four days later the ship eventually limped into Table Bay, at the aptly named Cape of Good Hope, under improvised 'jury' rigging. These details are known not only from the press accounts, but from three illustrations owned by Cleghorn – a lithograph by Thomas G. Dutton showing the magnificent *Sutlej* as launched, another Dutton lithograph showing the ship as imagined dismasted and adrift in a heaving Indian Ocean; and a charmingly naive painting by B. Jandrell, Marine Painter of Cape Town, showing the ship in Table Bay both as it arrived and as it left (Fig. 11).[3] It is sobering to think of the sufferings of the passengers in what they must certainly have viewed as their last hours on earth, but also of a minor miracle that Cleghorn's specimens and drawings survived without so much as a spot of water damage. The packers in Madras must have done their job well. According to a later account five 'able-bodied seamen' were lost,[4] though this was not reported in the contemporary press coverage, which stated only that several of the crew were 'much injured ... one with the loss of his arm'.

On reaching Cape Town it was found that, despite the battering sustained, the body of the ship was reparable. Rather than make other arrangements Cleghorn decided to wait until it was ready to resume its interrupted journey, which gave him two months for botanising in what is one of the richest floristic regions of the world, so that on returning to Edinburgh he was able to present at least 17 numbered specimens of Cape plants to the Botanical Society. It is not known exactly when the *Sutlej* resumed her voyage but by early July (by way of St Helena) London had safely been reached.

CHAPTER 3

Return to Britain & Evangelical Fervour

Religion played a central role in Cleghorn's life that is necessary to address. It might have been expected, as was the case for Alexander Gibson and Robert Wight (both of whom were Christian), that little evidence relating to such a personal matter would survive. But for Cleghorn there is much of relevance in letters written to him by his grandfather and father, in his own letters to J.H. Balfour but, especially, in a remarkable devotional common-place book in the Cleghorn papers at St Andrews.[5] This intriguing volume, used over a 40-year period, started out life very differently – to record medical case notes in Wards 11 and 13 of the Royal Infirmary, Edinburgh when Cleghorn was an undergraduate. Richard Grove must have seen this notebook as it is the only possible source for his view that Cleghorn had an 'almost fanatical adherence to a personal and idiosyncratic evangelical revivalism';[6] a view that cannot be upheld. Survival of such material is extremely rare for a botanist, but the contents are no more indicative of fanaticism or idiosyncrasy than the views shared by many at the time, both in Scotland and India – great popular evangelical revival was in full sway in the 1840s, and not only among the working class. Missionary activity was also rampant at the time in India: for example, in Madras in 1848 the Anglican establishment, under a bishop and archdeacon, had no fewer than 11 licensed Chaplains and 107 Assistant Chaplains; of other denominations there were Roman Catholics, Armenians and the Church of Scotland, to say nothing of a plethora of missionary societies: American, Baptist, Wesleyan, Basle Evangelicals, Dresden Lutherans, and various Anglican organisations (the Church Missionary Society (CMS), the London Missionary Society (LMS), the Society for the Promotion of Christian Knowledge, the British and Foreign Bible Society, and the Madras Tract & Book Society).[7] The names of many of Cleghorn's friends appear as subscribers and office bearers for these societies, so his interests were by no means unusual. There were, however, still (as in the early days of EIC, when missionaries had been forbidden), contrary views and Cleghorn would later (in 1859) write that 'the Anti-Missionary party are always glad to run down public Officers' who supported such activity.[8] The fact that Cleghorn's beliefs were not idiosyncratic is shown by much of the content of the 'devotional notebook' – extracts from popular, contemporary, evangelical texts (including works by the Rev. James Hamilton, Richard Marks and W.H. Dorman) and histories such as J.-H.M. d'Daubigné's 'History of the Reformation', but also from more 'orthodox' classics such as the 'Night Thoughts' of Edward Young (a believer in extra-terrestrial life), Archbishop [of Glasgow] Robert Leighton's 'Rules and Instructions for a Holy Life', and Thomas Babington Macaulay's account of the Church of England taken from his 'History of England'. There are also notes taken by Cleghorn from sermons, copied-out hymns (by, among others, Charles Wesley, Augustus Toplady and Amos Sutton), and many small tracts of the sort printed in their tens if not hundreds of thousands to satisfy huge popular demand. All belief is, of its very nature, 'personalised', and in Cleghorn's case it is possible that a life-threatening event may have led to an experience of personal revelation. When, over a decade later, Cleghorn got around to marriage, his wife was chosen from a leading evangelical Edinburgh family.

Typical of their class and of their intellectual interests, the Cleghorn family were traditionally conventional members of the Moderate wing of the Church of Scotland with leanings towards Episcopalianism. Serious, and clearly heartfelt, advice was given to the young man in epistolary form by both his grandfather and father – on the necessity for a reverence for a Creator God and His revelation in Christ, as described in the Gospel. They also held a

3. FURLOUGH & THE BRITISH ASSOCIATION, 1848–51

belief in an afterlife where social distinctions no longer existed, but in which an individual's status depended on conduct during their earthly life. In this is no hint of Calvinist predestination, and Cleghorn seems not to have been tempted into that narrow creed. Curiously his grandfather considered that Creation was most unlikely to be restricted to planet Earth and thought that sentient life probably existed on other stars (a view shared by Thomas Chalmers).

Not for Cleghorn, in his first period in India, the dissipated life of his unfortunate younger brother Allan; as a young man Hugh was clearly extremely pious, attending prayer meetings and being uplifted by the preaching of several notable evangelicals in Madras and Bangalore. Doubtless the way for this had been prepared during undergraduate days in Edinburgh, for which no evidence survives, though, of an older generation with whom he was interacting botanically, Greville and Balfour were certainly fervent evangelical Christians. In the evangelical tradition the Divinely-inspired writings of the Bible have a unique and absolute authority and Cleghorn wrote a note on the 'sufficiency of scripture':

> I confess myself little under the influence of human teachers, my being thrown exclusively on the Bible for a scheme of doctrine furnished me not only with a satisfactory one, but showed me so much of the inexhaustible riches of wisdom & knowledge hid in Christ – & of the Holy Spirit all sufficiency to take of those things & shew them to the humble diligent prayerful inquirer that in most cases of difficulty in stead of asking "What say the Commentators" I put the question "What says the Lord" For an answer I search the written word – for a Commentary – I study his visible work.[9]

Despite this he had an ear for a good sermon although, somewhat surprisingly, the preachers Cleghorn recorded as having heard in his first Indian period were nearly all Anglican, and it is curious that from this stage there are no references to the Church of Scotland (which had had a missionary in Madras since 1840), or to the Free Church to which he would later become attached and which (in the person of the Rev. John Anderson) was already active in Madras. Cleghorn was already in India during the final steps of the movement that led to the 1843 Disruption in Edinburgh (the 'Reel of Bogie', the confrontation of Chalmers with Lord Aberdeen and Lord Brougham's ruling against the anti-intrusionists – see Fig. 30), which will be discussed later in the context of the Cowan family, so it cannot have been direct experience of a surge of widespread emotion that inspired Cleghorn to go against what would have been expected from one of his caste. They may nearly all have been Anglican, but the preachers who inspired him were of several affiliations: the Rev. Alured Henry Alcock was a Company chaplain; two belonged to the CMS with its church in Popham's Broadway, Black Town whose incumbent from 1830 to 1846 was the Rev. John Tucker, followed by the Rev. Thomas Ragland (until 1857), both referred to by Cleghorn. But he also attended the LMS chapel at 10 Davidson Street, Black Town, of which the Rev. William Hoyles Drew was incumbent from 1843 to 1852. After his transfer to Mysore Cleghorn heard another LMS preacher, the Rev. Edmund Crisp, who was at Bangalore from 1840 to 1846.[10]

The last of the preachers encountered by Cleghorn in Madras was not Anglican: the remarkable dentist and missionary Anthony Norris Groves, who had been brought to India in 1833 by the engineer Arthur Cotton. Groves was one of the founders of the Plymouth Brethren but was also interested in agriculture and in 1850 won a prize from the Madras Agri-Horticultural Society for sugar made at his son's Astagram works near Seringapatam.[11] A letter written by

CHAPTER 3

Groves to Cleghorn in August 1848, shortly after the latter's return to Britain, shows that they were close, though Cleghorn seems primarily to have sought Groves's advice as a southerner as to where best to spend the winter – Hastings or Torquay. After the Mysore fever and a severe buffeting on the Indian Ocean, the returned surgeon was clearly feeling too fragile to face the chilly winter or the spring sea-mists ('*haars*') of the East Neuk of Fife. Appended to the letter is a touching note by Cleghorn on the influence of Harriet Groves (the missionary's second wife): 'The light & influence of her Xn. example have been seen & felt all around her. The spirit in her bosom has spread to many others',[12] which provides an example of the emotional language that Richard Grove may have considered fanatical. Despite this it appears that Cleghorn was not tempted by Groves's extreme Calvinism. Care is required as, like much of the highly personal expressions at the end of the devotional notebook, the writing is extremely difficult to read, nevertheless, what it *appears* to say is: 'Plymouth Brethren great sheep stealers only take the chosen lambs – not meddling with goats'![13]

After a period in the summer, staying in London with the Clarke family at 17 Kensington Square (the house next door was occupied at the time by John Stuart Mill, and later by the composer Sir Hubert Parry), the attractions of Torquay prevailed over Hastings.[14] His sisters Isa and Rachel must also have been keen to escape Fife and joined him, and the three siblings were in Torquay from December 1848 until April 1849. At this time the Devon resort was popular with the upper classes; recently connected to London by rail, Torquay was known as the 'Queen of Watering Places', but family finances must have ruled out one of the fashionable hotels and their address was 5 Torwood Row, perhaps a lodging house. At this time the family of Cleghorn's future wife Mabel Cowan also over-wintered in Torquay, so the connection between the families might have begun at this time. As if fever and shipwreck were not enough Cleghorn was now struck down with dysentery and it appears that in this period of extreme weakness he had a profound religious experience. The evidence is, once again, in the scrappy notes at the end of the devotional notebook, but seems to point to a dramatic intensification of already-held beliefs – on the importance of a living, loving relationship with God, through Christ and the Holy Spirit, who 'pour into our hearts His healing balm of His own divine comforting and sustaining sympathy'. At the same time, remembering that 'in our severest chastisements he is still chastening as a loving Father – the Child whom he loveth, only let us bear the Rod – & him who holdeth it'. In the following notes, which once again stress the fundamental importance of the Bible, the marine imagery suggests that the near-death experience on board *Sutlej* may, perhaps, have played a part in this episode:

> the sects of the true Church may be compared to the Billows of the ocean – separate yet one, so the Xn. living in the world may be compared to a ship floating in the ocean – the ship must be the sea – the Xn. must live in the world & it is only when the water gets into the ship that we have cause of alarm – so when the Spirit of the world enters the heart – the mischief has been effected.

> And so, the Xn. must guide his whole course by the Compass of Scripture – or he will strike upon the rock of presumption, or founder in the whirlpool of ignorance & infidelity – the thoughtful Christian ever consults his Bible as the Chart which is to point out his various dangers across this perilous ocean of life.

> Like a ship in the storm the soul loses one stay after another – The sails of love & gratitude are torn: the rudder of faith unshipped: the anchor of hope broken & the compass of the word too much neglected.[15]

Such extreme views were not to the taste of Cleghorn's father, Peter, who, clearly worried about his son's mental state, gave advice from a more rational, Enlightenment tradition:

> while we are in this world we must attend to the ordinary affairs of it otherwise we become useless to ourselves & others & become the laughing stock of all – this is a very different matter from being the slave of Mammon – neither will the human mind admit of being constantly upon the stretch upon deeply religious or metaphysical subjects – no ordinary mind can dwell too long upon these without being weakened so I advise you to remember the couplet of Sir William Jones, altered, I think from Milton
>
> > "Six hours to Law – to soothing slumbers seven
> > ten to the world & all to Heaven"
>
> you see there is still one hour to be accounted for – & this, one of the best of men, thought enough to be set apart for religious meditation.[16]

Peter was also concerned about his son's denominational affiliation, which must, by this time, have turned irrevocably to the Free Church of Scotland. When sending a Coutts banker's draft for £30, his father asked him to accept it, 'but only give as little of it as may be to the "Free Kirk" who are the sturdiest beggars in the world'.[17]

This Torquay experience stayed with Cleghorn for the rest of his life, and lies behind the work he took up on return to India, especially in his teaching of Indian and Eurasian medical students, and his advocacy for the need for *medical* missionaries to be sent out from Scotland. These roles (which surely had roots in St Andrews and Edinburgh – in the work of John Adamson, P.D. Handyside and W.P. Alison – as much as in Bible study), and the opportunities they provided, together with his eschewing of rigid Calvinism, suggests a warm humanity, and a concern for the well-being of the minds and bodies of the Indians with whom he interacted rather than an out-and-out desire to convert their souls to his own form of Christianity.

It appears that the company of his sisters, the mild climate, local expeditions to places like Sidmouth, and the clearing up of the dysentery, had a healing effect on Cleghorn by time they all left Torquay. Certainly he returned there the following spring of 1850 (when he stayed at Hearder's Hotel[18]) and again during the third winter of this furlough. A souvenir of this period is an album of seaweeds acquired in Torquay in March/April 1850.

Excursion to Northern Ireland

Unlike his father, his grandfather, or his 'fragile vessel' of a sister Isabella, Cleghorn seems not to have been tempted by Continental travel, but during his furlough he did take part in one longer expedition – to Northern Ireland, in August/September 1849.[19] Besides Cleghorn the party consisted of Dr John Merriman and his son Thomas. Dr Merriman was a neighbour of the Clarkes at 45 Kensington Square conveniently close to Kensington Palace where his father had been physician to Princess Victoria and her mother the Duchess of Kent, and jointly with his brother James, Apothecary Extraordinary to Queen Victoria. The excursion seems to have come about through two of Cleghorn's Mysore friends: Captain Gardiner Harvey one of the Assistants to the Commissioner of Mysore was married to Rosetta, daughter of the Rev. Robert Gage whose family had for a century owned the island of Rathlin, and Captain Richard Dobbs, who was brought up in Co. Antrim where his father the Rev. Richard Stewart Dobbs had lived at Bay Lodge, Glenariff and whose family was also connected by marriage with the Gage family.

CHAPTER 3

Fig. 12. Rathlin House, ink sketch by Mrs Catharine Gage. (Reproduced in Dickson, 1995).

The island of Rathlin, which on the map looks like the hook of a candle snuffer, is fascinating geologically – when seen approaching from the south a slab of brown basalt sitting on a basement of white chalk – and famous as the location for the story of Robert the Bruce and the spider. The Merriman party stayed at the Manor House (Fig. 12) with the Rev. Robert Gage, incumbent, as well as owner, of the island and his wife Catharine (Fig. 12).[20] In the 1840s two of the eight daughters of the family, Barbara (1817–1859) and Catherine (1815–1892), painted attractive studies of plants and of local birds and fish – three volumes containing 316 of these drawings are preserved in the Linen Hall Library, Belfast.[21] Catherine had also made a list of the plants of the island, of which Cleghorn took a copy back to Edinburgh.[22] The list included 204 dicots and 21 monocots – the most interesting of the latter being the pipewort (now *Eriocaulon aquaticum*), just about possible on distributional grounds, though on a recent visit the present author was unable to find it in any of the apparently suitable lochans, and it is possible that Miss Gage mistook non-flowering *Litorella* for it. On the nearby Antrim mainland the tourists visited two manifestations of the geological formation for which the region is best known: the columnar basalt cliffs of Fair Head (where they found the stonecrop *Sedum reflexum*) and the Piranesian Giant's Causeway.

Edinburgh & the Botanical Society (Part 2)

During his furlough Cleghorn spent appreciable amounts of time in Edinburgh, where he still had family – Aunt Campbell (reduced to living in Lauriston Place, following shady financial dealings on the part of her husband) and Aunt Wyllie (in Carlton Terrace), also Lady Foulis his childhood escort from India. In Edinburgh there were also friends from medical-student days including Andrew Halliday Douglas, the botanists R.K. Greville, J.H. Balfour and George Lawson, and also missionary ones including Dr John Coldstream. Greville and Balfour straddled these religio/botanical boundaries and Balfour in particular, an evangelical Episcopalian, became a close friend. On 20 July 1851 Cleghorn stood godfather to Balfour's eldest son, Andrew Francis, who was baptised by the Rev. D.T.K. Drummond at St Thomas' Church of England Chapel (a Neo-Norman structure by David

Cousin in Shandwick Place opposite Syme's consulting rooms).[23] In 1867 Andrew would join the Royal Navy as a midshipman, and he served as one of the Naval crew on the great *Challenger* Expedition of 1872–6, the scientific research of which was directed by his father's friend Charles Wyville Thomson.

There had, however, been changes. When Cleghorn left for India in 1842, Balfour had recently gone to the chair in Glasgow vacated on Hooker's appointment to the directorship of Kew. But after an unedifying rivalry with Joseph Hooker (more the fault of Edinburgh Town Councillors than either of the candidates), Balfour had returned to Edinburgh to succeed Robert Graham as Regius Keeper and Professor of Botany at Edinburgh University in 1845. With the support of the Commissioners of Woods and Forests, Balfour embarked on an ambitious programme to develop RBGE into a research as well as a teaching institution, with a major public profile (Sundays strictly excepted) and the formation of what was effectively a national herbarium and library. A larger classroom to designs by Robert Matheson was opened in May 1851, and the following month a Museum of Economic Botany in Graham's old teaching room (built by Robert Reid and converted by Matheson). Because Balfour used it as a major vehicle for communication and publishing, the Botanical Society of Edinburgh (BSE) was closely involved in these developments.

Since the time of Cleghorn's undergraduate involvement the BSE had grown into a national society every bit as prestigious as its London cousin (its junior by five months). It was significant not only as a focus for botanical collectors, but for the dissemination of botanical knowledge through its *Proceedings* and *Transactions*. Meetings were widely reported not only in the Society's own publications (which appeared only after long delays), but in popular periodicals including the *North British Agriculturist* and the *Botanical Gazette*. During his leave Cleghorn attended the Society's monthly meetings and was elected a Vice President for the year 1850/1.[24] Winter meetings (November to April) were by now held at 6 York Place, which, showing the close links between botany and godliness, also housed the offices of the Edinburgh Bible Society and the Sabbath Alliance (of which Greville was Secretary). Summer meetings (May to July) were still held at RBGE, and it was at one of these, on 13 June 1850, that Balfour was able to announce that Robert Wight (in the fourth volume of his *Icones* of 1848) had honoured their mutual friend with the genus *Cleghornia* (see Appendix 2).[25]

At seven BSE meetings between July 1849 and June 1851 Cleghorn is noted in the minutes as having read papers or exhibited drawings or herbarium/carpological specimens. The first was on 12 July 1849 when he announced his return by showing some of the 500 Shimoga drawings, suggesting how highly he valued them. The papers, all fairly short, reveal his particular interest in economic plants – those on the hedge-plants of South India, and on the 'grass cloth' of *Boehmeria*,[26] were dry runs for presentations at the 1850 British Association meeting, and with the former he showed more drawings, illustrating hedge-plants such as *Opuntia* and *Euphorbia*. On 9 January 1850 he read Catherine Gage's paper on the plants of Rathlin,[27] and Cleghorn's first interest in botanical history and biography was demonstrated on 9 January 1851 with a biographical note on J.P. Rottler, the Tranquebar Missionary who had taken part in his grandfather's Ceylon adventure in 1796.[28] Two of the papers have links with Robert Wight – one was on the planting of American cotton at Kadur,[29] and a taxonomic one on new plants described by Wight from Cleghorn's Mysore collections (of which he showed the related specimens).[30] However, it was the last of these papers, read on 19 June 1851, that proved the most prophetic of his future career, as it concerned the Government teak plantations of Mysore and Malabar.[31]

The cotton paper led Cleghorn to the herbarium to study the taxonomically tricky genus *Gossypium*, resulting in some astute remarks on the need for a broad species concept. While in Britain Cleghorn took the opportunity to study specimens in other herbaria – that of George Walker-Arnott (probably in Arnott's house at Arlary in Kinross-shire), and Paul Hermann's pioneering, seventeenth-century Ceylon collection in the British Museum. It was during this period that, with Thomas Anderson (then a medical undergraduate, later to be a Superintendent of the Calcutta Botanic Garden), Cleghorn recurated the Indian plants in the University herbarium,[32] which in 1838 had been amalgamated with the already much larger one of the BSE. Although it would later move to RBGE (to what had been the exhibition hall of the Caledonian Horticultural Society), at the time of Cleghorn's curation the herbarium was still housed in the University.[33] The collection was already rich in Indian material – including an early collection of William Roxburgh from his Coromandel days sent to the Royal Society of Edinburgh in 1789, Francis (Buchanan-) Hamilton's own beautifully mounted and well-documented set of specimens from his Bengal Survey of 1810–14, and Lady Dalhousie's Indian herbarium. Smaller Indian collections included Alexander Turnbull Christie's plants from the Southern Mahratta country, and recent specimens sent to the Botanical Society from southern India by Robert Wight and Cleghorn himself (from Mysore and the Cape of Good Hope).

St Andrews & Photography

Although Cleghorn spent time in Edinburgh during his furlough, he must have made frequent visits to St Andrews to visit his family. With the opening in 1847 of the railway between Burntisland and Tayport (with stops at Cupar and Leuchars) it was much easier to get there from Edinburgh, and in 1850 came Sir Thomas Bouch's roll-on roll-off train ferries between Granton Harbour and Burntisland, though passengers went on paddle-steamers such as the *Auld Reekie* or *Thane of Fife*. The small town itself had also changed during his absence, due largely to the efforts of several larger-than-life personalities.[34] Sir David Brewster, the physicist and encyclopaedist, had major interests in light and optics, but was also deeply religious and with interests in romanticism that he combined in what is a peculiarly Scottish way: for example in scientific explanation of phenomena he called 'natural magic', and in the cult of martyrdom, whether the result of scientific (Galileo), or (as in the Scottish Reformation) of religious, belief. It is hard not to see the latter lurking not far beneath the story of the Disruption. In 1838 Brewster had been appointed Principal of the United College and moved with his family to St Andrews. There they lived in part of St Leonards, the large house occupied by the Cleghorn family from 1800 to 1806, close to the dramatic ruins of the great cathedral, and today part of a well-known girls' school. The other part of St Leonards was occupied by the family of Hugh Lyon Playfair, a man with an Indian military background known as 'the Major'. Playfair became Provost (*Anglice*, Mayor) of the town in 1842, a position he would hold for the rest of his life. In the grounds of St Leonards Playfair developed a remarkable garden, full of inscriptions and automata, that has curious premonitions of the twentieth-century philosophical gardens of Ian Hamilton Finlay and Charles Jencks, which, as a major tourist attraction, must certainly have been known to Cleghorn. Though both prickly characters the two neighbours, Brewster and Playfair, were largely responsible for the renaissance of the small, run-down town and university that took place, ironically from the perspective of this biography, just as Cleghorn had left it. This period has recently been beautifully described by Robert Crawford – the foundation of the

3. FURLOUGH & THE BRITISH ASSOCIATION, 1848–51

Literary and Philosophical Society and the early history of photography, linked with the strange coincidence of the contemporary refuge in the town of Robert Chambers where he secretly wrote the scientific shocker *The Natural History of the Vestiges of Creation* at Abbey Park, just the other side of the Abbey precinct wall from St Leonards. Crawford also touches on the religious ferment of the Disruption, in which Brewster was heavily involved, and that led to his ostracisation by the conservative professoriate, and later to his appointment, in 1859 (aged 78), as Principal of Edinburgh University, where Free Church principles were more acceptable.

Almost as soon as he arrived in St Andrew Brewster had started to build a new, north range of the United College, and (at a meeting which took place in the University – now King James – Library on South Street on 6 April 1838), to start a Literary and Philosophical Society (he had been a founder of the British Association seven years earlier). The professors all signed up, but so did many of those old India hands already described – no fewer than seven of the 49 founder members were former EIC employees, including The Major and his brother George Playfair, Dr George Govan, and Patrick Cleghorn (interestingly using his formal Christian name) – others known to Hugh Cleghorn were his old French teacher Samuel Messieux and Dr John Adamson.[35] Adamson is the key character and retained the friendship of the Cleghorn family while Peter, like most people, had spats with Brewster, and also, it would appear, with Playfair.[36] Adamson was appointed curator of the new society's museum, which until 1912 was jointly run with the University and housed in the new north range of the College (the room now called Upper College Hall). The collections included both natural history and the arts – among the latter was John Knox's pulpit (now in St Salvator's Chapel), and, intriguingly, some 'Indian idols' doubtless brought back as souvenirs by one of the returned Company officials.

It was Adamson's expertise in chemistry, together with Brewster's friendship with William Henry Fox Talbot that led to the productive experimentation over the next few years during which St Andrews became the unquestionable 'Headquarters of the Calotype'.[37] The first of Fox Talbot's ethereal sepia images were shown at a meeting in 1839. Playfair also made many experiments though, despite dogged persistence, and the baroque complexity of the chemical processes, with poorer results than Adamson. The first great successes of the group came in 1842 – with both landscapes and portraits – which Crawford analyses in terms of content and meaning as records of mortality (geology and crumbling ruins), and as social documentaries related to what might be called Adamson's and Playfair's 'clean-up St Andrews' campaign based on Adamson's reports on sanitation and his recording of mortality statistics. But from a wider perspective it was the result of Adamson's teaching photography to his younger brother Robert that is better known. Brewster had suggested that photography should be used to record 457 of the key players who had taken part in the Disruption, as an aid to David Octavius Hill in the painting of an epic canvas of the scene at Tanfield (almost next-door to RBGE), the signing of the Act of Separation and Deed of Demission by which the Free Church was sundered from the Church of Scotland. A canvas of enormous size, twelve feet by four feet eight, described by Crawford as 'profoundly democratic' (see Fig. 30).[38] (Though the Disruption could perhaps more accurately be denoted as anti-establishment: against the ecclesiastical patronage wielded by landed interests and the legal system that backed this up). And despite the taint of martyrdom behind the act, this democratic element of the whole destructive process should never be forgotten, nor should it be forgotten that Cleghorn (though by this time in Mysore) enthusiastically, if metaphorically, 'signed up' to this new church going against class loyalties. As a result of Brewster's suggestion, and the technical expertise of the Adamson brothers, the partnership of 'Hill & Adamson', that lies behind this painting is world-famous, though Robert succumbed to tuberculosis only five years later, in 1848, aged 27, long before Hill had finished his painting, which took 23 years to complete.

Until recent work on early photographic history Dr John Adamson's major contribution had been almost entirely forgotten, at least outwith St Andrews; but he continued to take superb photographs (especially portraits) long after his more famous brother's death, and probably until his own, which took place in St Andrews in 1870. In October 1845 Peter Cleghorn had sat for Adamson and sent the resulting calotype image to Hugh in Mysore (Fig. 13)[39] – one can only imagine Cleghorn's sense of wonder at receiving a precise, if delicate, image of his father's features, five thousand miles away, examined under a brilliant Indian sun. A 'hard copy' of a piece of world-altering technology, sent from one of its key centres of development, a tiny, ruin-strewn town of learning, perched on a cliff above the waters of the North Sea, halfway across the world to one of the more remote, tropical corners of an empire approaching its zenith.

Cleghorn's own sitting for Adamson had to wait until he returned to Scotland on his first home leave. From the size of the plate, it was most probably in 1850, in St Andrews, that he sat for his old friend (Fig. 13).[40] That Cleghorn was closely involved in these networks is proved by the presence of his portrait in an album of calotypes assembled by Brewster's first wife Juliet,[41] with copies/variants in Adamson's own album in the National Museum of

Fig. 13. Calotypes by Dr John Adamson.
Peter Cleghorn, 1845. (Courtesy of the University of St Andrews Library. Alb-6-118-1) (left).
Hugh Cleghorn, c. 1851. (J. Paul Getty Museum, Malibu. Brewster Album (.7)) (right).

Scotland. This image shows a handsome young man of 30, wearing a jazzy chequed waistcoat, eyes modestly lowered and absorbed in the close examination of a plant specimen, seated on a leopard skin that had doubtless fallen to his gun in Mysore. Cleghorn's inclusion in this family album suggests that Brewster took the serious, younger man under his wing, which must, at least in part, lie behind Cleghorn's involvement in the 1850 British Association meeting in Edinburgh. On a later leave, in 1865, Cleghorn again sat for Adamson, resulting in a decidedly coy image of an older, bearded visage in the St Andrews University collection (Fig. 44).[42]

Cleghorn retained interests in photography throughout his career in India, though he seems never to have become a practitioner himself. Although he would become a member of the Madras Photographic Society, his main role both in southern and northern India was as a collector of photographs by others. These appear largely to have been 'work-related' – of landscapes, forests and economic botany – but unfortunately none of this collection has survived, although the author saw some of its remnants, large-format photographs of Himalayan forests, at Stravithie in 1998.

A Family Wedding

On 17 September 1850 an important family event took place at Stravithie/Wakefield House. Cleghorn's sister Rachel Jane was married to Alexander Sprot (1824–1854), heir to the estate of Garnkirk in Lanarkshire. Sprot was a grandson of the Rev. George Hill formerly Principal of St Mary's College, St Andrews, and they were married by the groom's uncle, Alexander Hill, Professor of Divinity at Glasgow.[43] At the time this marriage might not have appeared particularly significant dynastically, as there was still plenty of time for Hugh to marry and produce an heir for Stravithie, but it took Rachel off the family account book, as the Sprot family was wealthy. The Garnkirk estate of 1400 acres, had been purchased in the early nineteenth century by Alexander's great-uncle Mark, whose nephew Mark, Alexander's father developed its fireclay deposits. This fine white clay was used to make utilitarian objects such as chimney pots, but also handsome classical urns that were sold from a shop in Glasgow designed by Alexander 'Greek' Thomson.

Much can be told about the marriage from two portraits that survive in the Sprot family. The first, from 1851, is a chalk drawing by J.F. Thorpe made in Southampton showing Rachel, her handsome young husband and Janet Douglas, their first-born, draped in a shawl and sitting on her mother's lap as a tiny baby. Janet does not appear to have survived infancy, but she was not the only one to suffer ill health. That the drawing was made in the great southern port suggests that the family was on its way south for the winter, so perhaps Alexander was already suffering from something like consumption. The young couple had a second child, Alexander, born in 1853, whose health was fortunately robust, but his father succumbed in Pau, aged only 30, in 1854. Pau in the foothills of the Pyrenees in southern France was something akin to an Indian hill station for affluent Britons with weak health, where they could play golf, ride to hounds and admire the snowy mountain range.

That this marriage was a love match is suggested by the second of the portraits, a large oil, which shows Rachel, probably in her fifties, a commanding figure though with a kindly face, as if from the pages of Henry James, still dressed in widow's weeds. Letters in the Cleghorn archives reveal an extremely dashing hand, which suggests a confidence that allowed her to

CHAPTER 3

fight off the unwanted attentions of the fortune hunters whom she must surely have attracted. But, loyal to Alexander, she never remarried, which meant that both the Cleghorn and Sprot fortunes, and the lands both of Garnkirk and Stravithie, would devolve upon her only son Alexander. Nothing more is known about Rachel, though at the time of the 1861 Census she was living in St Andrews at 6 Playfair Terrace with her seven-year old son; she disappears from the censuses of 1871 and 1881, so may have been living abroad. In 1891 her home was in fashionable Mayfair at 5 Norfolk (now Dunraven) Street, just off Park Lane, though modestly so, with only a cook and a housemaid, and it was probably here that she died, aged 70, in 1893.

Fig. 14. The Twentieth British Association Meeting, Edinburgh 1850.
Sir David Brewster. Steel engraving by W.H. Mote, after a 'Talbotype' by Henneman & Malone (above).
Sir David Brewster delivering the opening address in The Music Hall, Assembly Rooms. Wood engraving, *Illustrated London News* 10 August 1850 (below).

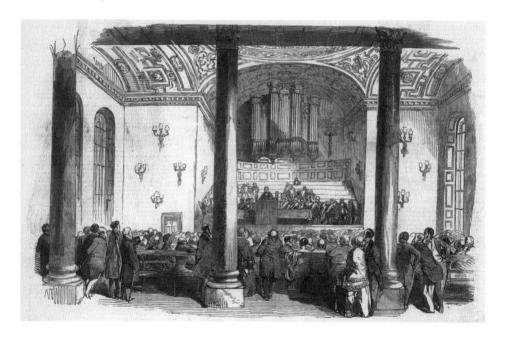

3. FURLOUGH & THE BRITISH ASSOCIATION, 1848–51

The British Association
for the Advancement of Science

The year 1851 proved to be one of the most important in Cleghorn's scientific life, as participator in two important events, one national, the other international – the annual meeting of the British Association for the Advancement of Science, and the Great Exhibition of the Works of Industry of All Nations.

Cleghorn's introduction to a meeting of the pioneering, interdisciplinary, scientific-umbrella group was in 1850.[44] It is probably not coincidental that in this year, its twentieth annual meeting, held in Edinburgh, was under the presidency of the veteran Sir David Brewster. Brewster had, in 1831, been one of the founders of the British Association, and was currently Principal of the United College of St Andrews University (Fig. 14). Although Brewster, due to his support of the Free Church, was (in the words of Allan Cleghorn) 'at loggerheads with the whole of the Professors',[45] this stance would doubtless have commended him in Cleghorn's eyes.

By reading two short papers to Section D ('Zoology & Botany, including Physiology') of the 1850 meeting – the first on the 'grass' cloth (botanically *Boehmeria nivea*, Urticaceae) of which he had purchased an example many years earlier, in Trichinopoly; the second on the hedge plants he had encountered in Mysore – Cleghorn's interest in matters economic botanical was declared.[46] Much more importantly it was on this occasion that the Association requested a report on tropical deforestation from a group of EIC employees comprising Cleghorn, Professor John Forbes Royle (effectively the EIC's economic botanist), and two members of the Bengal Engineers – Richard Baird Smith and Richard Strachey (the latter fresh from travels in Tibet and the Western Himalaya with J.E. Winterbottom). Details of how this commission came about are unknown; doubtless because it involved the Association in no expenditure – there is nothing in their unpublished minutes for 1850 other than the fact of the request, in which Cleghorn's name came first.[47] However, as several of the honorary office bearers for Section D for 1850 were personally known to Cleghorn, the President was John Goodsir, Professor of Anatomy, a long-standing friend of the Cleghorn family; R.K. Greville was a Vice-President, and the surgeon Douglas Maclagan one of the Secretaries, it was probably at least in part Greville's and Maclagan's informal conversations with Cleghorn over the previous two years at the Botanical Society that led to the realisation of the desirability for such a report. With recent experiences of teak (or, rather, its scarcity) in Shimoga, and conversations on the subject with W.C. Onslow, the matter was fresh in Cleghorn's mind, and due to commitments on the part of the other committee members (Baird Smith was working on a report on irrigation in northern Italy for the EIC at the time; Royle on the Great Exhibition), 'almost the entire labor of collecting and digesting the materials fell upon Dr Cleghorn'.[48]

The following year, from 2 to 8 July 1851, the British Association met in Ipswich, under the presidency of the Astronomer Royal, George Biddell Airy (Fig. 15). Section D meetings seem to have been held in the flourishing, three-year-old Ipswich Museum (Fig. 15) founded by George Ransome. Being within easy (rail) reach of London, the meeting was honoured with a visit from the Prince Consort, who was presented with a bound set of the Ipswich Museum Portraits. These depicted leading scientists (including many who attended the meeting), and had been commissioned by Ransome between 1847 and 1852 from T.H. Maguire, who both

CHAPTER 3

drew and lithographed them. It is uncertain if Cleghorn witnessed the royal visit as he is known with certainty to have been in Ipswich only from Sunday 6 July. On the following day he read the summary of his deforestation report to Section D, whose annual president was the Rev. John Stevens Henslow of Cambridge, the Museum's President who had curated its botanical element.[49] Assuming that the lectures took place in the museum's large upper room, Cleghorn would have passed on the staircase Benjamin Robert Haydon's huge canvas

Fig. 15. The Twenty-first British Association Meeting, Ipswich 1851.

The Ipswich Museum. Wood engraving *Illustrated London News* 18 December 1847 (left).

George Biddell Airy, Astronomer Royal. Woodburytype by Lock & Whitfield, from *Men of Mark*, 1877 (above).

Diorama of big cats in front of which Cleghorn probably read his address. Drawn and etched by J. Wolf, aquatinted by H. Adlard (above).

of the 1840 Anti-Slavery Convention,[50] in which his friend Greville is depicted standing next to Daniel O'Connell the Irish politician and campaigner for Catholic emancipation. It is intriguing to speculate that Cleghorn's talk was probably delivered in front of a spectacular and at least partly appropriate backdrop – one of the first museum dioramas, painted by the local artist Edward Robert Smythe, populated with stuffed tigers and leopards. The theme was feline and taxonomic not geographic (it also had lions and jaguars), but as symbols of arboreal tropicality four palm trees are included (Fig. 15).

Henslow himself read a paper on 'the vitality of seeds' and in the ensuing discussion Cleghorn stated that 'after the burning or clearing of a forest in India, invariably there sprung up a new set of plants which were not known in the spot before'. He was also present at a lecture by Richard Owen to Section C, which concerned fossil mammals from the Eocene deposits of the Freshwater Formation at Hordwell, Hampshire, based on specimens in the collection of the Marchioness of Hastings, herself a serious palaeontologist, and daughter-in-law of the 1st Marquess whose Plantation Committee Cleghorn had discussed in his own Report. Cleghorn was particularly impressed by the beautiful illustrations used by Owen.

The Report on Tropical Deforestation

The commission to write the report on tropical deforestation did not come out of thin air. By 1850 there was a general awareness of the subject, in India and in Britain, and a concern about the loss of forests. This was not only from the point of view of declining timber supplies, but (as a result of widely quoted desiccationist publications by Alexander von Humboldt and the French chemist and meteorologist Jean-Baptiste Boussingault) on environmental effects such as rain capture and run off, though in the mind of the EIC these were primarily causes of concern for their commercial knock-on effects, such as erosion and declining fertility of agricultural land. Another Scottish EIC surgeon, Alexander Gibson, had recently been appointed Forest Conservator in Bombay and three years earlier, in 1847, the Court of Directors had sent a request to the Indian Government for information on the occurrence (or not) of deleterious environmental effects, though it is certainly not true, as Richard Grove believed, that surgeons such as Gibson 'lobbied' the EIC on this issue.

Given the existing literature, Cleghorn's task was a review exercise, but geographically limited to India (in a broad sense, including Burma and the Malay Peninsula). Recent experiences gave him great advantages for undertaking the task – not only first-hand observation in affected areas, but in knowing where to find relevant information, both published and unpublished. The latter was available in manuscript reports (which he consulted in East India House in April 1851);[51] the former in Floras and travel books, in a few historical works on geography and forest management, and in periodicals both British and, more especially, Indian (including the *Madras Journal of Literature and Science*) – the majority of his sources were cited in a geographically arranged bibliography.

In addition to the literature and unpublished reports Cleghorn also sought anecdotal information from personal acquaintances by letter from the Madras surgeons Donald MacFarlane and Robert Wight, and from a Captain Gardiner Harvey (of the Madras Army, one of the Assistants to the Commissioner of Mysore). The report includes long quotations from these sources and brings clearly to light most of the major issues, including competing land use, the need for legislation, the difficulty of interpreting anecdotal information on the 'physical' (i.e., environmental) effects and a summary of the anecdotal evidence on negative climatic effects following deforestation elicited from the Court's 1847 enquiry. The needs (let alone rights) of local people to forest resources get scant mention – to them, rather, is ascribed a large portion of blame for the destruction of the forests. Cleghorn's own interests also emerge, with almost four pages devoted to the value of forests for non-timber tree-products as potential 'sources of wealth', including gutta percha, gums, resins, oils and pigments.

The seven conclusions reached were, using modern terminology, as follows:
1. That almost uncontrolled destruction of forests in India was widespread, and was to be blamed on the 'careless habits of the native population'.
2. It was possible to improve the situation by 'supervision', as had recently been demonstrated in Malabar, Burma and Sind.
3. These improvements should be extended by 'rigid enforcement' of existing regulations, and the bringing in of new ones, especially relating to better silvicultural practice.
4. Special attention should be given to preservation of forests in areas NOT suitable for cultivation.
5. Special care should be given to the preservation of woods on hills in watersheds.
6. Effort should be given to educating local inhabitants on the value of forests and their products.
7. The inaccessible manuscript reports of the Plantation Committee, full of 'practical utility' should be published.

According to a letter written to J.H. Balfour the following day, the reading and ensuing discussion occupied two hours, though it is hard to know what its wider impact was. There had certainly been no build up; in fact the report must have been completed only at the last minute as Airy had not seen it in time for his opening speech to the assembled conference in the Town Hall.[52] According to its author the report was received by a 'listening [i.e., attentive] audience' and was followed by an 'interesting discussion' participated in by Henslow, Charles Bunbury,[53] Lankester (with George Allman, Professor of Botany at Dublin, Secretary for the Section), Strachey, and two individuals unknown to Cleghorn – the *Athenaeum* magazine reported the comments of Strachey, Bunbury and Lankester.[54] Of these, Strachey might have been expected to be the best informed, having at this point spent 15 years in India (including work on irrigation projects), and was supposed to have helped in compiling the report, but he questioned whether forests had any effect on climate, especially in the tropics, and believed, pessimistically, that 'we had not the power to arrest the present destruction of forests in India'. In response to Lankester's remarks on the relations between forest distribution and rainfall, Cleghorn went off at a tangent, giving 'a short account of the destruction that is now going on in the forests of *Isonandra gutta*' the supply of which would probably become diminished due to the present process of collection. This evasion is perhaps significant, and could be taken to suggest that his interest in economically useful plants was always stronger than it was in the 'desiccation issue'.

The report was certainly of significance in drawing the subject to the attention of an influential body of British scientists. Few of these would have had access to the materials summarised – either the Indian periodicals, or, even less so, the manuscript reports lurking in Leadenhall Street. The question of its direct influence on directing Indian forest policy is hard to assess. In 1861 Cleghorn himself referred to it in the Preface to his *Forest & Gardens*,[55] and in 1865 it was noted that:

> it was only after the report was published ... and general attention thus prominently directed to the risks incurred by the reckless destruction of wood in India, that the question of forest management was seriously taken up by Government, and an attempt was made to establish a system of conservancy.[56]

In the opinion of Richard Grove the report represented 'a landmark in the history of the colonial response to environmental change ... and in the impact it had on government conservation policy'.[57] Grove's statement that the Report 'warned that a failure to set up an effective forest-protection system would result in ecological and social disaster'[58] is wide of the mark: the Report said no such thing, even in the more measured language of its time. The only emotive phrase, that the felling of trees on mountains prepares 'at once two calamities for future generations – the want of fuel and the scarcity of water', is an acknowledged quotation from Humboldt,[59] and Cleghorn's own improvement agenda makes his priorities crystal clear:

> In a single sentence, we would say that where human exigencies, whether for subsistence or for health, require the destruction of forests, let them be destroyed; but where neither life nor health is concerned [primarily on mountain slopes and ridges], then let a wise system of preservation be introduced and acted upon.[60]

The question of the influence of the Report would be raised two decades later, by Cleghorn himself, and by George Birdwood in 1884 and will be discussed in due course (see p. 245).

The Great Exhibition

The great public event of 1851 was not the British Association meeting in Ipswich, but the Great Exhibition, which had been opened a few months earlier (1 May) by the Queen in Joseph Paxton's epoch-making Crystal Palace in Hyde Park, London.[61] Cleghorn played a modest role in this great spectacle, in the Indian Section, which was supervised by John Forbes Royle and Royle's brother-in-law Edward Solly.[62] This Section occupied one of the most prominent sites, at the heart of the vast building, beside the south entrance flanking the west side of the great central transept on both the north and south sides of the grand central avenue. Over six million people visited and in the Indian Section they were dazzled by the Koh-i-Noor diamond (lent by the Queen, it had come to Britain with the recent defeat of the Sikhs), the gorgeous silks, cloth of gold, ivories, armour and thrones. But it was not only manufactured treasures that were on show. In the North-East Gallery the natural products and raw materials on which these were based (woods, gums, oils, resins, fibres etc.) were displayed and, by means of clay statuettes made at Krishnanagar in Bengal, miniature representations of the people – the craftsmen and farmers – who ultimately made it all possible.

Few details are known of what Cleghorn did other than to assist Royle, who had been confined to bed for ten weeks 'the joints of his fingers ... swollen from Chronic Rheumatism & his nervous system ... injured by overworking his strength'.[63] In later days (when his memory was not entirely reliable) Cleghorn stated that 'Professor Forbes Royle, of King's College, London, asked me to assist him in arranging the raw produce ... and [I] was occupied for several months in classifying the exhibits in the Indian Section'.[64] This material had been sent from the Subcontinent by specially convened Local Committees in each of the three Presidencies, of which Edward Green Balfour co-ordinated that for Madras, with contributors including Robert Wight and Walter Elliot. The fact that Cleghorn had to stay in a 'noisy Hotel with the Sappers and Miners' suggests that this was close to the building in Hyde Park, so at least some of it might have been practical work, supervising the installation of exhibits, though he had also recently spent time in East India House, which might have been in part for the Great Exhibition work in addition to the gathering materials for his British Association report.

Class IV of the exhibition was devoted to 'Vegetable and Animal Substances chiefly used in Manufactures', for which the jurors were Royle and Solly, along with Nathaniel Wallich – it was in this section that Cleghorn himself received a Prize Medal for a sample of gamboge that he must have collected in Shimoga.[65] Several other individuals from Madras known to him either at this point or later on, also won medals, including Walter Elliot (for resin from the succulent shrub *Euphorbia cattimandoo* – see *Cleghorn Collection* no. 65), G.F. Fischer of Salem (for indigo made from *Wrightia tinctoria* and cotton samples), Alexander Hunter (for vegetable fibre), Lieutenant Colonel Alexander Tulloch (samples of 'fixed oils' and wood) and Robert Wight (various cottons). In addition, five different collections of wood samples from Madras received an Honourable Mention. Another prize-winning exhibit was terracotta pottery made under Alexander Hunter's supervision at his recently launched School of Industrial Art. It is characteristic of the heterogeneous assemblage of the items displayed in the Great Exhibition to see this humble ware being given the same category of medal, in Class XXV 'Ceramic Manufacture', as the sumptuous productions of the royal porcelain factories of Denmark (Copenhagen), Bavaria (Nymphenburg) and Saxony (Meissen), or even that of Wedgwood.[66]

The Kew & Edinburgh Museums of Economic Botany

This experience at the Great Exhibition was important for Cleghorn, laying the foundation for similar work undertaken after his return to India – at the exhibitions held at Madras in 1855 and 1857, and the Punjab Exhibition at Lahore in 1864. But by this time he had also made his first donation to J.H. Balfour's museum of economic botany at RBGE. The story of the Great Exhibition, especially its visually spectacular side of arts and manufactures, and the role played by Prince Albert and Henry Cole, is well known but the displays of 'raw products' are less so. The latter were largely the province of J.F. Royle, but, due to the efforts of Balfour and Hooker, there was already a tradition of public display of such material. Among the increasingly urbanised population there was great curiosity, not only about the raw form and origins of their foodstuffs and beverages, and the raw materials on which the spectacular prosperity of the mid-nineteenth century was based (oils, timbers, dyes, resins, etc.), but also a nostalgic hankering towards an earlier tradition of the cabinet of curiosities – weirdly shaped fruits, objects associated with eminent people (secular relics), or with Biblical or religious associations, and also the need for reminders of the pre-Industrial world – objects from exotic ('primitive') cultures, or the sorts of medicine collected by herb-women and shamans.

3. FURLOUGH & THE BRITISH ASSOCIATION, 1848–51

Balfour had started his own private collection of 'economic botany' as early as 1840 to help to illustrate his lectures, and he greatly developed this during the five years during which he lectured in that great commercial centre and port, Glasgow.[67] After moving to RBGE Balfour had in 1849 been given permission to embark on a museum to house this collection, as part of major improvements to the teaching facilities. Robert Matheson, architect to the Board of Works, built him a large new classroom (now the Conference Room) and converted Robert Graham's 1820s classroom (where Cleghorn had attended lectures in the 1830s) into a museum (Fig. 16a).[68]

Like the 1820s greenhouse range this must have been shoddily built as the timber of both structures was already in need of major refurbishment within 30 years – the roof of the classroom was suffering from dry rot. This was not architecture of the grandeur of the palmhouse that Matheson would build for Balfour a few years later. In fact, the museum was not a great success, with large windows along both long walls that led to reflections on the glazed cases, which also suffered from too-small panes and heavy wooden glazing bars. But it was a cheap solution and a quick one, and ready to receive material from June 1851 (including material dispersed from the Great Exhibition). It opened to the public in January 1852, for it was to be a public attraction – 'useful instruction' for the working classes – as well as a teaching resource for undergraduates. In its first year Cleghorn donated specimens of Malacca cane (the rattan *Calamus scipionum*), fossil wood from Godavery (Madras) and the fruit of *Sterculia foetida* previously exhibited to the Botanical Society on which occasion he had observed that 'studying timber trees in the primeval forests ... [is] one of the most difficult departments of tropical botany'.[69]

Fig 16a. J.H. Balfour's museum (right), converted from Robert Graham's classroom; on the left is Balfour's own classroom. Anonymous photograph, 1870s (RBGE).

CHAPTER 3

Fig 16b. Interior of the Kew Museum of Economic Botany. Anonymous wood engraving, frontispiece of W.J. Hooker's *Popular Guide*, 1855.

This mixture of objects was typical, and a further 17 collections of Indian material would be sent to Balfour up to 1877 – wood samples, fruits and manufactured items including a cap made from the spathe of the *Areca* palm (November 1858); a 'packet of pilgrim's food, composed of dates and almonds enclosed in kid skin, from a Coptish convent at Mount Sinai' (December 1860) and the 'wood of *Heritiera minor* used by entomologists for lining cabinets' (July 1861). By 1855 at least the foreign parts of the University herbarium were kept in

Fig. 17. Cap (*muthalley*) made of leaf sheath of areca palm, from Nagar, Karnataka. A similar one was presented by Cleghorn to the RBGE Museum in November 1858. (Drawing by Claire Banks).

the Museum and adjacent rooms, before moving in around 1864 to the former exhibition hall of the Caledonian Horticultural Society where in 1873 it was joined by the Botanical Society's library. In an act of astonishing vandalism the RBGE Museum, by this time housed in two World War I huts on the site of the present herbarium, was destroyed in the early 1960s on the orders of a Regius Keeper who in the 1930s had been its curator, and was known as an aesthete and collector of contemporary art. The only one of Cleghorn's donations that appears to have survived is a coquila nut with the hole of a wood-boring beetle (*Bruchus ruficornis*), though this has nothing to do with India, being the fruit of the Brazilian palm *Attalea funifera*.

Balfour's Edinburgh Museum of Economic Botany was influenced by Sir William Hooker's similar one at Kew which had opened four years earlier, in 1848, aimed at complementing the living plants in the garden for public benefit.[70] It too was housed in a building converted from a previous function by an architect best known for a later palm-house. Decimus Burton adapted a fruit store that, with the royal kitchen gardens, had been given to Kew by Queen Victoria (Fig. 16b). Between 1854 and 1891 Cleghorn made ten donations to the Kew Museum, several of which survive. The batches of material are similar in scope to those sent to Balfour, except for a large collection of vegetable oils from the 1855 Madras Exhibition. Of particular interest in the donation of 25 February 1858 is material from Cleghorn's Burmese trip of the previous year, including 'torches filled with rotten wood saturated with the oil of *Dipterocarpus alatus*, and enclosed with leaves of Thaban ... (*Pandanus* sp.) commonly used at Rangoon', and strange-sounding 'cosmetic tubercles' – woody plant spines that Hooker thought might come from a species of *Aralia* (these would have been ground into a powder and made into paste known as 'thanaka', used for facial adornment by Burmese women, a custom that is still practised).

An intriguing late manifestation of Cleghorn's generosity in such matters is to be found in the *Dundee Courier* of 8 July 1880, which records his presentation to the Literary and Antiquarian Society of Perth of 'snow spectacles made of Yak's hair, cupping instruments (of Horn), &c, from Thibet', which he presumably obtained in Lahul in the 1860s.

CHAPTER 3

The Linnean Society & Metropolitan Scientists

During his time in London and Ipswich, Cleghorn met several of the leading scientists of the day. In 1835 a newly appointed Governor of Madras (Sir Fred Adam) on departing London had asked Robert Brown for recommendations for a good botanist, which had led to Robert Wight being assigned to botanical rather than military employment.[71] Cleghorn was as keen as Wight had been not to have to return to the duties of a military surgeon and asked Sir William Hooker for a reference attesting to his botanical qualifications to present to the Governor should a suitable post arise.[72] Links with London's botanical community also lie behind Cleghorn's fellowship of the Linnean Society, formal election to which took place in November, after he had left for India.[73] Chief among the scientists who signed the nomination were the Hookers (father and son), but there were other leading British botanists – George Bentham, Dr Charles Morgan Lemann, Dr R.C. Alexander (a founder member of the Botanical Society of Edinburgh), Arthur Henfrey, Professor Charles Cardale Babington of Cambridge and Robert Heward. Of the other proposers Nathaniel Wallich, J.F. Royle and Thomas Horsfield were closely connected with the EIC and Indian botany, the others being the Edinburgh-born Adam White, a zoologist at the British Museum, and the natural philosopher Edward Forbes. At this time Forbes was Professor of Natural History at the Government School of Mines in Jermyn Street, but Cleghorn had known him since attending his lectures in St Andrews in 1839, had renewed his acquaintance in London, and been with him in Ipswich. In London Cleghorn also met the zoologist E. Ray Lankester, as amusingly reported to Balfour: 'when getting into an Omnibus at Charing Cross, I found myself alongside of Lankester, who counts for two in any conveyance, he is getting so large– he desired to be kindly remembered to you, & said that his translations of Schleiden might be of use to you in the preparation of your new Manual [of Botany]'.[74]

The Bibliophile

Coming from a cultured and literary family, Cleghorn must have absorbed a love of books as a child from his grandfather's substantial library. Whereas most undergraduates are notorious for their unwillingness to purchase textbooks, it is a sign of the man that as a student he donated books to the library being assembled by the Botanical Society. His interest and collecting of books would develop through life, and he amassed a substantial library that will be discussed later. Within the scientific community intellectual friendships and patronage networks were important, but these interacted with and depended upon other networks, conspicuous among which (though often forgotten) were publishers and booksellers. Other than gifts from friends and patrons, botanists in India were reliant on London booksellers in order to keep up to date not only with the latest literature, but, in the case of taxonomists, with older works – some rare and antiquarian. Surviving documents reveal fascinating details of Cleghorn's interactions with the London bookseller William Pamplin.[75] In April 1851, while in London assembling information for his British Association report, Cleghorn reported to Balfour that he was 'purchasing largely for myself' at Pamplin's – books he intended to take back with him to India. Pamplin's shop was in Soho, at 45 Frith Street (now a Moroccan/Spanish restaurant, sandwiched between the Delhi Brasserie and Ronnie Scott's jazz club) and Cleghorn was interested to note that many books from Robert Graham's library (which had been auctioned in Edinburgh in 1846) were then on offer. This establishment was more than just a bookshop, as in 1839 Pamplin had taken over the successful business of his father-in-law John Hunneman, which dealt not only in books, but herbarium specimens and botanical illustrations (both prints and drawings). Hunneman had established a major network not only in Britain, but between Britain and the Continent, and Britain and India, and had, among many others, been Robert Wight's supplier and correspondent. Pamplin continued this link: 'I have several excellent Customers in India to whom I regularly send monthly packages of Books overland'.

The letters and bills from Pamplin cover the period 1851 to 1860 and add up to the substantial total of £160/1/5, most of which was for botanical prints. Rather few book titles are mentioned, which suggests either that the surviving bills represent only a fraction of the whole, or that most of Cleghorn's books came from other sources. In a letter of October 1855 Cleghorn told Balfour that he had spent £300 on his library since its destruction by fire in 1852. One other known source was Hooker's publisher, Lovell Reeve, as there are also some letters from him about sending *Illustrations of Himalayan Plants*, of which Cleghorn was one of the original subscribers, to Madras. Books sent by Pamplin included Burman's *Flora Indica*, Joseph Hooker's *Himalayan Journals*, and William Hooker's popular guides to *Kew Gardens* and to the *Museum of Economic Botany*. After Cleghorn's appointment as Forest Conservator in 1856 he also bought a number of related books, including Patrick Mathew's *On Naval Timber and Arboriculture* of 1831, with its appendix in which the theory of Natural Selection was proposed 28 years prior to Darwin. In one package Pamplin used 'a few seeds of some of those noble gigantic Umbels as *Heracleum gigant[eum]*' as an organic precursor of polystyrene beads – which he suggested that Cleghorn try to grow. Also mentioned as for sale by Pamplin in the letters are substantial collections of herbarium specimens – from Ernst Steudel of Esslingen, and a set of the EIC ('Wallich') collection – but it seems unlikely that Cleghorn bought any of these. He had been keen on, but was unfortunately not quick enough, to acquire a set of copies of the Roxburgh Icones that Pamplin had purchased from the defunct Medico-Botanical Society, now known to have been a set made for the distinguished Orientalist, Thomas Henry Colebrooke (their present location unknown). It was Pamplin who posted to Cleghorn a present from Wight, four volumes of the early eighteenth-century works of Leonard Plukenet now in the RBGE library. When back in Madras Cleghorn also acted as an intermediary between his botanically-minded Indian friends Walter Elliot, J.S. Law, Charles Drew and Pamplin.

The largest category of Cleghorn's purchases from Pamplin during the 1850s was of botanical prints for his collection of illustrations – almost 3500 are included in the bills, most of which survive at RBGE (if hard to distinguish from prints from other sources). These came from contemporary periodicals (the *Botanical Magazine* and the *Botanical Register*), from the folio medical works of Nees, Zorn and Plenck, and even from that great rarity Sibthorp's & Smith's *Flora Graeca*. The warmth of Pamplin's personality comes over in his letters – for example, in telling Cleghorn of his acquisition of the Nees plates of medicinal plants he wrote 'I will keep them back until return of post to prevent disappointment – I think they would suit you sir'. The letters also include European botanical news, including the shocking early deaths of Edward Forbes,[76] and of John Ellerton Stocks, a brilliant young Bombay botanist who died while on a visit to England.

CHAPTER 3

Return to India

Under Captain D. Robertson the ship *Trafalgar* set sail from the Thames on 31 August 1851 with Cleghorn on board. The voyage via the Cape was a slow one taking 104 days apparently with no stops on the way. Cleghorn took with him a Wardian case of plants – a reward from Kew for Lewis Manly, a correspondent of Hooker in Calcutta.[77] The time was used productively, instructing a fellow passenger in botany, writing an article on Indian agriculture, and reading Garcia da Orta's 'De aromatum'. His companions to Madras were mainly military couples, but Cleghorn was drawn to a group of soul-mates who met in his cabin twice a week for scripture readings and prayer. Services were led by the Rev. Robert Clark, recently graduated from Trinity College, Cambridge, on his way out as a missionary to the Punjab, with whom Cleghorn would renew acquaintance a decade later. Another passenger was Thomas Marden, headmaster of Bishop Corrie's Grammar School, an Anglican establishment in Madras. The surgeon and school-master appear to have made friends as when they arrived in Madras, on 12 December, Cleghorn moved into living accommodation adjacent to the school, and would serve on the school's committee until 1857. At this point (it later moved to 10 Semboodoss Street), the school was located on the top floor of a 'lofty and spacious', three-storey building, of which the lower two floors were occupied by Messrs Oakes, Partridge & Co., a firm of general merchants that dealt in hard- and soft-goods, wines and spirits, and acted as representative for several insurance and shipping companies (see below). The building's official name was 'Exchange Hall', but colloquially 'Waddell's Folly' – after its builder James Waddell who in 1819 had doubtless hoped to provide a replacement for the old Exchange in the Fort.[78] This stood on Popham's Broadway, the main north-south thoroughfare of Black Town, the city's major commercial street, conveniently close to lithographers such as Gantz and several printing houses (including the American Mission Press, the Scottish Press and the Oriental Press), the repository of the Madras Auxiliary Bible Society, and the Church Missionary Society's attractive Neo-Classical church ('Tucker's'), at the heart of what was to be Cleghorn's new 'parish'.

4
Madras Polymath, 1851–1856

> Without wishing to undervalue the labours of those whose attention has been chiefly devoted to the classification, nomenclature, structure and physiology of plants, we would always assign the highest meed of praise to the practical investigator, who employs his knowledge and his research in applying botanical science to the purposes of Agriculture, Medicine, and the Arts. This is evidently the leading aim of Dr. Cleghorn's labours. We have perused his writings with great interest, and doubt not that his name will rank hereafter, with those of Royle and McClelland as one of those pioneers of science, whose mission is to reclaim the peninsula of India from its pestiferous jungles, — to lay bare the innumerable vegetable treasures which lie hid not only in its forests, but in its hedges, in its brakes, and beside the stagnant water of its ditches and its swamps – antidotes mercifully placed there by the Creator against the mists of death which are there engendered; — and to introduce systems of scientific agriculture, the adoption of which, it may be safely presaged, would, in the course of a century or two, render British India the Garden of the World.[1]

On reaching Madras a new and exceptionally busy life opened up for Cleghorn – the most varied of his career. Doubtless inspired by the intellectual achievements of his recent leave, his work in the fields of botany, medicine and education was made possible by means of the patronage of the governors Sir Henry Pottinger and Lord Harris operating through bodies such as the Medical College, the Agri-Horticultural Society and the Madras Literary Society. This period also saw his involvement with a major exhibition and these activities were shared with a group of like-minded and talented Company officials, in a wide range of official postings – men such as Alexander Hunter and Walter Elliot, whose friendship Cleghorn enjoyed. Work in these areas was semi-official, additional to his official duties as a surgeon; but Cleghorn also undertook numerous voluntary occupations, including the support of missionary activities, and editorial work both for Madras publications and ones based in Calcutta. The period discussed here closes with Cleghorn's appointment as Conservator of Forests in 1856, during which time he was based entirely in Madras, living firstly in Black Town and later in St Thomé, but ending with what would be his first visit to Calcutta.

Before looking in detail at Cleghorn's major activities of the period, some of his more minor roles should at least briefly be noted. Although significant, relatively little is known about his editorial work, as much of it was anonymous, though hints are to be found in some of his letters to Balfour. One periodical on which he worked was the *Madras Christian Herald* and on 11 May 1853 he thanked Balfour for a packet of pamphlets noting:

> it is wonderful how much information it gives of what is doing in Edinburgh [probably referring to medical missionary work] – and it has afforded me some excellent "Selections" for the "Madras Xn. Herald" – which having 280 up Country subscribers requires much careful supervision and Catering to keep up general interest.[2]

A year later he reported that 'our little Herald' was edited by a 'Committee of three',[3] and by this time he was also undertaking editorial work (including book reviews) for the Calcutta-based *Indian Annals of Medical Science*, a periodical with a circulation of 200 in Bengal and Madras, edited by Alexander Grant 'Surgeon to the Governor General [Lord Dalhousie], an excellent man who has the true interests of Medical Science and philanthropy at heart'.[4]

CHAPTER 4

Cleghorn also wrote for the *Calcutta Review* and although these articles were anonymous it is known that he was the author of a review of Hooker and Thomson's *Flora Indica*.[5] In February 1856 Cleghorn was elected to the Managing Committee of the Madras Literary Society, and from a letter from his father in April 1858 it appears that he was acting as one of its journal's editors.[6] In 1856 Cleghorn performed the useful service for botanists of providing an index to the 2101 plates of Robert Wight's *Icones Plantarum Indiae Orientalis* – or, rather, two indexes, one alphabetical, the other systematic. He had compiled the work for his own use, and, when seeking approval for its publication at Government expense, he described Wight's labours as 'the most valuable contributions to Natural Science, which have ever been published in India'. The index was published by the Fort St George Press in an edition of 500 – 300 for the sets of the original edition of the *Icones*, and some extras 'useful to persons who do not possess the book, but who may have occasional access to it'.[7]

Cleghorn's membership of a wide variety of charitable committees is documented in the Madras Almanacs of the period. He was on the Madras Diocesan Committee of the Society for the Propagation of the Gospel in Foreign Parts (1854–7), the Madras Sailor's Home (1853–9) and the Madras School Book Society (1855–9). He also acted as Medical Officer for the Medical, Invalid and General Life Assurance Society (1857) and the East Indian Emigration Society (1858, though the society was in abeyance at the time). Strangely his name does not appear, at this point, on any of the missionary committees associated with either the Free, or the Established, Church of Scotland, but he was certainly actively involved with the former over the question of medical missionary work. The ethos behind this was 'with the one hand healing the body, and with the other healing the soul'.[8] In collaboration with the Edinburgh Medical Missionary Society, the Foreign Missions Committee of the Free Church was to send out its second Indian medical missionary to Madras. The first mention of this is in Cleghorn's letters to Balfour in 1855,[9] while bemoaning the deaths of two of the 'Fathers of Mission' – the Rev. John Anderson who had gone to Madras as a Church of Scotland missionary in 1837 and the Rev. Dr. John Scudder (1793–1855) the first American medical missionary in Asia. The following year Cleghorn was sending money to the Edinburgh Society's secretary Dr John Coldstream from himself and fellow medics William Evans and Ambrose Blacklock.[10] The medical missionary David Horn Paterson (1832–1871) arrived in September 1856, newly qualified with a diploma from the Royal College of Surgeons of Edinburgh, and the following March opened a dispensary in Black Town. Cleghorn took an active interest in Paterson's work, for which he continued to fund-raise. In 1859 Cleghorn was pleased that the Governor, Sir Charles Trevelyan, had himself taken out an annual subscription;[11] though at this very time there was outrage on the part of the Hindu merchant class by what it saw as the abandonment of the (former) Company's position of religious neutrality, and toleration of aggressive proselytising missionaries.[12]

Regarding his major professional medical duties, Cleghorn by April 1852 had been appointed Surgeon for the First Division of Madras, which included Black Town, the dense, urban area immediately to the north of the Fort, where he had responsibility for several institutions – the Leper Hospital (in which were 57 patients), the Orphan's Asylum, the Sailor's Home, and the House of Correction. Also in this part of town was the Monegar Choultry, a 'Native Hospital' that catered mainly for Hindu Tamils, for which Cleghorn had no responsibility, being staffed by Native Surgeons (whom Cleghorn would later be responsible for training at the Medical College). Cleghorn also spent:

much time every day in the Police Office, at Coroner's Inquests, or making surveys of Transport Ships, this comes within the duty of Port & Marine and Emigration Surgeon, by all of which titles I am accustomed to be addressed![13]

As early as January of this year Cleghorn had started to act as secretary to the Agri-Horticultural Society. But these labours were to be increased, and Hooker's letter of recommendation must have paid off as on 9 July the Governor Sir Henry Pottinger appointed Cleghorn as Professor of Botany at the Medical College. All, however, did not go entirely smoothly.

The Great Fire of Madras

Having survived ordeal by water four years previously, the early morning of Friday 16 July 1852 proved to be Cleghorn's time for trial by fire. According to a contemporary newspaper account he was woken by the whining of his dog at 3 a.m. to find his room full of smoke.[14] The premises of Messrs Oakes, Partridge & Co. beneath him had already been burning for some time when he raised the alarm, and four fire engines and a body of lascars were despatched from nearby Fort St George. The engines were faulty, the water supply inadequate, so that even a detachment of the 25th Kings Own Borderers, who eventually arrived on the scene some hours later, could not prevent the gutting of the building and the loss of stock worth Rs 300,000, and almost all of Cleghorn's property. The flames leapt 50 feet into the sky to a soundtrack of exploding kegs of gunpowder and spirits. It was 'probably the most extensive conflagration that has ever happened in the European quarters of any of the Indian Presidency Towns', though miraculously there was no loss of life. Truly the man of science Cleghorn chose to save his precious copies of the first six volumes of *Hortus Malabaricus* presented to him by Greville, and the notes made during his leave on the herbaria of Arnott and Hermann.[15] Despite the fact that his little dog had saved his life the favour was not returned, though Cleghorn did express regret for this which he concluded with the pious observation that 'a preserved life should be a sanctified one'. Also lost in the fire were many books, drawings and specimens, notes made for the botanical lecture course on which he was shortly to embark, and the nearly completed manuscript of a catalogue of the Agri-Horticultural Society's garden, his *Hortus Madraspatensis*, which he would have to start again from scratch. In 1853 Cleghorn received a touching letter from a fellow book-lover, the bookseller Pamplin who, on hearing of Cleghorn's misfortune, wrote:

> I beg to express my real sympathy and sorrow which I felt for you, when I heard some months ago that you have experienced very severe loss by fire and that among other property much of your Library &c had perished – Mrs Pamplin & I were quite grieved to think of it.[16]

Fire was not the only unpleasantness in the summer of 1852, and from May to December, under Dalhousie's aggressive policy, war was being waged in Burma. Two of Cleghorn's friends Alexander Christison and George Govan were among those ordered to Rangoon as medical officers and he commented: 'Horrida Bella! I am much mistaken if large reinforcements will not speedily be required',[17] and in October he was treating two officers wounded at Amherst. His view was that the annexation of Burma was 'the only course open to us', but that if this happened he hoped that the troops would behave 'so as to give a national testimony' and that the political officers 'may be instrumental in diffusing the blessings of peace where all has been anarchy and confusion'.[18] One minor blessing was the sending the following spring of some Burmese plants from Tavoy and Mergui to the Agri-Horticultural Society garden including a spectacular climbing *Thunbergia*,[19] which probably came from Christison (son of Cleghorn's old professor of Materia Medica). A much more significant outcome was the annexation of Pegu, the area of the lower Irrawaddy, with the important cities of Prome and Rangoon and its rich teak forests, which completed the Company's acquisition of Lower Burma – Arakan and Tenasserim having been ceded in 1826 after the first Burmese War (in which Alexander Gibson had served). The policies developed for the management of the teak forests acquired would have a major influence on the next phase of Cleghorn's career, and will be discussed in due course.

In this period a natural calamity also had to be faced, when the failure of the monsoon of 1852 led to famine in the Madras Presidency in 1853/4. Although the European population was largely immune to its effects, Cleghorn's humanity ensured that he at least noted it, even if with a fatalistic sense of inevitability, and no hint either of suggesting price controls or subsidies, or any sense of responsibility for the failure to import supplies of grain in time to avert catastrophe:

> We have had a scanty monsoon, and the price of rice has risen more than 100 per Cent – the prospect of the poor Hindoos is little short of absolute famine – which is specially alarming in a country so densely peopled, when Commerce from distant parts cannot assist the interim in sufficient time. We are yet hoping for the latter rains [i.e., the NE monsoon].[20]

Madras Medical College

On 9 July 1852 Cleghorn was appointed Professor of Botany, Materia Medica and Therapeutics at the Madras Medical College (initially in an acting capacity, confirmed in 1854) – a post he would hold for four academic years. This position gave him not only a platform from which to promote botany to young Indians, but, being the only Government post with 'botany' in its title or remit, made him, effectively, the Presidency's official botanist though such a post had been scrapped in 1828 never to be reinstated. Cleghorn would come to be consulted on any botanical (that is, economic-botanical) topic that arose, thereby succeeding to the semi-official role vacated by Wight on his departure from India in 1853.

The College had begun on a small scale as a Medical School attached to the Government Hospital, under the superintendence of Dr William Montgomery and the patronage of the Governor, Sir Frederick Adam.[21] Its first session began on 1 July 1835 and a year later a handsome building was opened, which looked (from the only known sketch, reproduced in the College's centenary *Souvenir Book*) like a Neo-Classical villa, of seven bays with a two-storey central section (Fig. 18). This must be the building known to Cleghorn as new, larger premises were not built until 1865/7. This Medical School had four rooms: museum, library, lecture room and theatre. As the theatre was described as 'a very elegant apartment, built nearly after the model of such rooms in Europe', this was probably the double-height, central space. The aim of the school was to train young men, from the age of 15 upwards, for the Subordinate Branch of the Madras Medical Service and from the start 'Natives' and 'Eurasians' – as well as Europeans – were admitted. Initially the course was of two years, but in 1847 a five-year course was started to train stipendiary students as Native Apothecaries (also called Native Surgeons) for the Civil Department. This was in addition to a four-year course to train Medical Apprentices for the position of Assistant Apothecaries in the army, and a three-year course for Medical Pupils to attain the rank of Native Second Dresser (also for the army). The previous year, 1846, the European staff (all members of the Madras

4. MADRAS POLYMATH, 1851–6

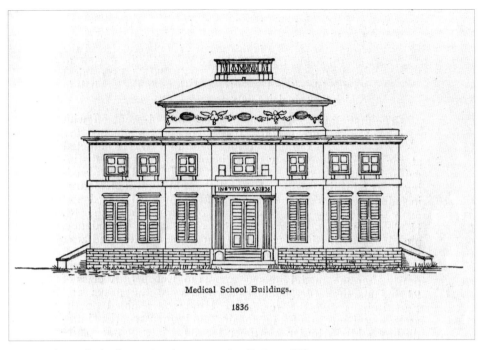

Fig. 18. Madras Medical College. Artist unknown, reproduced in Anon, 1935.

Medical Service) had increased from an original two to five, with chairs in Medicine & Clinical Medicine, Anatomy & Physiology, Surgery & Clinical Surgery, Midwifery, and Chemistry & Materia Medica.

In October 1850 the institution was relaunched as a Medical College under a governing council, of which the first President was the Edinburgh-trained Dr Thomas Key (1803–1880). It was at this point that the teaching of botany started; this was combined with Materia Medica which had been taken away from the responsibilities of the Professor of Chemistry. The first botanist was a Dr Currie,[22] who was replaced in the session 1851/2 by John Emilius Mayer. Cleghorn was the third holder of the post and would serve under four Presidents – Key, William Evans, John Lorraine Geddes and the Edinburgh-trained James Kellie. Although Cleghorn had brought back with him to Madras a letter of recommendation for a botanical post from Sir William Hooker it is interesting to find that Thomas Key was the son of Dr Patrick Key of Forfar, a favourite uncle of Sir David Brewster, so the Scottish patronage network seems also to have been in active operation. Cleghorn's colleagues in his first year were Ambrose Blacklock (Professor of Surgery, who also taught Medical Jurisprudence), William Evans (Medicine), James Shaw (acting for W.B. Thompson this year in teaching Midwifery and Ophthalmic Medicine & Surgery), George Smith (Professor of Anatomy & Physiology, who undertook the dissections and demonstrations), and J.E. Mayer (who ran a course in Practical Chemistry and the Toxicology department). The curator of the museum was Assistant Apothecary G.W. Flynn who was also assistant to the Professor of Anatomy.

At the time of his academic appointment Cleghorn was also made Surgeon of the Fourth District of Madras and moved from Black Town to the more salubrious St Thomé, close to the beach, some five kilometres south of the Fort, with which it had been linked by road only since 1846. It appears that he may have lodged there with General Alexander Tulloch. The villages of St Thomé and Mylapore formed a detached settlement, separated from Triplicane to the north by coconut groves. To the south lay the mangrove-fringed estuary of the Adyar River. The settlement was home to the large Shaivite Kapaleeswara Temple and a Roman Catholic Cathedral dedicated to the doubting Apostle, who is believed to have been martyred on nearby St Thomas Mount. Neither of these great ecclesiastical establishments is likely to have met with the approval of an evangelically-minded, Free-Church Presbyterian, and in fact Cleghorn considered that the main population of his district consisted of 'Mahomedan Masses' who must have worshipped at the mosque in Triplicane. It appears that Cleghorn, in common with many of his countrymen, got on better with Muslims than with Hindus – he described taking the medical missionary David Paterson to the 'houses of my Mussulmen friends',[23] which does not sound insincere. Earlier, however, he had described India as 'a country overspread with gross Gentilism, when the worship of Devils & the tyranny of Satan is constantly before our eyes'[24] – 'devil-worship' being an ancient trope by which the West identified the Hindu religion.

Having shed the responsibilities for the various prisons and hospitals of Black Town, the workload of Cleghorn's new academic duties must have more than replaced these, and it should be remembered that the teaching was on top of his district surgeon's work and the numerous other jobs that came his way. These included the secretaryship of the Agri-Horticultural Society, work on various reports commissioned by the Government (on the timbers used for gun carriages,[25] on pasture grasses as fodder for cavalry horses,[26] on lubricants for rolling stock,[27] and on plants for binding the mobile sand on the beach[28]). His father was pleased about the salaries that went with the medical duties and teased Cleghorn about 'getting rich in spite of yourself' from the 'full allowances of St Thomé – the College and your [army-medical] pay'.[29]

At the Medical College Cleghorn taught two courses each session, which ran from 1 August to 15 April – a junior one in Botany for first-year students, and a Materia Medica course for second-years.[30] Each year he gave between 68 and 80 two-hour botany lectures (two per week on Tuesdays and Thursdays at noon) and between 88 and 100 on Materia Medica (three per week on Tuesdays, Thursdays and Saturdays from 11 a.m. to noon). The number of students in each of the three different categories for the year 1853/4 were as follows:

	Botany	Materia Medica
Medical Apprentices	25	15
Medical Pupils	24	16
Students (Stipendiary, Scholarship or Private)	4	1

While on leave in Edinburgh Cleghorn must have been aware of (and possibly even attended) the botanical lectures given to medical students, and to extra-mural groups, by J.H. Balfour. These were almost certainly more lively affairs than those given by Graham in the late 1830s, though both professors were also keen promoters of the teaching of botany during field excursions. For his lectures Balfour used the full range of available resources of the garden, herbarium and museum, and drew on an extensive collection of large-format teaching diagrams, some of which had been painted for him by Greville and by the artist Neil Stewart, others by his wife. This use of illustrations, another tradition of Edinburgh botanical teaching, had been started by John Hope in the 1770s. Cleghorn clearly modelled

4. MADRAS POLYMATH, 1851–6

his own lectures on Balfour's and what he taught is known in some detail from two surviving syllabuses for his botanical lectures (for 1852 and 1854) and one for those on Materia Medica (for 1852).

In Materia Medica after an introduction, in which he gave definitions of allopathy and homoeopathy, he went through the major categories of substances – mineral, vegetable (family by family) and animal – teaching the actions, uses and commercial sources of medicines, and the native names of indigenous remedies. The textbooks he used were J.F. Royle's *A Manual of Materia Medica and Therapeutics* (London, 1847) and Edward Ballard's & A.B. Garrod's *Elements of Materia Medica and Therapeutics* (London, 1845).

In botany Cleghorn followed the natural system of De Candolle and gave among the reasons for its study as 'quickening the powers of observation, and training the mind to research' and the 'elevating influence of a study of the Works of Creation'; he also gave a brief history of Indian botany from Rheede to the contemporary work of Royle and Falconer. By 1854 the main part of the syllabus was in six parts: Structural, Physiological (carefully avoiding 'speculative theories'); Systematic (including the distribution of each family, noting members with medicinal or economic uses); Economical; Geographical (including the works of Alexander von Humboldt, Joakim Schouw and Franz Meyen, the 'Peculiarities of the Flora of India' and the acclimatization of plants) and Fossils. He ended with instructions on drying plants and how to form an herbarium, with a discussion of 'Botanical drawings and descriptions'. For the class textbook Cleghorn also relied on Balfour – in the first year, 1852/3 he had managed to find eight copies of Balfour's *Manual of Botany* (ed. 1 of 1849, or ed. 2 of 1851),[31] and asked the EIC to purchase 100 copies of the more elaborate *Classbook of Botany* (ed. 1 of 1854) of which 70 copies arrived in August 1855;[32] another 100 were ordered for the following session. Textbooks were sold to the students by the Madras Government at half the cost price, repayable in half-a-dozen monthly instalments.[33]

On several occasions in his Indian life Cleghorn expressed an intention of writing a major botanical work – as Professor this was to have been a Manual of Indian Medical and Economic Botany.[34] Later when Forest Conservator, at the request of the EIC, this had morphed into a 'Manual of South Indian Botany' of which he got as far as providing the Madras Government with a sample in 1858,[35] but which had become less necessary the following year with the appearance of Drury's *Useful Plants of India*.[36] He certainly assembled much material that would have enabled the preparation of such a work: a large collection of prints, copies of illustrations from books and periodicals, drawings of plants from nature, and some suggestive fragments that survive in the RBGE collection – extracts copied from books, and notes sent to him by correspondents, stitched together by species. No such volume, however, ever materialised.

In Cleghorn's first year 1852/3 a native draughtsman was appointed to the College to make teaching diagrams (both medical and botanical) – this may have been P. Mooroogasen Moodeliar, but the following year, when Wight left India, the post was taken over by Govindoo. Cleghorn also asked Balfour to send out diagrams, some of which were received in April 1853.[37] There was a strong practical element to the teaching, and a museum of 'vegetable productions' and items of materia medica was commenced. A garden in the College grounds also appears to have been started. Following the Edinburgh tradition of field excursions the 'lads' were taken on six Saturday afternoon outings each session – to the Agri-Horticultural Society garden

CHAPTER 4

and nearby localities, for which Cleghorn requested the hire of *hackeries* (two-wheeled carts, usually bullock-drawn) to enable the party to get to more interesting sites beyond walking distance. For assistance in his first year Cleghorn had the use of the 'indefatigable Native Assistant of the College', P.S. Mootoosawmy Modeliar, and of Chinnatumby, a student who held a Lane Scholarship.[38] For the next three years Cleghorn employed as a teaching assistant a Native Surgeon (of 'second dresser' rank) called Francis Appavoo, who had already been assisting him in his District Surgeon's role; the assistant's job was to prepare specimens and look out drawings. Cleghorn encouraged the young man, who continued teaching at the College 'distinguished by remarkable ability, and by a courteous and considerate though firm demeanour to the students'[39] until 1859 when he moved to join Cleghorn as office assistant in the Forest Department. In December 1860 Appavoo was elected an Associate of the Botanical Society of Edinburgh but he died in January 1863, when Cleghorn sent a short, but appreciative, obituary of him for publication in the society's *Transactions*:

> From the number and social position of those who followed his remains to the grave, he appears to have gained the esteem and regard of all those with whom he was brought in contact; and he affords a good example of the success which has been obtained by the Madras [Medical] College in educating natives for the medical profession.[40]

It is clear from this, and from other reports and letters, that Cleghorn regarded his native students with paternalistic pride and affection – he told Balfour 'we have some lads of wonderful quickness who w[oul]d pass, I believe, the required examinations in any of the Home Universities'.[41] In 1855 there had been a scheme to put all students (Native, European and Eurasian) on the same footing and while Cleghorn thought that this would be laudable in an ideal world, under present conditions it would put the Indians at an unfair disadvantage, as they had yet to reach the stage of what might be termed an even playing field.[42] This report (which he withdrew) was probably part of ongoing reforms within the College aimed at taking it away from control of the Medical Board with its heavy military influence. The result was its transfer to the Department of Public Instruction, and, after gaining recognition from the several British Royal Colleges, it later gained an affiliation with the University of Madras. But by March 1855 the College was already becoming more international, with three students preparing to go to England to take the exam required for the post of East Indian Assistant Surgeon.

In terms of 'achievement', at least when viewed from a Western perspective, Cleghorn's most successful Indian pupil was Senjee Pulney Andy (1831–1909), one of the earliest Indians to obtain a British MD, though obtaining this came at enormous personal cost. With the help of Andy's descendants his life has been studied as part of Anantanarayanan Raman's extensive researches on the history of medicine and science in Madras.[43] Pulney Andy, a Hindu descended from an ancestor who gave his life in the service of the Gingee (= Senjee) kings, was born at Trichy, and studied at the Madras Medical College under Cleghorn. After this he was one of those who wanted to join the 'East Indian' branch of the army medical service, despite the fact that travelling to Britain to take the required examination carried the risk of severe social stigma on returning home – the loss of caste by the crossing of the 'dark waters'. Pulney Andy arrived in England in May 1860, and though the army posting did not materialise, he on 28 September obtained an MD from the University of St Andrews, becoming a Member of the Royal College of [London] Surgeons the following year. This coincided with Cleghorn's second home leave and they corresponded.[44] Andy's family had cut him off for living too close to Smithfield Market (at 5 Noel Road, Islington) so Cleghorn lent him the graduation fee. Cleghorn must also have suggested that Andy join the Linnean Society, of which he became a Fellow in 1861. On returning to India Andy was appointed Superintendent of Vaccination in Calicut; he later moved to Trivandrum and in 1863 he was baptised by the Basel missionaries in Calicut. News of his conversion reached Cleghorn in the Himalaya and pleased him greatly.[45] Andy became involved with a new movement, the National Church of India, which aimed at breaking away from Western missionary influence and developing a form of Christianity more compatible with Indian traditions, though this movement had fizzled out by the end of the nineteenth century. In what seems like a final turning away from his roots Pulney Andy also became a Freemason, though he tried to reconcile this by publishing a theory that Indian temples were based on the Temple of Solomon, which itself had been built by Masons. He published on a variety of other subjects including medicine (vaccination, and the use of margosa leaves for treating smallpox),[46] and on Indian crafts. In the *Transactions of the Linnean Society* he published a paper on teratological specimens of palmyra and coconut palms that he had encountered in Travancore.[47] The beautiful original drawings (reworked for publication by Fitch) survive and, if they are by himself – which they may well be – then he appears to have benefitted from exposure to the botanical illustration that was being undertaken while he was a student in the Medical College. Perhaps the difficulties in reconciling the social tensions in his life became too much and around 1900 he emigrated to London, where he is said to have married an English woman and died in 1909.

Another Medical College student who did well was John Shortt, who went on a two-year leave to Britain in 1853 with £400 in his pocket and the hope of graduating. Cleghorn asked Balfour to give the hard-working young man 'facilities for enlarging his mind & progressing in education', adding 'He is an Anglo-Indian & I trust allowances may be made for his colour and manners'.[48] Shortt ended up with the rank of Deputy Surgeon General at the time of his death in Yercaud in 1889. He knew Pulney Andy and like him also published on branching palms but also on other scientific subjects including snakes, and in 1884 he edited the book *Forestry in Southern India* by Cleghorn's one-time assistant Henry Rhodes Morgan by this time a Major General.

With the arrival of his new Forest Conservator duties and its associated travels, Cleghorn had to give up the medical Professorship. In 1856 he was asked to give the annual end-of-Session graduation address,[49] which may have been his farewell to the College. By this time the term had shifted to run from 1 September to 30 April, with the annual examination in May 'when Diplomas &c, and prizes of books are given to the most deserving Students'. The prizes included the Johnstone Gold Medal (Rs 50) for the best Apprentice of the Senior Class and a Government Prize of Rs 30 for the best Native Medical Pupil. A fund for the book prizes had been set up by C.V. Conniah Chetty, though the sort of prize that Cleghorn gave to favoured pupils (including Balfour's *Phytotheology*),[50] probably came out of his own pocket.

The Madras Agri-Horticultural Society

The Madras Horticultural Society had been founded in 1835 as a means of improving the productivity of a country in which it was perceived by Westerners that 'immemorial customs and habits have for ages stood in the way of progress and amelioration'.[51] To show its practical function (which allowed of a zoological element) the Society was soon renamed in the form known to Cleghorn. It was closely associated with the Madras Government, which owned the land known as Namasavoya Puttadah, next to the Cathedral, on which the garden was established and at various points added to. From 1840 the Government provided a monthly grant of 150 Rs and financial premiums – cash rewards – of between 70 and 500 Rs to growers and raisers (both European and Indian) of crops or livestock deemed worthy, from coffee and cotton to bulls and sheep. The Society also issued its own silver medal for similar achievements. Another role was to supply seeds and plants to members – and on a large scale; in 1857, for example, it distributed 6000 trees and 500 packets of seed.[52]

The Society's Patron was the Governor and its Vice-Patrons included the Nawab of Arcot, the Commander-in-Chief of the Madras Army and the Bishop. The President was a senior member of the Government and the General Committee was drawn from horticulturally inclined members of the civil and military services, the most senior of which were sometimes raised to Vice Presidents, and one of whom acted as Secretary. After Wight the most influential of the Secretaries, from 1843, was Francis Archibald Reid, as and when his military postings permitted. Distinguished foreigners including Wallich and Royle were elected Honorary Members, and conscientious committee members were sometimes appointed Extraordinary Members on their departure. From about 1840 a Superintendent was appointed to take care of the garden. The first one may have been J. Stent, who certainly held this role in 1853 when he was said originally to have been a 'Bombardier of Artillery, untaught in Botany'.[53] The anonymous first garden catalogue of 1848 has been attributed to Stent,[54] but given his background this seems unlikely and the stress in its title on the use of De Candolle's natural arrangement strongly suggests Wight's hand. The 618 species and varieties grown at this date show an ornamental bias, including 11 passifloras, 11 roses, poinsettia, ten cacti and 20 orchids, though economic plants were also grown (including ten varieties of grape vine), and some of the exotics that survived into Cleghorn's time including *Guiacum*, *Bignonia jasminoides* and *Ochroma lagopoides*. The Society received samples and reports of agri-horticultural experiments being undertaken in the mofussil, but these would have been seen only by committee members prior to Wight's effort to print *Transactions* – of which a single volume was issued in 1842. From the point of view of the Ordinary Members, drawn from the upper echelons of the Madras public, but at least in theory open to 'gentlemen of every nation', the main role of the garden was to supply fruit trees and seed of useful and ornamental plants and to provide a venue for an annual Flower and Vegetable Show, which took place in February or March. For Membership the entrance fee was 10 Rs with a quarterly one of 7 Rs, or a single composition fee of 300 Rs. Responsibility for the running of the Society was vested in a General Committee of Management, with 12 members, which held a Special General Meeting on the second Wednesday each December and monthly meetings on the first Wednesday of every month. General Meetings for members were held on the second Wednesday of the months of January, April, July and October.

This is the body to which Cleghorn was appointed Acting Secretary on 28 January 1852, during one of Reid's periodic absences. Later in the year this was converted into a tandem tenure of the post jointly by the surgeon and the soldier when it was noted that:

> the Committee congratulated the Society on obtaining the benefits of Dr Cleghorn's services, his eminent attainments in Botany, particularly in its economic study, being vouched for by many of the leading men of science of the day. "Professors Forbes, Balfour, Arnott, Christison, and Anderson, besides Sir David Brewster, and Drs Wallich and Greville, all testify abundantly … to the value of his services during his former residence in India, but to the very valuable assistance he has afforded while in England in matters connected with Indian Botany, especially in arranging the Indian Botanical collection belonging to the Museum of the University of Edinburgh".[55]

4. MADRAS POLYMATH, 1851–6

At this point the Society's Patron was Sir Henry Pottinger and its President John Fryer Thomas, who, during the course of Cleghorn's six-year tenure as Secretary, were replaced by Lord Harris as Patron (in 1854), and Walter Elliot as President (in 1856). Cleghorn was fortunate with the two Presidents under whom he served, as both were sympathetic in outlook – Elliot was a keen naturalist and personal friend, and although Thomas was at one point Chief Secretary of the Madras Government, he held radical views and advocated state intervention in matters such as famine relief and forestry. His brother Edward Brown Thomas, another Madras Civil Servant, was also keen on horticultural and agricultural improvement: as Collector of Coimbatore he had been involved with Wight's cotton experiments, and he was also active in agricultural and forestry experiments in the Nilgiris (on which he would later correspond with Cleghorn) (see *Cleghorn Collection* p. 145). Although, as will be seen, record keeping and documentation were not Cleghorn's forte, he threw himself into the activities of the Society, and must have been largely responsible for several major developments.

Almost immediately he started to compile a new list of the plants in the garden, which under Reid's hand, with imports from 'Cape of Good Hope. Mauritius, Poonah [i.e., Dapuri], Calcutta, Rangoon, Edinburgh &c', had now reached over a thousand species and cultivars representing an immense range of taxonomic and geographical diversity. Published by the American Mission Press in 1853 this pamphlet hardly lives up to its ambitious title *Hortus Madraspatensis*,[56] being no more than a list annotated with references to descriptions and illustrations published elsewhere. That Cleghorn was not particularly proud of it is suggested by the fact that it is more or less anonymous, the preface signed merely with the initials 'H.C.' It is possible that something more substantial had been embarked upon, but the first draft had been one of the victims of the fire at Oakes, Partridge & Co. More significant was the appointment of a new garden supervisor. Although Stent had provided long and faithful service, it was felt that someone with botanical and horticultural training was now required, and it must have been Cleghorn who suggested applying to J.H. Balfour in Edinburgh to find a suitable young man.

From its foundation one of the Society's most active promoters had been Robert Wight,[57] but, on 11 March 1853, the day came for him, with his wife and 'four bairns', to leave India. He was described by Cleghorn as 'the acknowledged chief of the Corps Botanique in the East, a man who for 35 years (in this trying climate) has worked well, & has scarcely relaxed from labour for a single day'.[58] Cleghorn and Reid read a testimonial address paying credit to Wight's 'fostering care' of the Society, to his botanical research which had as its inspiration 'utility and application to the comfort and happiness of man' undertaken with 'unshaken perseverance and unwearied industry … amidst impediments and difficulties that required no common zeal to surmount'.[59] Cleghorn's hand in the text of the tribute is evident in the bibliography of Wight's works, of which he considered the illustrated works to be the most important: 'what Sowerby's English Botany is in Britain, your Illustrations and Figures have provided for the Student of the Indian Flora'. There was, however, a silver lining to the cloud of the departure in that it allowed Cleghorn to employ Wight's artist Govindoo and also his 'botanical writer' (though he did not take on his lithographer);[60] he also personally employed some of Wight's 'old [plant] Collectors' who by October 1855 had helped him to fill 'three immense chests with specimens dried during the last year'.[61]

Other developments in which Cleghorn must have had a hand included the development of the Society's library, and, in December 1854, a request for extra funding, which led to an extra 100 Rs per month as a grant from the Government of India to the Madras Government. This

CHAPTER 4

raised the monthly grant to 250 Rs, comparable to that received by the Calcutta and Bombay societies. In June 1855 the Madras Government also proved generous, allowing the purchase of additional land that increased the area of the garden to 12 acres, with a nursery area for trees, and an ornamental section where the Garrison Band played on Tuesday evenings. The annual flower shows continued, the Governor-General, Lord Canning, being expected at the one in February 1856. In May 1852 a show of agricultural products was held, including coffee, sugar, the wool of a hybrid Merino-Mysore sheep, and papers and fibres submitted by Alexander Hunter; the gums and resins (kino, guiac, gamboge and cattimandoo) were reminders of the Great Exhibition and a foretaste of the Madras Exhibition that would come three years later.

In July 1852 Cleghorn started to have the garden plants documented visually, with illustrations by some of Hunter's art-college students, and almost 150 such drawings survive in the Cleghorn collection at RBGE, dated up to April 1859 (see *Cleghorn Collection* nos 37–73). Many are attributable to Govindoo, but the work of Hunter's pupils is much harder to attribute. Some are tentatively attributable to P. Mooroogasen but others must almost certainly be the work of John Binny and the two Rungasawmys who are all known to have worked in the Garden. Some of the drawings are annotated by Andrew Jaffrey who must have had a hand in selecting the plants for the artists. The plants depicted and their annotations provide an additional source of information on the plants in cultivation. Many of the exotics from across the globe (including Australia, New Zealand, the Caribbean, West and South Africa, Texas, California, Peru and Chile), probably came through the botanic gardens mentioned by Cleghorn in the preface to his *Hortus Madraspatensis*; but on the drawings are also to be found the names of various Madras Company officials. Some of these were committee members who sent plants from their own gardens or native species with hoped-for horticultural potential collected in the wild (e.g., Naggary Hills, Ramamally Hills) – among these were Colonels Francis Reid and Henry Colbeck, the judge G.S. Hooper and the civil servant Walter Elliot. Lord Harris, the Governor, also took an interest and called upon a connection from his previous job as Governor of Trinidad when in October 1855 William Purdie (the Trinidad Government Botanist and Superintendent of the Botanic Garden) sent a box of plants that Harris forwarded to the Agri-Horticultural garden, including *Simaba cedron*, *Mammea americana*, *Dipladenia harrissii*, and various Malpighiaceae. There is also a drawing of a chili from the Caribbean that was supplied by Harris. In addition to visual documentation, Cleghorn also made specimens of many of the garden plants, which survive at RBGE and can often be correlated with drawings. Behind some of them, for example, *Erythrina* × *bidwillii* and *Bauhinia monandra*, lie interesting stories that have only recently come to light (see *Cleghorn Collection* nos 42 and 43).

After Cleghorn's appointment as Conservator of Forests his travels made holding the post impossible and in February 1857, after six years of service, he was replaced as Secretary by his assistant in the Forest Department, John Thomson Maclagan. Shaw, in his précis of the Society's *Proceedings*, noted that it was at this point that systematic recording of the Society's *Proceedings* was resumed, and it cannot be coincidental that Maclagan had started his working life in an Edinburgh insurance office. Cleghorn remained on the General Committee until his departure on sick leave in 1860, whereupon he was elected an Extraordinary Member. After his return to India and move to Northern India in 1861 he continued to send reports and specimens (including boxwood) from the Himalaya, and attended meetings on his periodic returns to Madras in 1866 and 1867.

4. MADRAS POLYMATH, 1851–6

Andrew Thomas Jaffrey (1824–1885)

Mr Stent may have provided long and loyal service but had fallen into habits of intemperance. In any case Cleghorn saw the need for a professional gardener, commenting that 'a Botanical garden minus a practical gardener is a joke'.[62] He therefore asked Balfour to send out 'the best you can procure' – the chosen candidate was to have his outward passage paid and to be provided with a house and a monthly salary of £5. The best that Balfour was able to procure was Andrew Jaffrey who arrived at Madras with his heavily pregnant wife Mary on the steamer *Queen of the South* on 21 April 1853 (the passage for the pair had cost £64).[63] With him Jaffrey brought a Wardian case of plants and some teaching diagrams from Balfour, and a pint of chloroform from the chemist Duncan & Flockhart requested by Cleghorn along with two of 'Symes' Abscess Lancets from Young the cutler'. Jaffrey came of a horticultural family: his gardener father had been born in St Andrews, and Andrew was born in Nursery Cottage, Govan on 23 October 1824.[64] At Glasgow High School Jaffrey was a prize-winning pupil, so it was presumably a limitation of financial resources that dictated his following in his father's footsteps into an horticultural apprenticeship. In 1851 he was foreman of the Warriston Nursery (belonging to the firm of Carstairs, Kelly & Co.), where James McNab spotted his talent and recruited him for the experimental garden of the Caledonian Horticultural Society that lay on the opposite side of Inverleith Row; he then obtained a post in the greenhouse department of the nursery from Dickson & Co., from where he was sent to Madras.

Despite problems with the '*anti-Horticultural climate* of the Carnatic',[65] which included the deleterious effects of a 'Tropical Monsoon upon his Dahlia-Beds',[66] dealing with 'the multitudinous minds' of a horticultural society, and appalling family tragedies, Jaffrey transformed the Society's garden. The first tragedy was the death of his wife in childbirth, aged 30, from faulty mitral valves. Their son, George, though premature was healthy and was fostered by a Mr and Mrs Fitzpatrick, but two months later he too died. Most of the time Cleghorn was delighted with Jaffrey's work, though he did not entirely escape censure when taking comfort in a little liquid relief, but it was Colonel Reid who acted 'almost like a father' to the bereaved and isolated young man. However, Cleghorn was certainly pleased enough with Jaffrey to send a photograph of him to Balfour in 1855.[67] This was taken by Andrew Scott, the Assistant Assay Master at the Madras Mint, with whom Cleghorn had earlier raised the possibility of sharing bachelor accommodation though it is not known if this happened.[68] Unfortunately this photograph is no longer extant, so there is no record of Jaffrey's appearance.

As already noted Jaffrey supervised some of the drawings made in the garden; he also became an author, publishing five pamphlets under the general title 'Hints to the Amateur Gardens of Southern India' (1855–60) – these are full of useful practical information and represent a significant contribution to the scanty early literature on South Indian horticulture. The titles of the pamphlets are 'Seeds', 'Flora Domestica', 'The Kitchen Garden', 'Indian Vegetables' and a 'Calendar of Operations' and were highly successful; the Madras Government purchased copies for the gardens they were establishing for retired soldiers and in 1860 they were published collectively by Higinbotham, being reprinted in 1874 and again in 1883. Doubtless under Cleghorn's direction or inspiration Jaffrey contributed a collection of wood specimens from the Agri-Horticultural garden to the 1855 Madras Exhibition; these won a second-class medal and ended up in the Government Central Museum. He also went on field trips and in January 1854 had 'been out for a fortnight in the Jungles', doubtless collecting material for cultivation in the garden.[69]

CHAPTER 4

Jaffrey appears to have had a radical streak and Cleghorn wrote of him: 'he is a capital Gardener but like Scotchmen, of that class, is marvellously free and easy – and speaks to Lord Harris in a manner which I would not venture to do'.[70] This led to an empathy with Indians and Jaffrey wrote to Balfour:

> it has truly been a wonder and amazement to me to see how an all wise Creator has provided wholesome food by waysides in the fields and ditches for this poor downtrodden race of people ... but wherever the hand of John Company rules supreme Barren wastes & misery with few exceptions is the general rule. Millions of acres of the finest land uncultivated through a bad revenue system & worse land tax.[71]

Although such views were kept largely to himself, they are likely to have been one of the reasons for the enemies he admitted to Balfour as having made. In the summer of 1856 Jaffrey was sent on an excursion to the Nilgiri Hills to visit W.G. McIvor at Ootacamund, and then on to Bangalore to advise Sir Mark Cubbon on whether the Lal Bagh might be revived as a botanical garden. Jaffrey published an account of this trip (the 'facts' retained, the 'reflections' duly pruned) in Hunter's *Indian Journal of Arts, Sciences and Manufactures*.[72] Cleghorn joined Jaffrey in Bangalore and endorsed his views, and the intention was that Jaffrey would get the post of superintendent when the garden was established. This, however, did not happen (perhaps Jaffrey's irreverence came back to haunt him), and the job went to William New who was sent out not from Edinburgh but from Kew. Early in 1857 Jaffrey was released shortly before the expiry of his four-year contract with the Agri-Horticultural Society, after which he went to superintend a coffee plantation in Wynad for two years.[73] This appears not to have worked out as in January 1859 Cleghorn described him as being 'reduced very low, he is now in the General Hospital and will require a short trip to sea, he has been humbled'.[74] He must have recovered as later in this year he reported on a disease of nutmeg in Penang and Singapore. Jaffrey next turned up in northern India, firstly at the Calcutta Botanic Garden (1862,) and then at Darjeeling where he worked on the Quinine plantations (sending back beautiful specimens of *Cinchona* species and varieties for the demonstration herbarium his old boss James McNab, still in the RBGE herbarium) and as curator of the Lloyd Botanic Garden. It was here that his third wife Phoebe Pearce supplemented the family income by making and selling attractive volumes of dried ferns. Jaffrey died in Darjeeling on 1 November 1885 after a useful and productive life that had been entirely forgotten until the researches of his great great grand-daughters Margaret Sargent and Mary Beresford.

Robert N. Brown

On Jaffrey's resignation Cleghorn and the Agri-Horticultural Committee again sought Balfour's help to find a replacement. Robert N. Brown arrived by August 1857,[75] on a five-year contract and monthly salary of 200 Rs. It has been impossible to find out much about him, but on his arrival Walter Elliot described him to Cleghorn (who was travelling in Calicut) as '"puer ingenui vultus" – speaks broad Scotch, and gives promise of doing well'. Two possible candidates can be found in contemporary Scottish censuses or parish registers:[76] Robert Nay Brown born at Kirkmabreck, Kirkcudbrightshire in 1831, and Robert Nicholson Brown who was staying with his grandfather at Kingincleuch, Mauchline, Ayrshire in 1851 aged 14, and therefore born around 1837. From the fact that Cleghorn asked for reports on Brown's satisfactory conduct to be sent to George Lawson and Lady Foulis it seems reasonable to deduce that he had worked in her garden and taken classes at RBGE where Lawson was a teaching Instructor, though Brown's name does not appear in any of the surviving class lists of the period. His catalogue of the Agri-Horticultural garden, first published in 1862,[77] effectively a Flora of the Madras area, is a huge advance on Cleghorn's 1853 list, though I have seen no copy of its first edition, and the second (1866) may or may not have been significantly added to by its editor J.J. Wood.

According to Alexander Hunter, who in 1863 was Secretary of the Agri-Horticultural Society and wrote to J.H. Balfour asking for help in providing another Superintendent, Brown's tenure of the Madras post had come to a sad end: 'Mr Brown got into low company & absconded at the expiration of his term after a fit of Delirium tremens– I hear he has returned to his friends in Edin''.[78] After which no more is known of him.

A Visit to Thomas Thomson in Calcutta

Between 12 December 1855 and 10 January 1856, Cleghorn made his first visit to Calcutta, to confer with Thomas Thomson, Superintendent of the Botanic Garden. Cleghorn travelled on the P & O steamer *Bombay* (obtaining favourable rates from the Marine Board: 160 Rs there and 120 back for his own fare, and a return one of 40 Rs for his servant).[79] Thomson followed in the footsteps of Roxburgh and Wallich, and more immediately in those of another Scottish surgeon, Hugh Falconer. As with Cleghorn's recent employment, the Calcutta job also went with a chair of botany at the Medical College. Although slightly ahead of itself in this narrative this visit is best discussed here – as it concerned two areas of interest that were effectively about to end for Cleghorn (botany and medical education), and a major one that was about to emerge (forestry). Of this intended meeting Cleghorn wrote in prospect: 'I trust the Garden at Madras and the progress of science in S. India may be benefitted by putting our heads together', and among the other medical and forestry matters discussed was 'the best organisation for Botanical distribution of useful plants'.[80] Despite their obvious benefits, such meetings were extremely unusual in India, due to inflexible Company rules and attitudes to cost, but this visit was supported by the enlightened Lord Harris and a deputy – Charles Drew – was put in place to cover Cleghorn's medical and teaching duties during his absence.[81] It was on the steamer returning from Calcutta to Madras that Cleghorn read the important introduction on botanical geography to Hooker & Thomson's recently published *Flora Indica*, of which he wrote a review for the *Calcutta Review*.[82]

The Madras Exhibition of 1855

20 February 1855 was declared a public holiday in Madras; the city was in a state of high excitement as this was the opening day of the great Exhibition of Raw Products, Arts and Manufactures of Southern India.[83] The brain-child of the Governor, Lord Harris, it had been six months in the planning in which Cleghorn, temporarily relieved of medical duties, had been heavily involved. This was the first of the 'special duties' with which Harris would entrust Cleghorn, and there seems to have been a personal friendship between them – certainly Cleghorn was appreciative of the 'great advantages and facilities which I enjoy under the Government of Lord Harris'. In fact family connections may go back further, as Cleghorn's uncle John was with the Madras Engineers at Seringapatam in 1799, and this was where Harris's grandfather had made his name as commander of the EIC forces that defeated Tipu Sultan.

CHAPTER 4

Fig. 19. The Madras Exhibition, 1855.
Firing of canons from Saluting Battery, Fort St George. Anonymous wood engraving, *Illustrated London News*, 7 July 1855 (above).
The Banqueting House, in which the Exhibition was held. Pencil drawing by Louisa Cursham, c. 1823 (below).

At the sea-front royal volleys blazed from the cannons of the Saluting Battery at Fort St George, and beside it, from the great flagstaff facing the Bay of Bengal, four strings of flags fluttered in the breeze beneath a large red ensign. It had been hoped that the Governor-General, Lord Dalhousie, would be in attendance, but though he sailed past Madras at this time, suffering from illness brought on by overwork and the loss of his wife less than two years earlier, he went straight to the Nilgiris via Calicut. So after singing a metrical version of Psalm 100, and some prayers from Vincent Shortland, Archdeacon of Madras, George Harris himself declared the exhibition open. The venue was in the grounds of Government House, the magnificent Banqueting House, a large Neo-Classical temple with an applied Tuscan order surrounded by a broad terrace on a high arcaded plinth, approached from the north by a grand staircase the breadth of the building (Fig. 19). The hall was designed by John Goldingham and, although built in 1802 to celebrate the victory of Seringapatam, it might have been built for the holding of exhibitions, its lofty, double-height interior was well lit, an aisled hall with two superposed orders and a first-floor gallery. For the exhibition the central space was filled with displays of artefacts and colourful textiles, carpets were underfoot and suspended from the galleries; Cleghorn's department, 'Raw Products', occupied the gallery, and the western aisle was filled with a display of goods imported and exported through Madras port – 'the Tariff'.[84] Photographs, paintings and drawings were attached to screens; on the walls were displays of animal heads and horns, and trophies of armour. Outside, on the verandas, were displayed the larger pieces of machinery.

The organisation, like that of the Great Exhibition from which it drew its inspiration, was efficient and minutely detailed – widely promoted through the *Fort St George Gazette* and pamphlets, as testified by a surviving folder of printed papers in Cleghorn's archive. A General and an Executive Committee were set up under the Governor, for both of which Edward Green Balfour acted as Secretary.[85] There were 28 Local Committees across the Presidency, including the territories of Goa and Pondicherry, the nitty-gritty of the work being undertaken by two Sub-Committees – one for 'Raw Products', whose secretary Cleghorn was; one for 'Machinery, Manufactures, Sculptures, Models and the Plastic Art' for which Drs George Smith and Alexander Hunter acted as secretaries. Noteworthy is the extent of Scottish involvement in the enterprise – of the 14 members of the General Committee nine certainly (and from their names probably 11) were Scottish, and of these medics played a major role. The aim, by means of pecuniary inducements and medals, was to attract as wide a range of exhibits as possible. Each of the Sub-Committees had 5000 Rs to award in prizes of between 50 and 300 Rs. Vast thought and effort went into the classification of the material received – there were 30 official Classes, virtually identical to those used for the Great Exhibition, but one of those for Raw Products was divided into eight Sections, and one of the arts into two, so there were actually 38 categories. For each of these a Jury was appointed whose job it was to judge the exhibits submitted. Cleghorn was delighted at the extent of the 'native' participation, some 640 Indian exhibitors;[86] and of its benefits he wrote 'it is universally acknowledged that this Public Exhibition admirably served its purpose of exciting attention to many novel products stimulating the Curiosity of the Hindoos, and directing public opinion'.[87]

Although Cleghorn's work on the exhibition was largely on behalf of others, he also exhibited material himself in three categories: 28 gums and resins from trees in the Agri-Horticultural garden, 43 wood specimens collected eight years earlier in Shimoga, and a collection of botanical diagrams and drawings. As well as the cash prizes and 'honorable mentions', the Juries awarded medals of first and second class.[88] The Reporter had the additional burden of writing up a report for his section. Cleghorn was a member or associate of 11 Juries and Reporter to eight of these: as usual he took the writing up extremely seriously, which lead to significant publications that reflect his existing medical interests, his embryonic forestry ones, and, to a lesser extent, those in the arts. To these reports he brought his extensive reading, citing historical literature from Roxburgh and Heyne to comparative information gleaned from contemporary journals such as the *Edinburgh New Philosophical Journal* and the *Transactions of the Royal Asiatic Society*. In addition to those awarded by the Juries, the General Committee issued medals of three classes – Lord Harris, Balfour, Cleghorn and Hunter received silver 'special medals of the first class'; ordinary 'medals of the first class' went largely to local rulers (but Cleghorn also got one); and 'medals of the second class' mainly to the (Western) members of the Local Committees.[89] The Jury awards were bestowed more widely than the Committee's, many (particularly in the textile categories) going to Indians, and four second-class medals, two honourable mentions and one pecuniary prize, were even given to ladies (all Western).

'Elite' contributions

The Exhibition was a major event in Cleghorn's life and reveals much about his interests, but also the interactions with others that enabled him to achieve so much. It is therefore worth looking in detail at some of the contributions to the exhibition from people known to him and also to his Reports, starting with contributions from those who might be termed his 'equals' – that is, other Company officials. At this point Cleghorn was Professor of Botany at the Madras Medical College, so it is unsurprising that he should have been Reporter for the Jury on 'Chemical and Pharmaceutical Processes & Products Generally'. This elicited contributions from the medical confraternity of the Presidency that, agreeably to Cleghorn, proved that:

> Southern India is abundantly supplied with simple, energetic, and appropriate remedies, well adapted for the treatment of tropical diseases … [and that] … many other indigenous drugs … might be brought into use, and improved by the operations of the Pharmaceutical Laboratory.

This shows a concern not only with reducing the need for costly imports of drugs, but a willingness to accept indigenous medical knowledge. The best of the contributions were 241 specimens from Edward Waring in Travancore, and 243 from James Kirkpatrick in Hyderabad. The former were accompanied by specimens (three of which are still in the RBGE herbarium, with related drawings that Cleghorn had made from them), and with the latter were drawings that have not survived. The other first-class medal-winner was Dr Andrew Scott, for crystallising an active principle from 'country sarsaparilla', *Hemidesmus indicus* (Asclepiadaceae), which had been collected for him at Courtallum by C. Appavoo Pillay, one of the new class of Indian being trained as a 'Dresser' at the Medical College, where some of the students continued to work as assistants after qualifying.

From the Northern Circars Walter Elliot sent samples of a 'Gentianaceous plant … largely exported [as a febrifuge to Northern India] from the Northern Circars – about 7000 Rs worth from Vizagapatam alone. It appears to be *Ophelia elegans*'. This kind of 'chiretta' (now *Swertia angustifolia*) resulted in a paper for the *Indian Annals of Medical Science*, published like most of Cleghorn's papers in several places and formats (in this case in Edinburgh and Calcutta).[90] It included illustrations by Govindoo of this plant and of another febrifuge gentian, and of the bundles of the sixteen-inch-long, dried stems wrapped in the leaves of *Bauhinia vahlii* by one of Alexander Hunter's Art School pupils, probably T. Rungasawmy.

Elliot, one of Cleghorn's closest friends, was also heavily involved with the Exhibition. He served on the General and Executive Committees, chaired two Juries and served on a third; he also sent samples of *Euphorbia cattimandoo* gum that had won him a medal at the Great Exhibition. Visually more spectacular were Elliot's contributions of mounted animal heads and horns including that of a narwhal (just what did these people take out to India in their luggage?). These were in the 'Animal Substances' section of Raw Products, for which Cleghorn was also Reporter; in this there were some altogether more arcane substances – oils made from the fat of peacocks from Tinnevelly and from Masulipatam alligators, feathers and down from the undertail coverts of storks used for making ladies' boas, and marine turtle-shells from Travancore. Some of the most striking exhibits, serving to remind one of the underlying military nature of the British-Indian encounter, were those of historical armour – to these Elliot also contributed, as did the Commander-in-Chief of the Madras Army, George Anson, and the Raja of Tanjore. Henry Valentine Conolly, Collector of Malabar, despite his 'green credentials' in setting up the pioneering teak plantations at Nilambur, also displayed his status with contributions of the products of 'shikar' – some 'Variegated Panther Skins', to the animal substances section. His 'War Knives from Calicut' to the armoury, however, proved to be a terrible omen as, later in the year, such knives were turned against him by fanatical Mopillas escaped from Calicut jail. Another old friend of Cleghorn, Captain R.S. Dobbs who was in charge of Chitaldroog District, contributed to a more pacific section, that for woollen goods – a check blanket or 'cumblie' of the sort for which the district had been known since the time of Buchanan.

The first section of the natural substance class used in manufactures was 'Gums and Resins', with Cleghorn as Reporter. There was great interest in these materials at the time, especially India rubber (from *Ficus elastica*) and gutta percha or caoutchouc (from the Malaysian tree *Palaquium gutta*), for use as an insulator in the burgeoning telecommunications industry. The supply of gutta percha was becoming exhausted (as noted four years earlier by Cleghorn in his British Association report) so merchants were anxiously seeking 'new sources of supply'. Of particular interest was the exudate of a tree newly reported by General Cullen, Resident at Travancore, and Frederick Cotton from the Nilgiris – both received second-class medals for their contribution, of which Cullen's was known locally as *pauconthee* (in fact a Kannada rather than a Malayalam name). At this point Cleghorn was unable to put a name on the tree-source, which was widespread in the Western Ghats, but it turned out that Nicholas Dalzell had described it from Maharasthra in 1851 as *Bassia elliptica* (now *Palaquium ellipticum*), a member of the same family as the true gutta percha, Sapotaceae.[91] This was the start of Cleghorn's interest in the tree, and over the next few years he accumulated a collection of notes and drawings on it from Cullen and from his young medical colleague Charles Drew (who made fine analytical drawings of the flower and fruit); he also commissioned an exquisite watercolour from P. Mooroogasen Moodeliar, one of Hunter's students who worked for Cleghorn both at the Agri-Horticultural garden and the Medical College. This material was eventually assembled in an extensively illustrated memorandum printed in 1858 by Henry Smith at the Fort St George Gazette Press.[92]

Another pharmaceutical substance in which Cleghorn had been interested since Mysore days was gamboge, the product of various species of *Garcinia* used both as a pigment and a purgative. An apothecary from Shimoga, George Wrightman, sent a specimen of gamboge 'collected with much care' for which he received a second-class medal.

'Subaltern' contributions

Gamboge was placed in the pharmaceutical class, but the greatest number of contributions from Indians were, hardly surprisingly, in those devoted to Raw Products – used either as food or for manufactures; but there were also notable contributions in the fine and applied arts (especially textiles, which cannot be discussed here). Several of these came from men closely associated with Cleghorn: Andrew Jaffrey, Francis Appavoo, Pulney Andy, P. Mooroogasen, Govindoo and P.S. Mootoosawmy Moodeliar, suggesting that he actively encouraged their participation. Class III, 'Substances used for Food', included plantation crops submitted by Europeans (coffee from the Nilgiris, cocoa from General Cullen, and sugar from the Astagram Sugar Company run by the sons of Anthony Norris Groves); but surprisingly it also included tobacco, and as a reminder of events that were happening in Europe at the time, 40,000 cheroots ('lunka segars') produced on islands in the Godavery River were purchased by the Government to send to soldiers fighting in the Crimean War. Heading the list of the second-class medals, however, were contributions from two Medical College employees – Mootoosawmy, the 'Native Assistant', sent 'samples in bottles' of 52 cereals with details of their local names and prices (including 36 rice varieties), and Appavoo, Cleghorn's own teaching assistant, submitted 24 pulses 'neatly arranged on a stand'. Pulney Andy's prize-winning contributions, a sheet of India rubber and a sample of camphor wood oil from *Dryobalanops sumatrensis*, came from further afield, from the island of Labuan off the Bornean coast. Why he went there is unknown, but presumably with some sort of military expedition.

Cleghorn was Reporter for the 'Dyes and Colours' section, in which indigo figured largely (30% of Indian production then came from the Madras Presidency) dominated by European-run firms. These, however, were dependent on local knowledge and this Jury had three Indian associates, one of whom, Bala Chetty, provided a fascinating appendix to Cleghorn's report on how to produce a range of colours from purple to sky-blue and crimson from plants.

It was not only Indians who occupied 'subaltern' positions in British India and Andrew Jaffrey also made medal-winning contributions to the exhibition. Section 5 of Class IV concerned 'Fibrous Substances', for which Alexander Hunter wrote the long Jury Report – and it is worth pointing out that these reports were of international interest, summarised not only in a lengthy article in the *Calcutta Review*,[93] but in both London,[94] and Edinburgh.[95] The first of Jaffrey's contributions, for which he received an honourable mention, was for a collection of fibres from plants such as *Abutilon polyandrum*, *Hibiscus vitifolia* and *Fourcroya gigantea*, doubtless all collected in the Agri-Horticultural garden, from which he also sent examples of ceramic flower pots. Jaffrey's medal, however, was for a collection of 88 wood samples from trees collected in the garden for which Cleghorn, in what, significantly, was the longest of all his exhibition reports, paid credit for the 'careful botanical nomenclature' in the identification and labelling of the specimens.

The timber report

Section 7 of Raw Products was devoted to 'Timber and Ornamental Woods', with Cleghorn as Reporter, a significant prefiguring of the work that Harris would ask Cleghorn to undertake later this same year, with confirmation as Conservator following in 1856. The report opened with the words:

> The importance of this Section of the Exhibition, can scarcely be overrated, in a country like this, for it must be remembered, that the value of wood and timber, here, is not to be measured by the estimation in which they are held, in temperate climates. Here, they are not only applied to those economic uses, with which we are familiar, but they also furnish fuel to all classes, supplying the place of ... coal, which has not yet been found in any quantity within the limits of the Presidency. Besides this, the influence of trees in climate is very considerable, tending as they do, to prevent the too rapid withdrawal of moisture from the soil, a point of great importance in a country, where the heat of the sun is intense, and the supply of water is dependent only upon periodical falls of rain.[96]

This is of considerable interest as the only place in the lead-up to the establishment of the forest conservancy where the forest/climate question is mentioned either by Cleghorn himself or any of the other individuals involved.[97] It is perhaps significant that this section was chaired by Lieutenant Colonel George Balfour, brother of Edward Green Balfour who had already taken a serious interest in the climate question and published significantly on the topic. As a result of poorly labelled samples, which did not allow an accurate assessment of the trees that might be suitable for exploitation, Cleghorn the economic botanist used the report to make a plea for the better collection of samples, paying heed to native names – to be recorded using their original script not English transliterations – commercial names, specimens of flowers, fruits and leaves, wood samples permanently numbered by branding, and labels written not on paper but on 'cadjan' leaves, the traditional writing material of the Tamil country – slips cut from leaves of the palmyra palm. The report ended with a Classified List of 155 Madras timbers and a separate one of 22 species authorised for use as railway sleepers, which resembles the list Robert Wight had produced in 1850 relating to the Great Exhibition. In Cleghorn's instructions for collectors is an interesting and eloquent warning about appropriate methodology for the collecting of ethnobotanical information: 'every effort should at the same time be made to test the intelligence given by one individual, by enquiries from others'.

Fine arts

With Class XXX, 'Fine Arts, including also Coins, Books &c', Cleghorn was only marginally involved. This was notable for the prominence given to photography, something of a craze in Madras at the time, which would lead to the setting up of the Madras Photographic Society the following year by Alexander Hunter and Walter Elliot. The greatest star was the wonderfully named Linnaeus Tripe, who exhibited photographs of the intricately ornamented Hoysala temples of Belur and Halebid, and who, partly at Hunter's instigation, would in October 1856 be appointed Government Photographer, which included teaching photography for the two wet months of the year at Hunter's art school. But many others contributed landscapes, portraits and architectural images, including three of Cleghorn's medical colleagues: Andrew Neill, William Pritchard and Andrew Scott. Cleghorn was an associate Jury member for the Section on 'Carvings in Sandal Wood, Ebony, Ivory, Stone, Metals and Inlaid Woods' – conceivably the result of an interest dating back to Shimoga days, as the principal carving was undertaken at Sorab. Of more relevance to the subject of this book, however, are the works submitted by Cleghorn, the work of two of the artists in his employment at the Garden and the Medical College, which received honorable mentions. These were 'two very good series of Botanical Diagrams', presumably large-scale teaching diagrams of the sort used in Edinburgh by J.H. Balfour, and '7 Vols of original Botanical plates ... executed under his superintendence from living plants by Native Artists, Govindoo and P. Mooragasen'. Another artist, T. Chengulroy, got an honourable mention for drawings of arms and ancient pottery, perhaps the 'Chengulvaraja' who signed at least one drawing made for Cleghorn. One of the Fonceca family was ambitious enough to submit works in oil that were deemed to 'require finish and taste' – this may well have been John Joseph Fonceca who acquired sufficient status to be allowed to paint a full-length portrait of Sir Charles Trevelyan that now hangs in the Foreign Office in London. There were, however, other artistic members of the family and Simon Fonceca drew a series of sketches of local scenery and people that were lithographed by J. Dumphy; for arboricultural reasons Cleghorn had two of these prints in his collection – both depict the banyan, one of which is strangling a palmyra palm.

CHAPTER 4

Significance

Of the 1855 Exhibition far more could be said. The machinery and manufactured goods have been entirely ignored, as it has been possible to concentrate here only on Cleghorn's role, which was with the 'Raw Products' on which these other categories were based or depended. In fact much more research needs to be undertaken generally into this well-documented and pioneering exhibition, which has received scant attention from historians. The little that Peter Hoffenberg,[98] the historian of colonial exhibitions, has to say about the Madras 1855 Exhibition is taken entirely from the anonymous account of the event in the *Calcutta Review*,[99] and he seems to have been altogether unaware of the 1857 Madras Exhibition (as also of the 1864 Punjab Exhibition). Of significance was its pioneering nature – it was one of the first generation of what might be termed 'colonial spin-offs' from the Great Exhibition. Planned at exactly the same time, and coming to fruition a few months earlier, there had been only three similar events in the Colonies – a little-known one in the Town Hall in Bombay and two in Australia. The latter pair were in Melbourne (for which a special building was erected) and Sydney (held in the Australian Museum), and were displays of material that was to be sent to the 1855 International Exhibition in Paris.

The Exhibition was reasonably popular with the public – during its nine-and-a-half-week course it received 26,563 visitors, but this was a relatively small proportion of the population of Madras which in 1857 was estimated at 720,000 and by comparison Balfour's Museum received 202,000 visitors in 1855. From Cleghorn's perspective one of its greatest benefits was the extent of the 'native' involvement. The contemporary Calcutta view of its importance was not only that 'novel products ... have [already] attracted the attention of merchants' and previously unknown labour-saving machines made known, but more widely that:

> a spirit of enquiry has been set on foot which may be expected to lead to important results. Distant provinces have been made acquainted with each others products, and have had an opportunity of observing their own dependencies. In no country is this likely to be fraught with good as in India, as there is no country in which there is so little intercourse between places far apart.[100]

A view of the role of such exhibitions that Hoffenberg eloquently glossed:

> exhibitions offered the intellectual, economic, and cultural sinews, or "intercourse", to tie together the vast leviathan of British India by testing and promoting new technologies, agricultural methods, and crops for internal economic development and oversees trade.[101]

Mention has already been made of the reports that appeared in learned periodicals in India and Britain, but the exhibition was also discussed in contemporary newspapers and the *Illustrated London News*. The event was not entirely ephemeral, parts of it lived on in specimens that were distributed both locally and internationally – to Edward Balfour's Government Museum in Madras, and to the museums of Economic Botany at Edinburgh and Kew. The greatest sign of its success, however, is that soon after it ended it was announced by Lord Harris that a similar exhibition would be held only two years later, in February 1857.

Photography in Madras

Photography took off in India at an early date, first the daguerreotype, then paper-based techniques – initially the calotype followed by the wet collodion process. The work was initially undertaken by 'amateurs', often members of the military, for recording topography, but very early on came portraiture, and studies of archaeological subjects. It is only possible

4. MADRAS POLYMATH, 1851–6

to discuss the subject briefly here, with respect to Cleghorn's minor involvement, and further information on early Madras photography can be found in the work of Roger Taylor,[102] and Christopher Penn.[103]

Photography had made a deep impression at the 1855 Madras Exhibition, which led Alexander Hunter to repeat his 1853 request to the Government for support for a photographic teacher at his Art School. This time it fell on the receptive ears of Lord Harris, so the proposal succeeded and Linnaeus Tripe was appointed Government Photographer in October 1856, with a remit both for practical photography and teaching. Tripe started work in March 1857 and continued until the post was axed under Sir Charles Trevelyan's pruning of the Madras administration two years later. As part of this photographic enthusiasm a Photographic Club was formed in Madras by two of Cleghorn's friends, Alexander Hunter and Walter Elliot; meetings were being held by September 1856 that were reported in the *Madras Journal of Literature and Science*. It is noteworthy that while a similar society had been established in Bombay two years earlier, and in Bengal earlier the same year, this was a year before Brewster started the Photographic Society of Scotland. That Cleghorn took part in this early period of Madras photography is known from references in Hunter's *Indian Journal* and tantalising mentions in his letters to Balfour. For example, at a meeting of the Photographic Society on Thursday, 25 September 1856 (with Findlay Anderson in the chair, and Murray, Clarke, Cleghorn, Jesse Mitchell, Alexander Hunter, Dr Cochrane, Dr John Pearson Nash and Dr Andrew Scott present) Cleghorn exhibited:

> photographs on albumenised paper of the Medical College Hospital, Calcutta, by Newlands, and of the Medical College, Bombay; he also exhibited photographs of Professor Balfour of Edinburgh, and of Dr. R. Wight, late of Madras, which were pronounced to be not only excellent pictures but striking likenesses.[104]

At the same meeting botanical photographs were also exhibited: of the Sago Palm by Andrew Scott and of *Yucca gloriosa* by Jesse Mitchell.

Of photographs sent by Cleghorn to Balfour, Andrew Scott's photograph of himself and Jaffrey in 1855 has already been noted, and in 1859 he sent nine landscape photographs taken in the Nilgiri Hills by Captain James Buchanan of the 4th Madras Light Cavalry.[105] This was for use in Balfour's and Greville's anticipated publication 'Plant Scenery of the World', which, had it been completed or published, would have been a ground-breaking series of chromolithographed illustrations of what would now be called major world biomes.

The Madras Exhibition of 1857

Although out of chronological sequence, the 1857 Exhibition is appropriately discussed here.[106] From a Cleghornian perspective less is known about it than its predecessor because he was less heavily involved; by time it happened his employment and priorities had changed, as in the meanwhile he had been appointed Conservator of Forests. The structure and mechanisms were exactly as in 1855, with the same 30 Classes of exhibits under the presidency of Lord Harris; there were Local Committees, and a General and an Executive Committee with Edward Balfour as secretary to both who doubtless, once again, was the presiding genius of the whole event. This time Cleghorn sat on the General but not on the Executive Committee, was a Juror for six of the Classes and two of the Sections of the 'Manufactures from Animal and Vegetable Substances' Class, but was Reporter for none, submitted no exhibits and was awarded no medals. He appears largely to have delegated responsibility to his assistant in

CHAPTER 4

the Forest Department, John Maclagan, who was Secretary of the Madras Local Committee. There are few papers in the Cleghorn archive relating to this Exhibition, and knowledge of it derives mainly from a bound volume in the RBGE library containing the Jury Reports and those from eight of the Local Committees.[107] In July 1856, before forestry duties took over his life, Cleghorn had got together with Elliot to publish a pamphlet with the hierarchically arranged list of articles required for the raw products, but when the time came, in February 1857, Cleghorn's membership of the juries of Classes XXI and XXII ('Cutlery and Edge Tools' and 'Iron and General Hardware') were indicative of his new responsibilities, though these classes included objects like handcuffs and locks as well as arboricidal tools, though many were manufactured by that great establishment of deforestation, the Railway Workshop at Palghat.

This time Cleghorn was not even on the Timber and Ornamental Woods jury, though for it his assistant Richard Beddome provided a catalogue of timber trees similar to Cleghorn's 1855 one, listing the Telugu names for the trees of the Godavery Forests in the Northern Circars.[108] Other individuals who came under Cleghorn's sway, if not exactly employees, contributed – from the Ooty Garden William McIvor sent samples of woods, fibres, and a bundle of porcupine quills. Jaffrey was still active in Madras and won a general medal 'for services' to the exhibition. The report for Class III 'Substances used for food' is the lengthiest and most elaborate of all, thanks to John Mayer, Professor of Chemistry at the Medical College, who made an apologetic nod for his lack of botanical knowledge to an (un-named) Cleghorn, but more than made up for this in his chemical analyses of food grains. P. Mooroogasen Moodeliar contributed ten large portrait plates of the major cereal crops to this report, but the original drawings for these did not find their way into Cleghorn's collection.

One of the other of the reports was highly illustrated, that on 'Mining, quarrying, metallurgical operations and mineral products', the work of Alexander Hunter, who took the opportunity to reprint a series of previously issued circulars – two on 'Blue Mountain [i.e., Carboniferous] Limestone' and seven on 'Practical Geology'. These had been written with the laudable aim of stimulating interest in fossil collecting throughout the Presidency (as indicators to potential coal deposits), and were illustrated with beautiful woodcuts and etchings, some copied (unacknowledged) from the works of Hugh Miller. Sadly Hunter's geological knowledge was flawed and his good intentions brought down the vicious wrath of an anonymous reviewer of the Jury Reports in the *Calcutta Review*, almost certainly the geologist Thomas Oldham.[109] The review is fair, but one squirms for poor Hunter, and wonders why on earth he had not sent the circulars to a more competent geologist for vetting and correction before publication. What enraged Oldham was that they were issued in an official Government publication, and thereby invested with authority, and he used it as a call for better scientific education. In fact Hunter's education in such matters was more highly developed than that of many of his colleagues, he was merely out of his depth in geology and palaeontology – Oldham, whose major concern was also to use fossils to find coal, also mocked Hunter's attempts to equate stratigraphy with the Mosaic days of creation.

Class XXX, Fine Arts, under Walter Elliot, was largely devoted to photography, which had continued to grow in popularity since 1855, new exhibitors being Alexander Greenlaw, Jesse Mitchell and Henry Otway Mayne from Madras, and from Bengal Drs John Murray,[110] Alfred Mantell, and Josiah Rowe. Unlike its predecessor the 1857 Exhibition was itself recorded photographically, the first commission by the Madras Government from its newly appointed photographer Captain Tripe. The 35 photographs mainly depict groups of objects forming individual exhibits, but one image survives showing a crowded section of the central display, and is reproduced in Janet Dewan's catalogue of Tripe's work.[111] In the Report is a fascinating and minutely detailed list of artisans, which specifies not only how many 'moochies and painters' there were in Madras at the time, but where they lived. Out of 17,666 artisans, 142 were so classified and their geographical distribution was as follows: Peddonaik pettah (in Black Town) 105; Triplicane 17; Teeroovatasem pettah 5; Mylapore 12. However, it is two exhibits of paintings that call for comment; first a group of drawings of 'native figures on talc [i.e., mica]' that Elliot praised as 'remarkable for careful execution and finish'. The name of their artist was 'G. Mooragasen Moodeliar'. It is possible that the 'G' is a misprint of 'P', in which case they would be the work of one of Cleghorn's painters, but even if not, then he was surely a close relation, and of interest in proving that painters from a traditional background could 'cross over' to produce scientific work.

Margaret Read Brown (1816–1868)

A second group of paintings shown at the 1857 Exhibition is relevant to the Cleghorn story, and, though almost unknown until recently, more widely to that of the history of botanical art in India. Hidden away in Walter Elliot's jury report for Class XXX is the following entry: 'A series of Drawings from Flowers by Mrs Col. J.R. Brown are worthy of notice; they are excellent as botanical studies, very characteristic, and true to nature, being also well drawn but requiring more finish in colouring, especially in the larger leaves'.[112] Despite her perceived technical deficiencies the jury gave Mrs Brown a First Class medal and it emerges that Alexander Hunter had wanted examples of her work for the 1855 Exhibition – a request that shows her to have been a friend of Cleghorn: 'I wish you could persuade your other Lady friend [later in the letter given as 'Mrs Browne'] to exhibit 5 or 6 of her best Botanical drawings. I have seen nothing to surpass them in India and I think it a pity that such talent should not be represented'.[113]

Almost nothing was known about this obscure colonel's wife, though her sole publication (see below) did not escape the eagle eye of Claus Nissen,[114] when on 11 August 2009 I received an email from David Simonds, which began:

> I picked up your 2005 lecture on Robert Wight in the Linnean Society recently and read it with great interest, because my great grandmother Mrs Margaret Read Brown was painting the wild flowers of Southern and West India at the same time.

The links between these obscure facts are a manifestation of the resonances in time and space that have emerged time and again in the course of research into the RBGE Illustrations Collection. These extend beyond the realm of botanical art, as Mrs Colonel John Read Brown – denied her own name by the conventions of the time – turns out to have been born Margaret Mary Inverarity,[115] of an Angus family whose grandmother Margaret Duncan was sister of John Duncan of Rosemount and Alexander Duncan of Parkhill. The latter introduced the tree peony to Scotland, of which a direct descendant of the original plant still grows at RBGE. Margaret's father Captain David Inverarity was Secretary to the Bengal Marine Board who died prematurely at Penang in 1818, so that she and her sister Anne Lilly returned with their French mother Anne de Sorel to Paris where Margaret learned to paint. The two sisters (aged only 18 and 16) went back to India to be married to two soldiers of the Madras Army in a joint wedding ceremony in Madras in 1834. John Read Brown was in the 6th Madras Light

CHAPTER 4

Cavalry with whom Margaret would lead a peripatetic life around India for the next 27 years, painting plants as she went, especially in hill stations such as Mahabaleshwar and Ooty – it was presumably at the latter, or in Madras, that she met Cleghorn in the early 1850s. Margaret's composition, colouring and style were bold and confident and expressed on a large scale, frequently grouping several species together, while including no botanical details. In addition to Cleghorn, another player in this story is Arabella Roupell, wife of a Madras civil servant, who painted flowers in a similar style to Margaret and published a volume of drawings of Cape of Good Hope plants in 1849 to which Mrs Read Brown subscribed.

The Read Browns (he by now a Major-General) returned to London in 1864, where they were clearly ambitious both socially and in her case artistically. Her paintings were exhibited in the prestigious surroundings of Burlington House on at least four occasions between March and May 1866 at soirées both of the Royal Society and the Linnean Society, where they were highly praised in the contemporary press.[116] While not wanting to diminish Mrs Brown's hard-earned achievement, it is hard not to feel the poignancy that it should have been her work, rather than the much more highly finished work of Govindoo or one of the other Indian artists, that got to be shown in such surroundings. The Royal Society conversaziones, under the presidency of General Sir Edward Sabine, were glittering social occasions. Exhibits included the scientific (such as Frank Buckland's illustrations of the life-cycle of the salmon), but also the fine arts, including drawings by Nicolas Poussin lent by the Queen and contemporary sculpture by Thomas Woolner and Joseph Durham. The Prince of Wales attended the second of the three Royal Society conversaziones, which is probably what led to the patronage of his wife Alexandra when Margaret decided to follow Arabella Roupell's example and have her work published by subscription. Originally 20 parts, each of five plates priced at two guineas, were to be issued, but only two were, in 1868, as Margaret died the same year (Fig. 20). The dramatic paintings (the originals of which survive with her descendants, and a few that were donated to Kew in the 1960s and a second batch in 2015), were faithfully reproduced using the most advanced techniques of chromolithography available, for which Margaret had to go to Brussels, to the firm of François de Tollenaere.

The conclusion to this digression came with David Simond's generous presentation of a copy of the first fascicle of this extremely rare work *The Wild Flowers of Southern and Western India* to the RBGE library, where it joined Cleghorn's own copy of Mrs Roupell's similar elephant folio, and represents a colourful reincarnation of one of the exhibits from the 1857 Madras Exhibition.[117]

4. MADRAS POLYMATH, 1851–6

Fig. 20. *Beaumontia jerdoniana* and *Loranthus longiflorus* (now *Dendrophthoe falcata*). Chromolithograph by F. de Tollenaere after painting by Margaret Read Brown, from Brown's *Wild Flowers of Southern & Western India*, 1868. (RBGE).

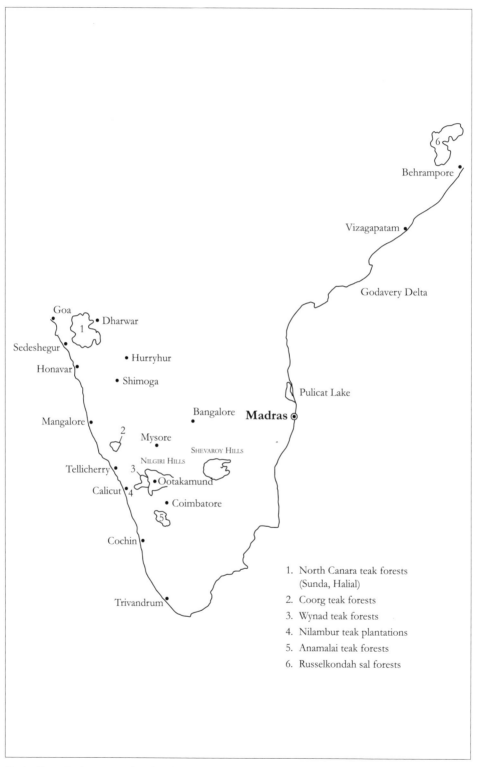

Fig. 21. The major forest areas of South India, simplified from the map in Cleghorn's *Forests & Gardens of South India* (1861).

5
Forests of South India, 1856–1860

'Applications [for land by coffee planters] should be liberally responded to'[1]

Three months after the end of the first Madras Exhibition, on 23 July 1855, at the suggestion of Walter Elliot, Cleghorn was summoned to an initial meeting with Lord Harris and the engineer Arthur Cotton, about the setting up of a forest conservancy for the Madras Presidency. Cleghorn's own description of this meeting was related in a letter to Balfour:

> The Governor informed me in his private room that the greatly increased price of Timber for Railways, Ordnance, Commissariat and Shipbuilding purposes had of late been so frequently brought before him that he considered a thorough exploration of the Forests necessary– He asked me if I was willing to organize a system of conservation, and desired me to state my views generally – I explained my opinions on the Timber question, and requested three days to consider the offer he had made to me.
>
> At the end of three days, I again breakfasted with Lord Harris, & told him that after mature deliberation I was prepared to assent to his proposal, & to undertake the management of the department – that I had weighed the magnitude and responsibility of the undertaking, wch. would compel me to give up my District and Educational duties at the Presidency.– I said that a long and interesting Journey on the Western Coast would occupy 4 or 5 months and the necessary organisation with the drawing up of a Report wd. occupy the first year entirely.
>
> I made no allusion to the allowance but I am sure that I may safely leave my remuneration in the hands of Govt. and be no loser.
>
> The whole matter is now under reference to the Court of Directors, in 8 or 9 months a reply may be expected, till then, we must stand and wait quietly.[2]

This document has been quoted at length, because, contra the view of Richard Grove,[3] it shows that it was neither the Court of Directors, nor the Company Surgeons (least of all Cleghorn) prodding the Government on any, least of all on environmental, grounds that led to the establishment of the post. There are in fact suggestions that its origin goes further back than Lord Harris, and ultimately to Arthur Cotton's brother Frederick, an engineer concerned with timber supplies, who had been behind the initiation of conservancy work of the Anamalai Teak forests as much as seven years earlier. George Birdwood, in his brief summary of the history of Indian forestry, wrote 'the Court of Directors ... sanctioned the extension of General Cotton's [Anamalai] scheme to the remaining forests in the Madras Presidency'.[4]

As explained in Cleghorn's letter nothing was to be rushed. After the discussions in Calcutta at the start of the year with Thomas Thomson, whom Cleghorn had asked to collect together Government of India papers on forest management, it was not until August 1856 that Cleghorn submitted firm proposals. This led to his appointment as Conservator on 19 December 1856 – a post he would hold officially for nearly eleven years, though actively so only for the first four, with short periods of activity between his periods of Government of India employment between 1865 and 1867. Though Cleghorn was unworldly about his salary (clearly having failed to learn from his grandfather's experience of the generosity of governments), his father was persistent with enquiries on the matter (concerned about how Stravithie was to be maintained),[5] and eventually the answer came back that his monthly salary

CHAPTER 5

was now Rs 1000 (which Peter took to include his son's army pay).[6] While not ungenerous (£1200 a year, the same as the salary of the Superintendent of the Calcutta Botanic Garden) this would clearly not allow Cleghorn to amass a fortune anything like the one his father had been able to as a lawyer in 13 years in the 1820s.

This was by no means the first time that forest conservancy had been attempted in Peninsular India.[7] In 1806 the Bombay Government had imposed control over the teak forests of Malabar, appointing a Forest Conservator (Captain Watson) who maintained a monopoly on teak for the Government, but in 1823 this had been stopped by Sir Thomas Munro, Governor of Madras, on the grounds that it was an unwarranted infringement on private property and that market forces were sufficient to ensure preservation. As early as 1837/8 the far-seeing John Sullivan, Collector of Coimbatore, who played such an important role in the development of the Nilgiris, had forbidden felling of trees in the mountains without permission, on the grounds that such felling was 'highly prejudicial to the springs that take their rise there and upon the irrigation on which the low country depends'.[8] The first steps towards a more modern system, with a slightly more holistic approach (and an allowance, if a limited one, for villagers' needs), came in the Bombay Presidency where from 1840 onwards Alexander Gibson made forest surveys, culminating in his formal appointment as Forest Conservator in 1847 with the primary aim of conserving teak. In Sind a Forest Ranger was appointed in 1847 and in the Western Himalaya a timber agency had been started in the Chenab Valley in the early 1850s. Even in the Madras Presidency there had been earlier forestry activity and as an alternative to the complex questions raised by managing multi-owned natural forests, the Madras Government had supported an initiative by its Collector of Malabar, Henry Valentine Conolly in his setting up of the Nilambur teak plantations in 1843. And, as noted above, in 1847 the engineer Frederick Cotton had drawn attention to the importance of the Anamalai teak forests, which led the following year to the appointment of Lieutenant James Michael to manage them. The Collector of Canara Thomas Law Blane had also drawn up and implemented some rules for the forests of North Canara (another teak district). But the teak question remained a fraught issue, with differences of opinion over private rights and public needs – both between different departments within the Bombay and Madras Governments, and between those and the Government of India in Calcutta.

The Teak Forests of Burma

At this point a digression on the pre-eminent teak forests of Burma is called for, as their management (issues including state ownership, the selling of leases to private merchants, over-exploitation and exhaustion of supplies) was to play a key role in the development of forest policy for India. The coastal territories of Arakan and Tennaserim had fallen to the Company in 1826 and the latter, southern, province was renowned as a source of teak for naval and other purposes, and one that was conveniently close to Calcutta. In 1827 Nathaniel Wallich, Superintendent of the Calcutta Botanic Garden, was sent as a scientist to report on the Tennaserim forests. The timber had historically been regarded as royal property, on which duty had to be paid for felling by timber merchants on a licence system, but Wallich pointed out that this would almost inevitably lead to over-cutting and depletion of the resource, which is exactly what happened over the next 25 years, despite attempts at control (the appointment of Captain Tremenheere as a superintendent in 1841) and further critical reports by two later

5. FORESTS OF SOUTH INDIA, 1856–60

Calcutta botanists – J.W. Helfer in 1837 and Hugh Falconer in 1851. It was the acquisition of Pegu in 1853, under Dalhousie's expansionist administration, and his appointment of John McClelland[9] as Superintendent of the Pegu forests that proved to be a turning point. Pegu included the lower reaches of the Irrawaddy and its delta, abundant teak forests and the cities of Prome and Rangoon, and completed the gap in the British territories of Lower Burma. In 1854 McClelland made a lengthy and detailed report on the forests, and it was this that drew attention to the need for a much stricter governmental control – if the forests were to be preserved and remain profitable. It was, however, the appointment in January 1856 of Dietrich Brandis as McClelland's successor in Pegu (with responsibility for the forests of Tenasserim added the following year) that marked the beginning of 'scientific forestry', and the setting up of a national Indian Forest Department to be discussed later.

Before that it is necessary to explode a myth about Dalhousie's personal role in this, which has been widely misunderstood, partly because of ambiguous wording in the account in E.P. Stebbing's influential history of Indian forestry, but not least because few of those who have quoted Stebbing have queried him or gone back to his original source. This is indeed hard to find: but a copy of the relevant letter can be found in the India Office collections at APAC F/4/2659 ff 131–138.[10] This was written to Captain Arthur Phayre, Commissioner of Pegu, on 3 August 1855 by the Government Secretary, Cecil Beadon in the name of the 'President in Council' also referred to in the letter as 'His Honour in Council', which immediately raises suspicions as to authorship and authority. Had it been written directly on behalf of Dalhousie (as Stebbing claimed) the titles used would have been 'Governor-General in Council' (or 'His Excellency in Council') and 'His Lordship in Council' respectively. Although the letter was an official one, and therefore under the overall aegis of the Governor-General, it was, in fact written on behalf of the Honourable Joseph Alexander Dorin, who had previously served as financial secretary to the Indian Government but whose role in 1854 as First Member of the *Supreme* Council gave him the title 'President in Council'[11] – Dalhousie was President of the *Legislative* Council. The letter was in response to McClelland's report, and its covering letter by Phayre,[12] which in Dorin's view was badly worded and flawed. The letter was first drawn attention to by Ribbentrop,[13] who described it as a 'memorable reply ... in which Lord Dalhousie laid down, for the first time, the outline of a permanent policy for forest administration'. Stebbing vastly inflated this claim,[14] calling the document a 'famous Memorandum' amounting to a 'Charter of the Indian Forests', a view that has been uncritically accepted by all subsequent writers on Indian forest history including Grove.[15] This distorted view was further inflated by Gregory Barton,[16] who seems to have taken Stebbing's ambiguous wording 'the ... letter ... communicating Lord Dalhousie's famous Minute on a future Forest Policy' to imply that there was a separate minute, when there was no such thing – had Dalhousie wanted to issue a minute on the subject in his own name he would surely have done so. Beadon's letter is rambling and not particularly clear, but the major claim it makes concerns the absolute rights of the Government to *teak timber*, which, by the annexation of 1853, it had inherited from the previous ruling dynasty. Its major concern is the question of how to get the highest price for timber on behalf of the Government, which should be an absolute price (at open auction) rather than by claiming a duty; it supports McClelland's view over Phayre's (who favoured selling the timber, by the payment of a duty, to a limited number of 'approved purchasers'), and in safeguarding the forests by the Superintendent marking trees for felling by Government contractors and not letting merchants into the forest. The letter, however, refers *only* to Government forests, and furthermore to only a *single species*, so can hardly be said to bear the enormous claims made on its *indirect* author's (Dalhousie's) behalf by Stebbing and others.

The reason for pointing this out is not to diminish Dalhousie (more than enough opprobrium has been heaped on his head, both by some of his own contemporaries on account of the annexation of Oude and the origins of the 1857 Mutiny, and more recently by historians of the post-colonialist school) – but rather to correct Stebbing's misinterpretation; to show Dalhousie's true views; and to give the credit for the real steps towards forest protection, which lie elsewhere, with another Scottish peer. There is no question that Dalhousie was seriously interested in the matter of protecting (and exploiting) forests, and *had* himself written a Minute on Arboriculture in the Punjab in 1851, which shows his concerns about deforestation and its climatic consequences,[17] and this same year he had appointed Captain Longden to survey the Himalayan forests. However, his Government's track record in such matters was far from unexceptionable, for in 1849, on receiving papers on teak requested from Bombay and Madras, it had objected to interference with private property believing unrealistically (as earlier had others, notably Sir Thomas Munro) that the free market would lead to protection:

> His Honour in Council finds it difficult to believe that proprietors of forests will be found, as a body, to pursue to an excessive degree the unwise and improvident course, the existence of which is assumed as sufficient proof of the necessity for the interference of the law to restrain them in the management of their own property.[18]

Stebbing quoted this passage noting that it demonstrated a 'surprising lack of statesmanship and want of foresight', but what he failed to notice was that it, like the letter to Phayre on Burma, was written under the authority of the Governor-General whom on forest matters in Burma he credited as 'a far-sighted statesman … far ahead of his times'.[19] To be sure, one response was with reference to *Government* property (Burma), the other to what was regarded as *private* (Madras), but a far-sighted statesman would have realised that the two were incompatible, and that some sort of resolution of the competing claims was required. The Scottish peer who actually did so was Lord Elgin, but that is to anticipate.

Cleghorn visits Burma

It was a logical step that, immediately after his official appointment, and before the first tour of his own new fiefdom, Cleghorn should make a trip to observe the teak forests of Burma and study their management. This he did, in January 1857, in the company of Walter Elliot and Sir Patrick Grant, Commander-in-Chief of the Madras Army.[20] This was Cleghorn's first meeting with Brandis, and though they could not have known it, it was a portentous one – a prefiguring of their close collaboration in working out national policy in Simla and Calcutta from 1863 onwards, following Brandis's summons from Burma by Elgin to organise a national Forest Department. Minor benefits from this trip were some ethno-botanical specimens sent to Kew that have already been noted. Seeds and plants were also taken back for the Madras Agri-Horticultural Society garden, including a handsome climber that Cleghorn and Elliot found 'widely diffused on the banks of the Salween River for fifty miles from Moulmein'.[21] This had already been sent to Madras during the Burmese war and was named by Sir William Hooker for Lord Harris as *Thunbergia harrisii*, but it is no longer regarded as distinct from *T. laurifolia*.[22]

5. FORESTS OF SOUTH INDIA, 1856–60

First Days as Madras Conservator

While not entirely novel, as already outlined, the Madras department was certainly a significant development in terms of co-ordination and extension, and its rapid success in trying to exert some degree of control of over-exploitation and wastage is greatly to Cleghorn's credit. In 1901 George Birdwood wrote of Cleghorn's work that he 'carried out the organisation of the new department in Madras with ... astonishing energy and success'.[23] E.P. Stebbing attributed this success primarily to Cleghorn's personality, local knowledge and respect for the 'natives':

> As a medical man his name was widely known and ... [this] gave him great influence amongst the people. They trusted him and believed in the disinterested [sic!] nature of his work and proposals, and were aware that he had an intimate knowledge of their mode of life and system of agriculture ... Cleghorn's popularity with the people and his known keenness for their welfare, so universally acknowledged, was naturally common knowledge amongst the higher officials whose confidence he enjoyed; and to this factor, more especially in the light of the subsequent retrograde policy introduced [under Beddome], may be attributed the signal initial success secured by the Conservator in this important matter of protection.[24]

In fact this is a more or less a quotation, if reworded and unacknowledged, of a statement in Brandis's article on 'Dr Cleghorn's services to Indian Botany',[25] but although contemporary and written by a friend it is purely anecdotal, and Brandis cannot have observed Cleghorn at first-hand in Madras, the period to which this empathetic picture must largely refer.[26] It is certainly a rose-tinted view and at the time Cleghorn's own father was far more perceptive: 'The natives probably consider the woods as their own & you as the Great Thief & entitled to no mercy'.[27] In any case the practical good sense expressed in Cleghorn's reports are probably sufficient explanation in themselves to account for what he was able to achieve – seeking to balance potentially conflicting needs of the Government, private owners and investors, and – at the bottom of the heap – the *ryots*. Cleghorn's prime aim was for the 'authorities to economise public property for the public good',[28] whilst being acutely aware of the 'difficulties in blending the interests of the State with those of private enterprise'.[29]

As a result of writing the British Association report Cleghorn was well aware of the nature of such competing needs, and that the result of absence of Government control in the past had been large-scale forest destruction, with deleterious effects not only on supplies of timber (especially teak) but also on other useful forest products and resultant environmental damage, perhaps including climate change. But to be a Conservator was not to be a modern 'conservationist' and Deborah Sutton has aptly denoted conservancy (in the case of the Nilgiri sholas) as 'the selective, and taxed, destruction of a barely known and unquantified resource'.[30] Cleghorn believed it a duty of the colonial state that forest should be cleared for agriculture for the development of new cash crops (especially coffee) and foreign investment. On the other hand, shifting cultivation had to be severely controlled or stopped, there should be no deforestation of watersheds, and certain forest areas should be protected for the production of teak, other useful timbers and non-timber products. As is also clear from his reports Cleghorn had a utilitarian hatred of wastage – whether from inefficient agricultural methods (kumri), bad harvesting or timber-processing techniques (axe rather than saw), or from natural causes such as wood-boring insects. These were views formed, and firmly held, *before* the commencement of his official employment in the nascent forest service.

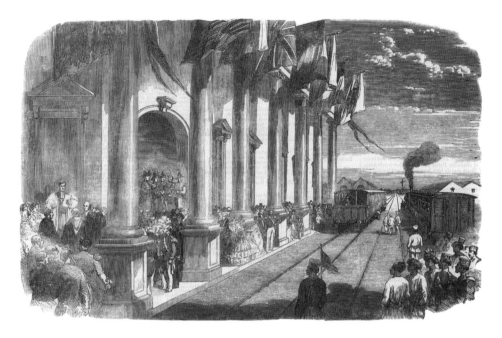

Fig. 22. The sources of two of the greatest pressures on timber.

'Opening of the Madras [to Arcot] Railway' by Lord Harris, showing the terminus at Royapuram, 1856. Anonymous wood engraving, *Illustrated London News* 6 September 1856 (above).

'Constructing the Jackatalla [later Wellington] Barracks', near Coonoor, Nilgiri Hills; the standing figure (right) is probably Captain John Campbell who superintended the work, and who also planted the Australian plantations. Anonymous wood engraving, *Illustrated London News*, 10 April 1858 (below).

5. FORESTS OF SOUTH INDIA, 1856–60

What he set up was a Forest Department, with its headquarters in the Presidency town of Madras, an office at Nungumbakum, and a small staff consisting initially of two European Assistant Conservators (also called Travelling Assistants, stationed in the important and *already managed* teak forests of the Anamalais and North Canara) and a European office manager.[31]

As his first Office Assistant in Madras in 1856 Cleghorn employed John Thomson Maclagan (1828–1897), who had trained as a clerk in an Edinburgh insurance office, so was ideally suited to write a Manual of Accounts for the use of the Forest Department.[32] In 1855 Maclagan married Euphemia Scott Parker who died in 1864 after their return to Edinburgh, suffering from what Cleghorn called 'Pthisis', presumably tuberculosis; two years later, Maclagan would marry Margaret Dalziel Pearson (1841–1877), one of Mabel Cleghorn's numerous second cousins.[33]

The Assistants managed establishments with European 'Overseers' (often retired soldiers) in their forest areas, the work of cutting and trimming being undertaken by hired Indian labour, usually recruited from local tribes (Wuddurs and Malsars in the Anamalais), haulage being by means of elephant. In the managed forests substantial amounts of timber extraction (e.g., for supplying the railway companies) were let out to private contractors, the part being undertaken by Government employees being termed 'amani'; the latter, being under closer control, was the more efficient if not necessarily more profitable. The annual budget for the first year was a modest Rs 28,000, of which Rs 17,000 was the salary for Cleghorn and his Office Assistant. When the Assistant and Overseer posts in the new department were advertised there were 200 applicants, but only a handful of these had any relevant experience, though in this, as in his Medical College work, Cleghorn was aware of the potential and need for training his staff. In this case in silviculture and account keeping, not only with the help of books (Brown's *The Forester*, though dealing with temperate conditions being the nearest approach to anything useful), but by personal tuition. Botanical instruction was also necessary, to avoid wastage through misidentification of tree species, so that valuable timber was not used in low-grade construction work through ignorance. He therefore wanted his staff to collect flowering and fruiting specimens of trees, and to keep a herbarium of such material in Madras (to which was added a collection of named vouchers sent from Kew). Another idea to help with identification was the circulating of nature-printed representations of 30 of the most useful timber trees to the Overseers.[34] It is not known if such a work was ever produced, but it may lie behind some of the nature-printed tree leaves in the Cleghorn collection (see *Cleghorn Collection* nos 123–5).

The qualities looked for in the Overseers were robust health, integrity, knowledge of basic forestry, accounting and local languages. The jobs, however, were far from attractive as they exposed individuals not only to the danger of disease, but loneliness and the temptation to corruption. Another disincentive was the salary, which was unattractive compared with similar posts in the police or revenue departments. Though nowhere mentioned by Cleghorn, there may also have been personal physical danger from those whose livelihoods forest-control was likely to interfere with. It is worth reproducing here the rules Cleghorn wrote to guide the Assistants as reproduced annually in the Madras Almanacs, one of the vehicles through which the new department was announced to the Madras public (much information was also provided in the local press including the *Fort St George Gazette* and *The Athenaeum*):

CHAPTER 5

> GENERAL INSTRUCTIONS TO ASSISTANTS
> 1. To keep a Diary of work done.
> 2. To obtain a complete knowledge of the quantity and quality of Timber in each Forest of the District, and to prepare a Forest Chart according to a fixed scale, indicating as far as possible the number and size of the more valuable Forest tees within their respective ranges.
> 3. To prevent any kind of depredation or damage being committed in the Forests, to aid in which the Civil Authorities will give every support in their power. It will be the duty of the Assistants also to make circuits of the Government Forests, and to prevent private individuals cutting or damaging trees of any description in them, and to complain to the nearest Police official against all who violate the order – to be dealt with according to the offence.
> 4. To improve the Forests by clearing, planting, and by unremitting attention to young trees. To see that Teak and other fine Timber is carefully seasoned, and to take care that no trees are felled except under the Orders of the Forest Department, and when felled that the Timber is properly protected. If Forests are being worked by Contract to watch against injury to seedlings and undersized trees, also to see that trees are cut near the root (always within two feet) and to guard against wastage of timber from any cause whatsoever.
> 5. To present Monthly returns and Half yearly Statements of the work done.
> 6. To supply Tabular Statements of the quantity and description of the seasoned and green Timber contained in their respective Districts. Personally to pay the Establishment, and to take every care of the health of the employees. Also to report every instance of neglect on the part of their Subordinates.
> 7. To give details of all sales effected whether by notification or otherwise.
> 8. To transmit to the Office of the Department as opportunity offers, any new or remarkable production of the Forests.
> 9. The services of the Assistants belong entirely to Government, and they are strictly prohibited from engaging in private transactions of any kind whatever.
>
> H. CLEGHORN.[35]

Cleghorn, it must be stressed, had no formal training in forestry himself: 'I had had no training in my youth specially to fit me for such a duty',[36] and in his *Encyclopaedia Britannica* article on 'Forests' he explained (doubtless autobiographically) that 'Many of the forest officers first appointed [in India] were chosen because of local knowledge and love for natural history rather than their knowledge of practical forestry'.[37] Even his scientific botanical education was, as will be seen, nothing like as extensive as that of Brandis, and in forestry he was largely book-learned (though with unusually wide reading in both English and French), with a healthy dose of common sense. Temperate forestry had, however, been absorbed in childhood – at the age of 14 his grandfather's North Plantings at Stravithie had, after 25 years, been thinned – the poles to make palings for fences, the branches as wind breaks and for field drainage. It had been the same story with Wight and his cotton experiments – a slightly naive belief on the part of the EIC that the scientific training of its medical staff was sufficient to qualify them for practical work in the field of what would now be termed agro-forestry.

The First Forest Report, 1857–8

In his first year, as in his other employments and activities, Cleghorn applied himself with marked enthusiasm, making the first of what would be three annual forest tours of the Madras Presidency.[38] He covered a 400-mile section of the Western Ghats – from the Anamalai Hills in the south to North Canara almost as far as Goa. During the year he wrote individual reports on topics including the forests of particular regions (the Nilgiri and Anamalai Hills, and Canara); on the Ooty garden;[39] on the Jackatalla plantations near Ooty; on the Striharikota forests (on

5. FORESTS OF SOUTH INDIA, 1856–60

the coast between Madras and Pulicat, the main source of firewood for Madras);[40] on timber prices; and on a substance he believed would be a substitute for gutta percha as an insulator for telegraph wires. The accounting year ended on 30 April 1858 and the following day, from Mangalore, he punctually submitted his first Annual Report, summarising his travels, with observations and recommendations on a variety of topics.

The tour had taken eight months, between March and November, and on it he took Govindoo, so that the itinerary can be pieced together by the dates and localities on the drawings as much as by the rather unspecific itinerary and chronology given in the Report. They started from Bangalore, via Chikmagalur and the Bhoon (or Bun) Ghat to Mangalore, then up the coast of Canara visiting the timber depots at Mangalore, Calicut, Tuddru, Kumta, Honavar and Sidashegur. In North Canara he met with his Assistant Sebastian Müller and visited the forests around Soonksul, Soondah and Yellapur from which timber was floated by river to the west coast, though this timber was no longer of substantial size. Cleghorn also looked at the Gund Forest, which retained much larger teak trees, but with problems of transport to the coast, for which either roads needed to be built or the banks of the Kalinada ('Black') River blasted to improve its flow. While in this area Cleghorn visited the Southern Mahratta country where he was 'delighted with the vast sheets of Cotton Cultivation, which is now provided with a great No. of admirable gins supplied from a Factory which is under the able management of Dr Forbes of Dharwar'.[41] Cleghorn then travelled back south, and from July to October seems to have been based largely in the Nilgiris where in July he wrote a report on McIvor's work at the Ooty garden (see Chapter 6). From Ooty Cleghorn made excursions to the forests of Waynad and Coorg (probably including Sampagee and the ghats of Perambady and Tambacherry mentioned in the report). In August he visited Conolly's 15-year-old teak plantation at Nilambur, and in July and September made excursions to Coimbatore and to the two nearby forest areas – Waliar, and the Anamalais where Richard Beddome was by now Assistant Conservator. During this period Cleghorn also visited the coastal timber depots of Ponany and Cochin.

The main concern of the Forest Department was to stop wasteful felling and to bring the forests into the 'control or regulation of authority'. The primary interest was teak, which occupied two-thirds of Cleghorn's time and (with the poon spar, *Calophyllum*) was required by the naval dockyard in Bombay. The main areas that retained teak timber of usable size were all visited on this first tour: North Canara, Wynad/Coorg and Anamalai. But the other major concern developing at this time was wood for railways and telegraph posts, and firewood for the settlements of Ooty and Madras. At this time the South-West railway from Madras to Ponany on the Malabar Coast was making rapid progress and had reached Salem – wood was needed for sleepers, but much was also cut for firewood for the construction workers and for charcoal for smelting, and there had been extensive felling for such purposes around Palghat, in the Shevaroy Hills, and in North Arcot. If not as thorough in terms of quantification and statistics as the contemporary German methods of 'scientific forestry', Cleghorn's methods were extremely practical: to survey and classify the timber into four classes and to divide forests into divisions, with rotation so that a proportion of the trees of first class (6 feet and above in girth) would be taken out of one division, allowing the second-class (4½ to 6 feet in girth) ones to develop as the cutting moved to other sections of the forest.[42] Another timber required by the government was blackwood (*Dalbergia latifolia*) for gun carriages. Some natural products also merited some sort of protection including red sappan (a red dye, *Caesalpinia sappan*, planted by Mophlas – Malabar Muslims as a dowry for their daughters), giant bamboo, sandalwood, *Acacia catechu*, kino resin (*Pterocarpus marsupium*) and the 'pauchontee' (*Palaquium ellipticum*) that had been discovered by Frederick Cotton in Waynad in 1855, for which

CHAPTER 5

Cleghorn had hopes as a gutta percha substitute.[43] He later published a lavishly illustrated paper on this supplemented with information and drawings supplied by William Cullen in Travancore and by Charles Drew,[44] but disappointingly to Cleghorn the substance was not highly rated when it came to be examined in England.

At this stage the idea was developing of what came to be known as 'Reserved forest', for the sole use of Government as the best way of protecting the Navy's interests, by 'consolidating ... that part belonging to Government as well as that rented ... into one Government forest, which should be a *reserve* in perpetuity for the Indian navy'.[45] It was also suggested that an officer should be put in charge of the Waynad forests that were currently under dispute between Mysore, Coorg and Malabar; Cleghorn also wanted overseers for controlling the over-exploited forests at Putur (South Canara), Waliar and in the hills west of Vellore. Another activity that had to be watched was the timber auctions that took place in the Anamalais and Mysore – where merchants ('sowcars') and others could legitimately purchase timber required for construction purposes (priced at 1 Rs per cubic foot). Cleghorn did not want to interfere with the needs of *ryots* for wood for their agricultural implements, or the branches and leaves that were burnt for fertiliser.

In addition to the management of natural forests, attention was also paid to plantations for commercial exploitation. The plantations at Jackatalla in the Nilgiris had been started in March 1856 by Capt. John Campbell, an interesting experiment of planting mixtures of Australian acacias for firewood, with nurse trees and native hardwoods that would become useful for timber; these were planted in a scheme derived from one in Brown's *The Forester* (Fig. 23).[46] The Nilambur plantations at the western foot of the Nilgiris, were of native teak, established by

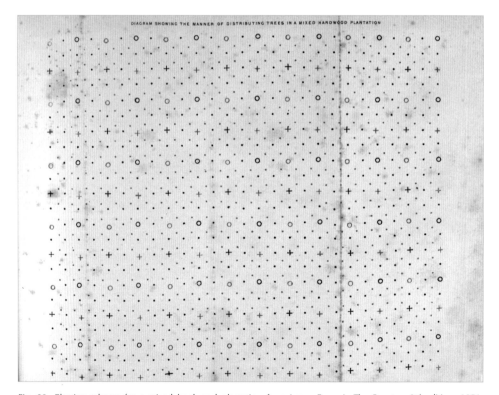

Fig. 23. Planting scheme for a mixed hardwood plantation from James Brown's *The Forester*, 2nd edition, 1851. This scheme was adopted for the Nilgiri Australian Plantations by John Campbell. (RBGE: Cleghorn's copy).

5. FORESTS OF SOUTH INDIA, 1856–60

H.V. Conolly, Collector of Malabar in 1843, which after Conolly's murder continued to be run by his assistant Chatu Menon. Cleghorn visited Nilambur and admired its running and wanted to establish similar plantations in North Canara and the Anamalais.

On the other hand, being keen on the investment of foreign capital, Cleghorn was also enthusiastic about the clearance of forest for coffee plantations, which he believed best suited to the wet climate of the western slopes of the Nilgiris ('Kundahs'), in contrast to the drier Eastern Ghats. At this time E.B. Thomas, Collector of Coimbatore was becoming worried about the clearing of shola forest for coffee plantations and asked Cleghorn for advice. His reply of 2 October 1857 is summarised in the First Report,[47] in which Cleghorn regarded three important issues to be considered when making over Government land to planters – the value of timber that would be destroyed, the protection of springs, and, least important, but still worthy of mention 'the ornamental appearance of the hills'. Cleghorn attempted a balance and recommended strict preservation of forest over 4500 feet for the sake of climate and springs, which was above the level of coffee cultivation and therefore should not cause controversy. However, he was wrong. This was unsatisfactory to the planters and brought him into conflict with W.G. McIvor, who believed that preserving this higher forest would kill the start of the tea industry (tea thriving at higher altitudes than coffee). In 1860, in his Third Report, Cleghorn would again use the same reasons, 'their beauty, but also from fear of injuring the water springs', for the absolute reservation of sholas close to the hill stations.[48]

The First Annual Report was (re-)published in *Forests & Gardens* followed by a minute written on it by the Government Secretary, Thomas Pycroft. The minute largely summarised the key points, but is noteworthy in that even the points that called for extra expenditure were agreed to, on which topics Cleghorn was asked to provide firm proposals.

Novara *calls*

Previously regarded as the 'benighted Presidency' this period of Cleghorn's life coincided with a flourishing of intellectual and cultural activities in Madras – it may or may not have been coincidence that this should have come to fruition under the Governorship of Lord Harris, as most of the strands had begun before his incumbency. It similarly remains a moot point as to the degree to which these were contributory factors for the revolution of 1857 scarcely touching the southern Presidency. Supporting evidence for the cultural renaissance comes from a rather surprising external source – the visit of a shipful of *savants* from the land-locked country of Austria – sent on the frigate *Novara* by the Archduke Ferdinand Maximilian, under the command of the dashing Commodore Bernhard von Wüllerstorf-Urbair, who visited Madras from 30 January to 10 February 1858 during a three-year circumnavigation. Although Cleghorn is not mentioned in the text of the popular write-up of the voyage by Karl von Scherzer,[49] in the official account his name is acknowledged in a list of those who helped the expedition in Madras. From letters home it is known that he accompanied the party on a trip to Pulicat Lake, and his sister Isabella read about her brother's participation in a German newspaper.[50] The Austrian savants (including the botanist Edward Schwarz, and horticulturist Anthony Tellinek, with a geologist, a zoologist, an ethnographer and an artist) were struck that in a town of 700,000, but with only 1600 Europeans, there was a thriving museum and zoo (though Balfour was not mentioned by name), a school of art (Hunter), a medical college, a literary society and a thriving interest in archaeology (Elliot). The officers and naturalists were hosted at a banquet at the Madras Club by Sir Christopher Rawlinson, and by Lord Harris at his annual children's fête at the Governor's country house at Guindy.

CHAPTER 5

The Second Forest Report, 1858–9

Cleghorn was eight months late in writing his Second Forest Report – although the reporting year had ended on 30 April, it was not dated until the last day of 1859.[51] In the same format as its predecessor, this outlined his travels, had sections on the major forest districts and on topics including plantations, auctions and sandalwood, with comments on events of the year, results, problems, and improvements actual or desirable. It is as hard to construct an itinerary from it as it was from the First Report; it is also notably incomplete, with nothing about most of the events of the year 1859 (though some of these, including a major trip up the east coast to Orissa, would later be included as appendices to the Third Report). The mileage clocked up on this year's tour was astonishing, exceeding that even of the previous year. Before embarking upon it Cleghorn had made several local excursions from Madras to Pulicat, including the one with the Austrian scientists and another to look again at the firewood supplies of the Striharikota forest (which led to the establishment of a set of rules on the cutting of firewood, which had previously been untaxed and uncontrolled).

The middle of April 1858 found Cleghorn in the hills of the Salem District, to the south-west of Vellore, from where he went to Coorg, then to Mangalore and up the coast to North Canara where he went inland as far as Bellary. By mid-June he was back in Madras, and the following month he returned to the Shevaroy Hills in Salem District, this time in the company of Lord Harris who was embarking on a major tour of his Presidency. Harris, however, was taken ill in Ooty so did not head south to Tinevelly and Madura as intended, though Cleghorn did so and visited the Coimbatore and Waliar forests followed by a week-long expedition with a large party to the Anamalai Hills in September (see p. 95). From here Cleghorn probably returned to Ooty where a large expansion of the Jackatalla (later Wellington) Barracks must by now have been nearing completion – the enormous quantity of timber (probably largely teak) required for this project is dramatically shown in a contemporary woodcut (based on a photograph) in the *Illustrated London News*, showing that there was interest in these large colonial projects (which included Cleghorn's work, if indirectly) even in Britain (Fig. 22).

A state visit to Travancore

Following his recovery, Lord Harris had been invited from Ooty to visit the State of Travancore; Cleghorn accompanied him, starting with a visit to the Nilambur Plantations on 19 November.[52] In view of the state of Harris's health he had decided to return from the Nilgiris to Madras by sea, for which the Indian Navy's steam frigate *Ferooz* had been sent to Calicut by the Bombay Government. This provided the opportunity for a cruise down the Malabar coast, though there seems to have been an 'incident' as in a reply by his father to one of Cleghorn's letters is a reference to a 'narrow escape you & all on board the *Ferooze* made from getting into Davy's Locker' – Cleghorn's second narrow escape from drowning.[53] No reference to this is found in any contemporary newspaper: perhaps the near-drowning of a Governor was considered best hushed up.

The Governor's party reached Calicut on 20 November where they stayed at the Collector's bungalow. The newspapers record a festive atmosphere: triumphal arches had been thrown across the Governor's route, and the doors and gates of the bazar were adorned with 'plantain trees studded with Areca nuts and dressed in garlands of flowers', the:

> Mophla sword men performed their feats of Tomfoolery, the Native music struck up nobody knows what, and cannonading commenced without reference to rank or the corresponding age of the personage in whose honour the salute was offered.[55]

5. FORESTS OF SOUTH INDIA, 1856–60

Rani Hayashabe, Bibi of Cannanore, the Muslim ruler, was there to visit Lord Harris, as was 'Kakassara Darasow, an opulent Parsee of Tellicherry'. A visit was made to Beypore and on 24 November the party reboarded the *Ferooz*, to sail to Cochin, accompanied by the *Beemah*, a lateen-rigged wooden sailing ship of the kind known as a *pattimar*. Lord Harris and his party disembarked and went to the British Residency at Balghatty where they were met by General William Cullen and his ADC Major Heber Drury. At home in Fife Peter Cleghorn was pleased to hear of this meeting as he and his brother John had known Cullen as a promising young officer in the 1820s.[56] Cullen, grandson of the notable eponymous Edinburgh Enlightenment medic, had since 1840 been Resident to the Courts both of Cochin and Travancore and was a keen naturalist who provided Cleghorn with botanical specimens, drawings and notes. So too did Drury, who shared Cleghorn's interest in economic plants and wrote the *Useful Plants of India*. More festivities awaited the party at Cochin where, on the night of 24 November, boats or *jagars* (a pair of canoes linked by planks) moored along the water-front from Mattancherry and Jew Town to Ernakulam were illuminated with blue lights and fireworks; paper fire balloons were let off, though the weather somewhat spoiled the effect. At Cochin the party remained for several days and on the 26th Cleghorn and his Madras medical colleague Dr James Sanderson visited the hospital at Ernakulam before attending a grand durbar in the Cochin palace where entertainment was provided by bejewelled dancing girls. The following day the party sailed on down the coast to Quilon, disembarking at Thangasseri, where they were met by the Dewan (Prime Minster) of Travancore, T. Mahava Rao. At Trivandrum, which they reached on 30 November, they stayed four days, with hospitality including another firework party. On 4 December Raja Uthram Thirunal and the princes, curious to see the steam mechanism, went for a short voyage on the *Ferooz*; the Governor's party then returned to Madras round Cape Comorin and Ceylon, but Cleghorn stayed on in Malabar.

Fig. 24. View of East Hill, Calicut by moonlight. Watercolour by Georgina Philips, December 1856.[54]

CHAPTER 5

In December the Madras Government sent Cleghorn to investigate, and recommend a solution to, a boundary dispute between Malabar and Mysore, in an area of thick forest between Sultan's Battery and Berambadi. The report of this expedition provides a uniquely surviving vignette showing Cleghorn interacting with locals, riding on elephants, examining boundary stones – one of which bore a Kannada inscription and figure of Vishnu holding an umbrella, locating the storage place of temple jewels, distinguishing between different castes, and possibly speaking Kannada; but also trying to reconcile this indigenous knowledge with demarcations made in three British surveys, going back to Colin Mackenzie in 1809.[57]

As in the previous year the Second Annual Report was only a summary of proceedings, and during the course of the year Cleghorn had also submitted reports on the Ooty garden (again), the Canara forests, on his Anamalai expedition, on fuel and firewood, and on the railway requirements of the Salem and Bellary Districts. Government demand for timber was ever-increasing, no longer so much from the navy, but from the Public Works Department and the military (for the building of barracks, gun carriages etc.), and from the Telegraph Department, which by now had lines that 'stretch across our Peninsula', with an experimental line in the Nilgiri Hills between Ooty and Coonoor. These lines were no longer supported (as previously) on small trees, but on sawn timber of teak, sal or *Pentaptera*. By far the greatest demand now came from the railways – in addition to the South-West line, a North-West one linking Madras to Bombay was now under construction, as were parts of the Great Southern Railway to Tuticorin (with a branch from Negapatam to Trichy) in which Cleghorn had personally invested money, purchasing 50 shares in January 1859. With 1760 sleepers required for each mile of track, each sleeper of three cubic feet, the railways simply gobbled up wood – Cleghorn estimated the annual demand at 253,000 sleepers, the equivalent of 35,000 trees. Only some of this could be met by imports (including *Eucalyptus* from the Swan River in Western Australia) – most had to be supplied locally, the best timber being teak or sal, but most coming from *Terminalia* and *Conocarpus* in the family Combretaceae, or from the leguminous *Pterocarpus*, *Inga* and *Hardwickia*. Even if treated, as Cleghorn recommended, with metallic oxides (using Bouchière's process), wooden sleepers had a limited lifespan, succumbing to the tropical climate and the depredations of the white ant.

It was this railway demand that led to the visits to the Salem District, where in August 1858 Cleghorn opened a further forest department establishment under a third Assistant Conservator, Louis Blenkinsop.[58] Railways led to another change in his staff: John Maclagan, who had performed such useful service in ordering the affairs of the Madras office, left in early 1859 to become Agent of the Southern Railway.[59] The accounting system having been set on a sound footing allowed Cleghorn to replace Maclagan with his former assistant from the Medical College, Francis Appavoo, whose primary function was to run the herbarium (though Cleghorn had still made no progress on his Manual of Botany). The running costs of the department had risen to Rs 41,000, but each of the three regional departments was running well and making a profit – that of the Anamalais Rs 97,000, Canara Rs 67,000 and Salem Rs 5000. Problems, however, arose through the lack of authority held by the staff of the department, so that assistants could be over-ruled by officers of the Public Works Department demanding timber at low rates. Cleghorn commented that his work was only possible through the goodwill of the Revenue Department and its key officers, the District Collectors.

5. FORESTS OF SOUTH INDIA, 1856–60

In this year control of forests was extended to include those of his old stamping ground Nugger and also to those of Wynad, and of Sigur (to the north of the Nilgiris, added to the Mudumalai forests under the responsibility of Captain Henry Rhodes Morgan); in addition Overseers were placed at Puttur and Hallihal in South and North Canara respectively. Seven new species were added to the list of reserved species (*Pentaphora, Cedrela, Inga, Pterocarpus, Mimusops* and two of *Artocarpus*).

Cleghorn was also attempting to increase awareness of the importance of tree-planting, both by private individuals (along water-courses and beside wells), and along roads ('avenue trees') by the PWD. This was helped by the fact that the gardens of Ooty (under William McIvor), the Lal Bagh in Bangalore (William New) and the Madras Agri-Horticultural Society (Robert Brown) were now viewed as adjuncts to the Forest Department and could provide seed and seedlings. McIvor was doing his bit in and around Ooty, planting Australian acacias (*A. robusta* and *A. melanoxylon*), cypresses, willows and Himalayan deodars some of which can still be seen today. Always keen to develop the economic potential of native species, Cleghorn's subject of investigation this year was a species of wild madder, munjit (*Rubia cordifolia* – see Cleghorn Collection no. 110), a common plant on the Nilgiris, which had been sent to England and Calcutta for testing its dye properties.[60]

The Second Report ended with reports by W. Fisher, the Collector of Canara based at Mangalore, reporting on Sebastian Müller's work on what were the most extensive and usually most profitable of the Madras forests, and by Beddome on the Anamalai forests. To give an example of the scale of the Anamalai operations, in this year 319 teak trees were felled and converted into 1565 dressed pieces. Of these a third were rejected and another third were of small dimensions. Only a third were usable for naval and other purposes (ranked into three classes), cut into 2424 planks for the Bombay dockyard. Beddome was having constant problems with his native work-force, and was unable to supply the blackwood required by the army (which he therefore had to purchase from an Indian proprietor), and there was no spare wood for a public auction this year.

During the year 1858/9 Cleghorn made two other major excursions that were not covered in his Second Report – the first to the Anamalai Hills in September 1858, the second to the Northern Circars in February and March 1859.

A week in the Anamalais

The expedition to the upper parts of the Anamalai (= Elephant Mountain) in Coimbatore District is known in detail from a lavishly illustrated report, of which Cleghorn was clearly particularly proud – it was read to the Royal Society of Edinburgh on 29 April 1861 and printed the same year both in the society's quarto *Transactions* and in *Forests & Gardens* (in which six of the seven the illustrations had to be concertina-ed into duodecimo format) (Fig. 25).[61]

Although lasting only a week (15 to 22 September 1858) this was a major expedition in the old 'statistical' tradition, to an area of great potential, which, despite its dramatically rugged terrain, had for a decade been exploited for its teak forests and for coffee planting (the latter, interestingly, by an Indian, Ramasami Mudeliar). The party consisted of Revenue Department officials (the Collector of Coimbatore and his assistant), four military personnel (including three engineers), two from the medical service, the foresters Cleghorn and Beddome, with Douglas Hamilton as artist – though this classification is an overlapping one as all the men

CHAPTER 5

except for the Revenue officials held military rank. To carry the gear seven elephants were taken: 'the most trying work for laden elephants is crossing the mountain streams, as the sloping boulders offer a precarious footing for these heavy animals', evoking a vision of Hannibal's crossing of the Alps. The reason for the diverse team lay in the expedition's aims and the range of subjects to be investigated, starting with mapping, and proceeding (in no particular order) to:

Forests (valuation of the standing crop of teak and other timber species).

Archaeology (an ancient cromlech was discovered).

Ethnobotany (the building material of the huts erected for the party's accommodation)

Ethnography (three tribal groups were ranked – the Koders 'lord of the hills', hunters who were 'truthful, trustworthy and obliging', collectors of honey from bees that nested on sheer cliffs reached by an eighty-foot long chain made of rattan links; the Paliars, herdsmen and merchants; and, lowest of all, the Malsars – shifting cultivators.

Colonisation opportunities (especially for coffee plantations, and the possibility of setting up another hill station or 'sanatarium').

Natural History: (geology, botany and zoology, particularly game animals: bison = gaur and ibex = Nilgiri tahr).

For botany Beddome's contribution was particularly valuable as he had already embarked on his extensive taxonomic study of the South Indian flora. It was Beddome who was the true successor to Robert Wight in a field to which Cleghorn, somewhat surprisingly, made virtually no contribution. The expedition did, however, give Cleghorn the opportunity to describe habitats and plants in a way unique among his other reports, but the most unusual feature of the trip was the employment of a (military) artist to record the landscapes. Douglas Hamilton already knew the area well, having briefly (before being recalled to military service in 1857) preceded Beddome as Forest Assistant in the Anamalai teak forests. Hamilton made large numbers of delicate ink sketches in South India (see *Cleghorn Collection* Fig. 28),

Fig. 25. Expedition to the Anamalai Hills: 'Foot of Punachi Pass, crossing the Torakudu River'. Lithograph by W.H. McFarlane after drawing by Douglas Hamilton, reproduced in Cleghorn, 1861a & b. (RBGE).

5. FORESTS OF SOUTH INDIA, 1856–60

and used some of his drawings in his memoirs of life as a *shikari*. Many of these, including the originals (and more that were unpublished) of the Anamalai drawings, survive in the British Library. Through J.H. Balfour Cleghorn obtained a grant from the Royal Society of Edinburgh to have seven of Hamilton's drawings turned into lithographs by W.H. McFarlane for illustrating his paper, though these chalky images scarcely do justice to the delicate lines of the originals.[62]

The route taken by the party was from Pollachi southwards to the village of 'Erular' (probably modern Aliyar), up the 'Torakudu' river towards the twin peaks of Tangachi (2380 m) and Akka (2483 m). This area is still preserved, as the Indira Gandhi Wildlife Sanctuary (since 2008 a Project Tiger Reserve), though the Aliyar River has been dammed at three points for the purposes of irrigation and the creation of hydro-electricity.

Excursion to the Northern Circars

On 31 January 1859 Cleghorn boarded H.M. steamer *Dalhousie* for the start of what would be a two-month excursion up the Coromandel Coast, with the primary aim of investigating the sal forests of the Gumsur District (now in the State of Odisha) – 'the most valuable tract of wood on the east coast'. Details of this excursion are known not only from Cleghorn's report,[63] but from one of only a handful of his notebooks to have survived.[64] On the way there and on the way back to Madras he stopped at various points noting much of interest, including not only timber and forest resources, but gardens, sand-binding plants and, at Masulipatam, an interesting experiment by the local judge, John Rohde, of planting several species of mangrove for use as firewood. From Vizagapatam a visit was made to the garden of the late Goday Surya Praksa Rao (d. 1841), a zamindar, at Ankapilly, in which he had planted exotic trees including mahogany and *Melaleuca cajaputi*, some of which had been supplied by Wallich from the Calcutta Botanic Garden in 1828.[65] Cleghorn was so impressed that he got the Madras Agri-Horticultural Society to send Rao's heirs a silver medal. He must also have visited the site for a new sanitarium at Galiparvatam in the Vindyan Hills inland from Vizagapatam about which he later wrote a report on its coffee-growing potential.[66]

Cleghorn reached Behrampore on 12 February, from where he investigated what are the most southerly sal forests, around Russelkondah and Aska on the Duha River. He confirmed the identity of the tree as the same as that of the Sub-Himalayan tracts (now *Shorea robusta*), was impressed by its prolific and rapid germination leading to the formation of almost pure stands, and its rapid and straight growth (see Fig. 76). In the past the felling of timber had been 'systematically discouraged' by the local rajas, on the grounds of the dense forests rendering the country 'less accessible to a military force' – an interesting example of indigenous forest conservation practice. The local tribe the 'Kunds' (Khonds = Kandhas) were more interested in other useful forest species, including wild date and mango, wood-apple, the dye plant *Rottlera tinctoria*, and especially in two liquor-yielding species: toddy palm and *Bassia latifolia* (which, in Cleghorn's disapproving view, were in danger of making them a 'dissipated people'). However, there were increasing demands from Europeans – the need to open up the country and build roads, and timber demands from a factory at Aska. Cleghorn recommended to the local Collector that the forest should be reserved and put under an Overseer, who could control felling and allow a supply of timber that could be exported from Ganjam to Madras for railway and telegraph use. For the felling and carting of timber Cleghorn was particularly interested in employing 80 Meriahs who had been saved from a form of ritual sacrifice by the Government (they called themselves 'Sirkar ki Bucha'). These unfortunates were purchased for the purpose by the Khonds, tied to a post, plied with hopefully anaesthetic toddy and killed with a ritual knife; their body parts were then strewn on the earth to appease the earth goddess Tari Pennu.[67] On the return journey down the coast Cleghorn visited Arthur Cotton's great irrigation operations in the Godavery Delta, and the works at Dowlaishwaram, returning by boat to Madras from Roxburgh's old stamping ground of Cocinada.

Further Travels of 1859

Lord Harris's five-year term of office had come to an end in April 1859 and he was presented with a testimonial on 18 March to which Cleghorn had subscribed Rs 50 (£5).[68] Harris's replacement as Governor of Madras was Sir Charles Trevelyan. Trevelyan's early career had been spent in India; as a leading Anglicist on the Committee of Public Instruction he had already started the reforms that were formalised in Macaulay's (in)famous 1835 Minute on Education. Thomas Babington Macaulay had arrived the previous year as Law Member of the Governor-General's Council with his sister Hannah on whom he was emotionally dependent. Hannah fell in love with Trevelyan, so by the time of the Minute the two were brothers-in-law, and living in a (platonic) *ménage à trois*. After their return to London Trevelyan had made a name for himself as a radical civil servant as Assistant Secretary in the Treasury, and had been sent to Madras to cut costs though the thought of Lady Trevelyan's defection to India hastened Macaulay's death.[69] Trevelyan accordingly used 'the pruning knife most freely', with a ten per-cent cut in government salaries, and the raising of heavy import duties on European goods.

Cleghorn regarded the new Governor as 'an able but most eccentric man',[70] but they had a remote point of contact in that Trevelyan's cousin Sir Walter Calverley Trevelyan was an early and active member of the Botanical Society of Edinburgh. At this time Cleghorn was pursuing his interests in medical missionary work and asking Balfour to campaign for funds in Edinburgh but asked for the request not to be put 'prominently forward' in the press as 'the Anti-Missionary party are always glad to run down Public Officers' such as himself.[71] In May and June he was in the northern part of his territory, attending the annual auction of sandalwood in the Nugger District then going on to the timber (primarily teak) auction in North Canara – Cleghorn supported the sale of timber by such auctions of timber not required by Government (mainly of second and third class), on the grounds that if there were no legitimate way of purchasing timber required for purposes such as construction then illegal felling would be the inevitable result. After the sales he went to the Nilgiris via Mysore and was based at Ooty in July and August, investigating and writing reports on the Mudumalai Forest and on a new teak plantation at Coonoor. But a new scheme was in view, with potentially enormous medical benefits – the introduction of *Cinchona* to India. While the Dutch in Java had made great strides in cultivation, British efforts in India had so far been unsuccessful. Clements Markham had been despatched to Peru to collect plants to be taken back to Kew, and preparations had to be made to attempt to grow them in Ceylon and the Nilgiris, the latter under the care of William McIvor. Due to his sick leave Cleghorn would miss out on this first introduction, but he continued to take an interest in it and would meet with Markham on the latter's second Indian visit.

From the Nilgiris Cleghorn made a trip to the south, and on 1 September was in Palghat, discussing timber requirements with his former office assistant John Maclagan in the latter's new capacity as Agent for the Southern Railway. It was on this trip that Cleghorn appears to have met the German missionary Johann Friedrich Metz (1819–1885) of the Basel Mission at Mangalore.[72] Metz was a keen botanist, and had made extensive collections in the Nilgiris and Canara which were sent to R.F. Hohenacker in Esslingen who distributed them through a subscription society called the *Unio Itineraria*. From Palghat Cleghorn went to the Anamalais and then further south to look at the forests of the Palni Hills, the Cumbum Valley and Madura, before returning to the Nilgiris for October and November. There was a report to be written on the Malsar Farm in the Anamalais (which was closed as a result, probably because the Malsars didn't take to fixed agriculture). In November came a visit to Bangalore, where Cleghorn wrote a report on the Lal Bagh, now under the hand of William New. From Bangalore Trevelyan asked Cleghorn to go to Goa to settle a dispute over the boundary between the Presidencies of Madras and Bombay – 'an unusual occupation for an M.D.' was his father's comment,[73] though Cleghorn already had experience in such matters. At this point, for the first time since Shimoga, illness struck, an ague that he called 'Wynad Jungle Fever', and it appears that the boundary expedition was postponed until the following January. The first attack of malaria was severe but was 'cut short apparently by Warburg's Drops' – the fever recurred every two weeks, inclining Cleghorn to believe in the theory of Lunar influence.[74] After an exhausting year of travel and report-writing, and still incapacitated by the revisitations of fever, December found Cleghorn back at home in Madras, writing up reports on familiar subjects relating to the Nilgiris (the Australian plantations and the ever pressing problem of fuel supply) and his long overdue Second Annual Report. He also met Sir Bartle Frere, the Governor of Bombay, on his way to Calcutta to take up a seat on the Governor-General's Legislative Council, who took the opportunity to compare notes with Cleghorn on the very different forests that Nicol Dalzell was supervising as Forest Ranger in Sind.[75]

The Third Forest Report, 1859–60

Cleghorn's Third Annual Report is the longest of the three,[76] but, given the state of his health, his forest tour of this year was by far the shortest, covering the months January to July 1860. Despite suffering from malarial attacks he continued to function, starting the tour in January with his established pattern of first heading north. He attended the sandalwood sales in Shimoga then continued into North Canara to examine the forests with Müller – those of Sirci, Yellapur, Hallial, and the Gund Forests that (with improved transport routes thanks to the engineer Captain G.W. Walker) were now in production. An excursion was made to the 'Sidh Temple on the Belgaum frontier ... the subject of an old standing controversy' (doubtless the one he had asked him to investigate the previous November). Cleghorn then went to the timber depot at the port of Sideshegur, from whence he was summoned to Ooty to attend Trevelyan who was at the hill station from 13 to 27 February. Almost never did Cleghorn refer to his mode of transport (there is one mention of reading matter being put in his 'Palankeen drawer', and one to some 'rough riding'), but it is likely that from Sideshegur he took a boat to Calicut as he took the opportunity to visit the Nilambur Plantations on the way to the Nilgiris. In the hills (where he must have been based from February to July) he reported on familiar subjects: the Australian Plantations, and the forests of Sigur and Mudumalai. Excursions were made to Wynad and Coimbatore and by 7 July he was in Bangalore about to return to Madras on sick certificate having been prostrated with dysentery. By time he got back to Madras a major political upheaval had occurred. Trevelyan (and his Council) had disagreed with a new policy on Income Tax by the Government in Calcutta, and Trevelyan had leaked to the press the Madras Government's criticism of the policy – or as Edward Balfour tactfully described it had acted 'with less dependence on higher authority than was deemed expedient'.[77] As a result Trevelyan was recalled to Britain on 10 May 1860 (though this did not damage his career permanently; he returned to India as finance minister two years later, when the by now married Cleghorn would renew his acquaintance in Simla). In view of the state of Cleghorn's health, dysentery following closely on the heels of malaria, the only option was a sick leave to Britain, but this could not be arranged immediately. First he had to write a final annual report, submitted on 31 August 1860, and make arrangements for the running of the Forest Department in his absence by Beddome.[78]

The Third Report is similar in content and format to its two predecessors, the major issue remaining the insatiable demand for timber for the railways (that for telegraph posts having declined), but with demands from a new source – from the British Navy to supply timber of *Artocarpus hirsutus*. The question of wastage was still of major concern to Cleghorn, for example the fact that wooden sleepers which ought to last five or six years often survived only two, because felling had taken place in the wrong season or the wood had been poorly seasoned; there were also problems from wood-boring insects (carpenter bees and white ant). The recommendations on the problems both of seasoning and insect damage show Cleghorn's wide reading, both modern and historical, in both English and French. One of the sources mentioned is a beautifully illustrated work on insects by Samuel Neville Ward (1813–1897). Ward, whom Cleghorn must have met in Sirci where he was based, is better known for his exquisite bird paintings and there is a single lithograph by him (of a bee-eater) in the Cleghorn collection and also a watercolour of *Curcuma aurantiaca* from Canara by one of Ward's daughters (*Cleghorn Collection* Fig. 21).[79] Wastage from trimming with axes was on the wane with a rise in the use of saws, and at Wellington in the Nilgiris a steam saw mill was now in operation. At the other end of the tree life-cycle there were problems with seedlings in the Nilgiri nurseries from climate (frost), over-crowding and predation by the Nilgiri rat, grubs and ('on clear moonlit nights') by hares.

One of Cleghorn's oldest forest bug-bears, dating back to Shimoga days, was partly resolved this year. In August 1859 he was asked by the Government to consider reports on the matter from those with a wide variety of interests,[80] and the result was that the practice of *kumri*[81] was banned in Canara on 23 May 1860 in all places other than where there was no significant timber remaining. His objections were not only to the destruction of valuable timber, but, despite the good grain crop gained in the first year, that this did not even benefit the shifting cultivators significantly, the profits largely going to 'soucars' on the coast (bankers, to whom the cultivators owed money), and that the resulting scrub was more unhealthy than the forest it replaced.[82] Further forests were added this year to those under Forest Department control – Overseers were appointed to Ossur (near Bangalore) and Madura, and an Assistant stationed at Cuddapah. The Nallamala Hills to the east of Kurnool were investigated and Beddome's skills as an 'excellent explorer' were to be used to investigate them further. The Nilgiri shola forests were also to be reserved 'not only for their beauty … but also from fear of injuring the water springs'. A new topic this year was a consideration of 'fancy

5. FORESTS OF SOUTH INDIA, 1856–60

woods', notably ebony and satinwood; blackwood (*Dalbergia latifolia*) from plantations in North Canara and Nugger was also now being put to far better use by Bombay furniture makers than for the making of gun carriages, its previous fate. The three public gardens were doing well distributing seeds and plants – the numbers supplied to the public for the year 1859/60 being 17,000 from Ooty, 4000 from Madras and 2000 from the Lal Bagh. Cleghorn had still made no progress on his Manual of Botany, and with the new generation of colonial handbooks starting to come out of Kew (*Flora Capensis* and *Flora Hongkongensis*) as daunting items of comparison, he was put off from the idea.

Müller's report on North Canara for this year was a substantial piece of analysis,[83] and Cleghorn's comment that allowance had to be made for the author's 'difficulty of writing in a foreign language' is ungenerous and inaccurate. In it Müller was at pains to point out the need for meaningful comparisons of profits before and after the setting up of the Forest Department, but the difficulty of drawing correct conclusions due to the fact that the apparently more profitable early results were from felling in the more easily accessible lower regions that had been exploited by Contractors. The Government, or *amani*, extraction allowed, by contrast, for a much less wasteful and more profitable exploitation of the upper regions. The profit from his operations this year was Rs 143,000, with wood worth Rs 133,000 in storage. This year 7087 teak trees had been felled by the Government, turned into 14,000 logs and 16,000 candies weight of smaller timber; substantial numbers were also being taken by four contractors – 5747 teak trees and 4500 'jungle trees'.

As Cleghorn was about to go home on sick leave he made some general remarks about his Department, which had now been in operation for almost four years. In many ways it was doing well – the major forests had been surveyed, and the most important of them were under a reasonable degree of control. But there were major problems, one of which was lack of knowledge, for example on the basic question of the growth-rate of teak they were 'groping in the dark' (the best figures available coming from Brandis in Burma). The major problem, however, was one of scale and limited resources. The area covered was far too large for the 'means placed at our disposal', and the 'field of operation [had] gone beyond the original scheme'. The need for policing in particular had been completely underestimated both in terms of the number of *peons* required and who should pay them. There was also the difficulty of getting good Overseers, and the need for ones for additional areas. However, the profits for this year were Rs 400,000, of which four-fifths of the total derived from a combination of the Anamalai and North Canara Forests (the other areas contributing very small amounts financially). The Department also had a growing reputation and copies of its rules had been requested by Ceylon, Hyderabad and Nagpur. The importance of his work to the Madras Government was acknowledged by the fact that it commissioned him to 'publish in England a "Selection of Forest and Horticultural Records" connected with [the] Department'.

What is not included in his report is anything about Cleghorn's his own role in the enterprise. Quite apart from the official side it is worth drawing attention to the human aspect, and the remarkable nature of Cleghorn's forest tours (which are comparable to those of Alexander Gibson in Bombay, of which he must have been aware). Before his appointment he had seen them, in a visionary vein, as 'the wandering life of an exploring pilgrim',[84] but one should not forget the considerable physical discomfort of such tours, given the climate of southern India. These must have been planned with the physical constraints in mind – the three tours started going up the west coast in the first few months of the year, when it was at its hottest, but before the breaking of the South-West Monsoon. By the time of the heaviest rains (June, July August) he was usually in Ooty, or back in Madras where the wettest season comes with the North-West Monsoon in October and November.

Beddome was left in Madras as Cleghorn's *locum tenens* but by late 1860 Cleghorn heard that he too had been sent home on sick leave, 'prostrated by the June fever',[85] which Cleghorn worried might 'require me to hurry back much sooner than intended'.[86] This proved unnecessary and Beddome was replaced by Captain Morgan, to whom it fell to write the annual report as Officiating Conservator in September 1861.[87]

Forest Department Assistants & Christianity

Richard Grove noted that 'German foresters had been employed as early as 1856 by the Madras Forest Conservancy',[88] the implication being an early link with German 'scientific forestry'. The reference is to Sebastian Müller, who was appointed as assistant for North Canara in 1858. Although Cleghorn believed that Müller 'had been early trained in the Black Forest of Germany',[89] this turns out not to be the case. From the records of the Evangelisches Missionswerk, Basel, it is found that Müller in fact went to Mangalore as a missionary and had originally been apprenticed as a clock-maker.[90] He was born in 1825 in the village of Villingendorf on the eastern edge of the Black Forest in Baden-Württemberg. As his father was a farmer, Sebastian, like Cleghorn himself, may have had at least some early informal training in rural pursuits. Despite the fact that the population of Villingendorf is (still) primarily Catholic, Müller joined the Protestant Basel Mission in 1846, travelling to India in 1848. He resigned from the Mission in 1854 though clearly kept links with it as his marriage to Rosa Wanderer in Mangalore in 1856 was performed by the missionary Hermann Gundert (who was Herman Hesse's grandfather). After Rosa's death he married for a second time, a woman called Margaret Millar in Dharwar in 1862. It was doubtless in his missionary days that Müller became a 'good Canarese scholar',[91] one of his greatest strengths as a forest assistant.

It is noteworthy that several of Cleghorn's other forest assistants or staff had strong religious links or leanings. Louis (Lewis) Blenkinsop (1827–1861) was the son of an EIC Church of England chaplain, the Rev. William Blenkinsop (1802–1871) of Cuddalore. He was appointed forest assistant based in Salem in August 1858 to supervise the provision of wood for railways on a monthly salary of 200 Rs,[92] but 'was suddenly carried off by cholera at Hurroor in the 34th year of his age' and is buried at Yercaud. Cleghorn's first Office Assistant in Madras in 1856 was John Thomson Maclagan, from a family well known to Cleghorn in Edinburgh and notable for its piety; on returning to Scotland in 1863 Maclagan 'gave up the latter portion of his life to Church [of Scotland] work' based around South Leith Parish Church and from 1873 to 1877 was Secretary and Home Agent for the Church's Foreign Mission Committee and involved with other charitable work.[93] Maclagan was succeeded by Francis Appavoo, whom Cleghorn had trained as an assistant in the Medical College. Nothing is known of his religious affiliations, but Cleghorn's account of the large numbers following him 'to his grave', suggests that he was Christian, and from his forename perhaps of Catholic Portuguese descent. From his later devotion to theosophy, and as a friend of Madame Blavatsky, Captain Henry Rhodes Morgan (1822–1909), clearly also had spiritual interests – if less conventional ones. He started as a soldier in the 13th Madras Native Infantry and was put by Cleghorn in charge of the Sigur and Mudumalai forests; following Beddome's illness Morgan then acted for Cleghorn during leave of 1860/1.

Cleghorn's forestry therefore had links not with German scientific forestry but with his own religious beliefs and outlook. Blenkinsop was son of an EIC chaplain; Maclagan was deeply religious with a brother who became the Archbishop of York; Müller was a lapsed missionary and Morgan who became a Theosophist. Nothing is known of the background of Richard Beddome, who had succeeded Hamilton as assistant in the Anamalais in 1856. He was appointed by Cleghorn 'solely on account of his powers of observation and description',[94] but it could perhaps be significant that his parents were married at Holy Trinity, Clapham, stronghold of the 'Clapham Sect'. Such links were to continue into Cleghorn's Punjab period: Brandis's first wife Rachel was a daughter of Joshua Marshman one of the best known of all Indian missionaries, and in the single example of nepotism in which Cleghorn appears to have been involved he, in 1863, recommended his wife's distant cousin Lieutenant John Chalmers (1821–1883) as Deputy Conservator of Forests in the Chenab Division. Chalmers had gone to India (under some sort of cloud) as an engineer, became involved in the Mutiny and transferred to the military. Like his uncle, the Rev. Thomas Chalmers, he held strong evangelical beliefs, and his holding of Christian services for his soldiers led to his unpopularity with Punjab officials.[95]

Was this simply, or primarily, because Cleghorn needed staff of high moral probity? Or that this was his 'circle' (friends and relations) from which he had no need to go beyond for recruitment? It seems likely that there was more to it and that underpinning Cleghorn's philosophy and career was an unshakeable belief in man's supremacy within the Created order. In the Old Testament, which not only justified, but demanded, his exploitation of natural resources, God's instructions to Adam and Eve in the first chapter of the Book of Genesis (verses 28–29) are explicit:

> Be fruitful and multiply, and replenish the earth, and subdue it: and have dominion over the fish of the sea, and over the fowl of the air, and over every living thing that moveth upon the earth. And God said, Behold, I have given you every herb bearing seed, which is upon the face of all the earth, and every tree, in the which is the fruit of a tree yielding seed: to you it shall be for meat.

Despite the underlying Christian faith of most of those involved in nineteenth-century colonialism this is a subject that has been fought shy of by most environmental historians, with the notable exception of Richard Drayton who explained that 'The knowledge of nature had a central place in one Judaeo-Christian myth of the origins of human sovereignty, within which power over and within nature was a right which had been passed from God to Adam … to the Christian princes and lords of modern Europe'.

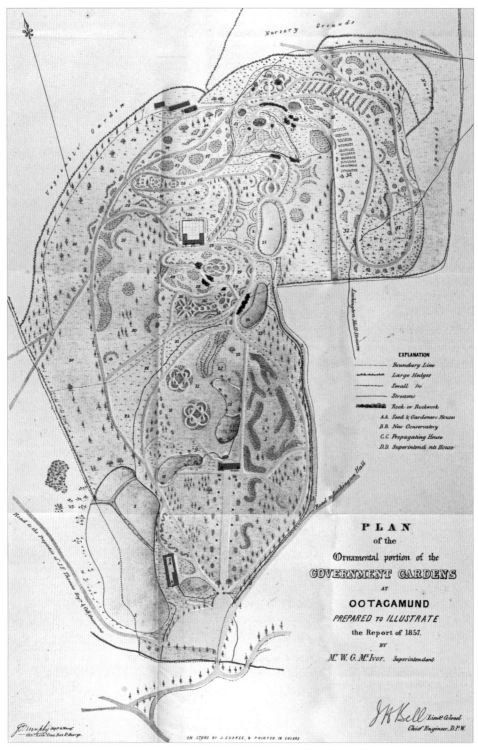

Fig. 26. Plan of the Government Garden, Ootacamund, 1857. Lithograph 'drawn on stone' by J. Suares, and printed 'in colors' by J. Dumphy, Government Lithographic Press, Fort St George, from Cleghorn & McIvor, 1859. (RBGE).

6
Gardens of South India

The garden of the Madras Agri-Horticultural Society, its role in plant introduction and Cleghorn's role as its Secretary, has already been discussed. Although, following his appointment as Conservator of Forests in 1856, this involvement had largely had to be given up, it was compensated for by his connection with two publicly funded gardens, one in a supervisory role, the other merely advisory. The owners of private gardens, many of them known to Cleghorn, also played a significant part in the introduction of new plants – especially of ornamentals – and this merits a brief discussion, as does a new type of horticultural enterprise in Madras, a 'People's Park'.

The Government Garden, Ootacamund

The garden at Ooty is still by some way the most picturesque and attractive botanical garden in India. That this is so is due largely to the vision and energy of William Graham McIvor who arrived to run it in March 1848. Following an unsatisfactory period reporting to a succession of committees, it was Lord Dalhousie who, during his period of convalescence in the Nilgiris in 1855, recommended that responsibility for the garden be taken over by the Madras Government. This happened in September 1857 when it was placed under Cleghorn as part of his duties as Conservator of Forests,[1] having visited the garden for the first time three months earlier.

The story of the garden has been told by Frederick Price in his history of Ooty,[2] but it is worth summarising here, being a place of some importance to Cleghorn – not only for its major role in the introduction of Australian trees, cinchona and ornamental plants, but also as a source of subjects for his artists. Its origins were in 1845 as a subscription garden, on a six-acre plot, for the growing of fruit and vegetables that were provided free to members. In 1847 more ambitious plans were hatched for a Horticultural Society and Public Garden and on the new committee were several people known to Cleghorn: George Hooper (a horticulturally-minded judge), the engineer Frederick Cotton and the missionary the Rev. Bernhard Schmid. The Governor, the Marquess of Tweeddale, agreed to provide 100 Rs a month for running costs and a one-off grant of 1000 Rs that allowed the recruitment of a gardener from Kew on a monthly salary of 150 Rs.

The McIvor brothers

Rather little is known about William McIvor and he certainly merits further biographical research as he was clearly a man of exceptional ability. Born at Dollar in Clackmannanshire on 12 November 1824, the son of a gardener, he was educated at the Dollar Institution, that as a local boy he attended free of charge.[3] From 1838 McIvor would have been taught botany and horticulture by John Westwood (himself a former pupil and gardening apprentice of the school before spending time at the Royal Botanic Garden Edinburgh). After completing his schooling McIvor stayed on as an apprentice from 1839 to 1843, and at this time took lessons with the school's drawing master Patrick Syme. Following this McIvor 'obtained a thorough horticultural training at the Royal Botanic Garden, Edinburgh', after which, from the data on a collection of lichens, he seems to have been based somewhere around Ripon in Yorkshire (perhaps at a garden such as Grantley Hall) in the year 1844/5 before continuing a southern migration to Kew.[4] From Kew he issued an exsiccata of liverworts,[5] which must have attracted

CHAPTER 6

the eye of Sir William Hooker and J.F. Royle and that led to his Indian appointment. When he went to India in 1848 he was accompanied by a brother James and a sister Catherine, both of whom would die there young.[6] In 1850 William married Anne Edwards, daughter of a Welsh colonel and James married Susan Charlotte Higgins. Anne's supportive role in her husband's career was stressed in the *Gardeners' Chronicle* obituary published on McIvor's premature death, aged only 51, on 8 June 1876.[7] This support was probably both social and financial, for someone who exemplified the Victorian ideal of self-improvement from humble origins; somehow he acquired capital to invest in the burgeoning plantation industry, though some of this was sent back as annual donations for the poor of Dollar. James was also involved in plantations, but in these activities both brothers sailed close to the wind. James came unstuck – after being given a grant of land of shola forest for an 'experimental Scotch farm', what he actually started in 1861, against the wishes of the Forest Department, was a coffee plantation. He sold off the felled timber and charcoal, which the Collector E.B. Thomas deemed 'an unscrupulous, unparalleled, and unjustifiable destruction of government property'. James seems never to have mended his ways: 'his speculations and profiteering caught up with him' and he ended up being arrested for debt.[8] William, by contrast, escaped largely unscathed, though was himself censured for selling for his own advantage timber that he had cleared for the Government quinine plantations at Dodabetta.[9] As noted earlier he was one of the many planters who disagreed with Cleghorn's attempts to preserve (on climatic grounds) the higher forests, which made the job of Cleghorn, and his foresters including Henry Morgan, an impossible balancing act. But the situation was complex and conflicted, as many of Cleghorn's friends including Rohde, Morgan (and even Thomas himself) also owned tea/coffee plantations.

After William's death his widow Anne paid to rebuild the chancel of St Stephen's church at Ooty in his memory and gave money for a prize at the Dollar Board School. McIvor's funeral at St Stephen's was a demonstration of just how far he had risen socially;[10] it was taken by the Bishop of Madras and the pall-bearers were the upper echelons of the Madras Civil Service led by Sir William Rose Robinson.[11] Also at the funeral were two of Cleghorn's former forester colleagues, his successor Richard Beddome and Colonel James Michael. McIvor's social standing doubtless owed more to his business interests in various estates and plantations, and his role in the establishment and running of the hugely successful government quinine plantations (discussed elsewhere), than to his garden work. The development of the quinine plantations took place largely after Cleghorn's departure for northern India but were built upon McIvor's remarkable achievements at the Government Garden in the period 1847 to around 1860.

The Government Garden

Much information is to be found in a fascinating Garden Report for 1857 written by McIvor after a decade in office, with Cleghorn's comments thereon,[12] accompanied by a fine map showing the garden in its heyday, prepared by McIvor, that Cleghorn ordered to be lithographed in colour (Fig. 26). It is hard to believe that this creation was achieved on a small budget, in a mere ten years, out of a ravine through some shola woodland, resulting in a 30-acre garden with the full range of features expected in a mid-Victorian garden, one that met with the approval of the discerning eye of Lady Canning when she visited Ooty in 1858. There were lakes (no fewer than five), lawns with specimen trees and beds of 'florist's plants' of which 30,000 were raised annually, including verbenas, geraniums, calceolarias

6. GARDENS OF SOUTH INDIA

and the quaintly named pikotees (*Dianthus caryophyllus*), parterres, rose gardens and sections with geographical themes (Turkey, Australia, the Cape and Italy). There were two octagonal bowers and a fountain, and even some areas with native plants – one area of the original shola woodland had been left with its native shrubs of *Rhododendron arboreum, Ilex wightiana, Michelia* and '*Andromeda*' (probably *Gaultheria fragrantissima*), and in other parts there were rocks covered with native lycopods, sonerilas and orchids. The creation of this ornamental garden was made possible through McIvor's thrift (of which he was proud) and his imaginative use of labour-saving devices (e.g., damming the stream to move earth to form terraces without the need for manual labour).[13] However, one of the garden's major aims remained from its early days – the production of seeds and plants – fruit, vegetables, medicinal plants and timber trees – that were sold to offset running costs.[14] Another stated aim was the propagation of some of the most ornamental native Nilgiri species such as the 'Sispara creeper', *Torenia cordifolia*, for sale or exchange with British nursery firms such as Veitch.[15]

Although McIvor seems never to have produced the catalogue of garden plants that Cleghorn asked for (a 'Hortus Neilgherrensis' like his own 'Hortus Madraspatensis'), a large number of species are listed in the 1857 Report, so that it is possible to identify the precise garden area where the plants painted for Cleghorn were grown. The garden was divided into two sections, of which the upper, of 18 acres, was the first to be developed: Cleghorn noted that 'the view from the higher terrace is romantic' and this was where McIvor chose to build his Superintendent's House. To the east of this was the Fountain; the 'Italian parterre' with an Octagonal bower covered with fuchsias (including 'Princeps', see *Cleghorn Collection* no. 81), below which were 'two Medici vases and garden seats' and gardens with ornamentals including hydrangeas, heartsease and gesnerias; further east again was the 'natural forest'; a sloping bank with camellias, fuchsias, roses and berberis; and a lawn with eucalyptus, *Acacia pulchella* and an *Edwardsia* (now *Sophora tetraptera*, see *Cleghorn Collection* no. 77) from New Zealand. On the slope between the Superintendent's House and a substantial Propagating House were Cape and Australian gardens, a 'Rosery' and 'Turkish walk', to the west of which lay lawns with Australian acacias and walks lined with Cape heaths. Below the Propagating House were two large specimens of *Brugmansia suaveolens*, and on the lower slopes of the upper garden were flower beds with statices, cupheas, stocks, dahlias, roses and hollyhocks; this is also where the Australian legume *Sutherlandia frutescens* (now *Lessertia frutescens*, see *Cleghorn Collection* no. 78) grew. To the west of the Propagating House was a lawn with more Australian trees (some of which had reached almost 50 feet in only eight years) and also Himalayan trees including an avenue of *Cupressus torulosa*; and to its east lay the highest of the ponds (used for irrigation).

The lower garden, originally a swamp and susceptible to frost, had taken McIvor longer to create, but was substantially complete by 1858 (Fig. 27). From the entrance gate a formal walk flanked by six pairs of urns led to a Conservatory. The Conservatory was built between 1856 and 1858 to designs by the engineer Peregrine Madgwick Francis.[16] Costing an initial 4300 Rs, then a further 1000 Rs for completion (authorised by Sir Charles Trevelyan in 1860), the design was not a success: Cleghorn described it as an 'unsightly building' and McIvor likened it to '3 propagating houses placed together' – it was demolished in 1887.[17] Despite its appearance, Cleghorn approved of its role not only as an attraction (in it were grown orchids), but as being 'of material aid to Naturalists visiting the Hills'. Along similar lines Cleghorn wanted, generally in the garden, 'permanent plant labels to designate the different species [which] would be a useful means of instruction and attraction, and tend to correct errors of

Fig. 27. Government Garden, Ooty. Photographs of the 1870s by Albert Thomas Watson Penn. (Courtesy of Christopher Penn).
Octagonal bower and pond in the lower garden (above).
The lower garden looking south-east, showing Peregrine Francis's conservatory (below).

nomenclature'. Above and to the south-west of the Conservatory lay a series of ponds, the lowest being the largest, beside which was the Seed House and the gardeners' huts; McIvor was particularly pleased with the picturesque effect of fringing borders of 'German daisies' around the ponds. To either side of the main axis were lawns where there was the second of the octagonal bowers, and some thirty-foot high trees of the native *Rhododendron arboreum* that had been preserved for their scarlet flowers, which provided a magnificent sight at the end and beginning of each year. Here also were specimen trees including more Himalayan conifers and yet more Australian acacias, including the 'melanoxylon' (known to McIvor as *Acacia longifolia*, see *Cleghorn Collection* no. 76) and 'wattle' (*Acacia dealbata*, see *Cleghorn Collection* no. 75) that, with the blue gum, became the longest lasting, if pernicious, legacy of the garden to the Nilgiri landscape, while providing a temporary solution to fuel demands. The disadvantage of these sclerophyllous Australian aliens was not only their devastating effect on ecology, but an aesthetic one. Eugenia Herbert in her history of Indian gardens aptly commented that 'Ootacamund was being Australianised as fast as it was being Anglicized ... an unhappy example of arboreal globalization'.[18] And as early as 1856 Andrew Jaffrey had warned:

> Ootacamund is really a pretty place, rich in natural scenery; but I am sorry to observe that its beauty is becoming deteriorated by injudicious planting of trees. There does not appear (with a few laudable exceptions) to be any taste or forethought displayed in this matter. Wherever the eye turns it is met by the dark towering forms of Australian Acacias the majority of which is *A. robusta* reminding one of a Nicropolis [sic].[19]

McIvor had reason to be grateful not only to one Governor-General and several Madras Governors (especially Tweeddale and Trevelyan) but also to botanists. It was Robert Wight who in 1853 had recommended the scrapping of the original garden committee, which had been interested only in vegetables for giving away, and gave McIvor constant hassle. This resulted in the name changing from the Horticultural to the Government Gardens, with McIvor responsible to a much smaller committee consisting of the Collector of Coimbatore,[20] and the heads of the Nilgiri military and medical establishments. In the 1857 Report McIvor paid tribute to Cleghorn for the 'warm interest he has evinced in the welfare of these Gardens', and Cleghorn was responsible for getting him a pay-rise from 120 to 200 Rs per month. This is not to say that they always agreed – for example Cleghorn thought that the prices McIvor charged for seeds and plants were too high, and would cause private individuals to turn for supplies to European nurseries. McIvor, not only from professional pride, but on economic principles, tried to hold out for higher prices partly on the grounds that he considered that people valued more expensive produce more highly, that there was no real danger of individuals going to British nurseries, and that by charging reasonable prices, he would stimulate the establishment of commercial nurseries through a spirit of competition. Despite these philosophical/practical differences, Cleghorn provided support for McIvor's struggle with philistine visitors, whose attitudes and expectations can be inferred from the rules he wrote for the garden (based on those of Calcutta):

1. To keep on the walks, especially near the shrubberies and flower beds.
2. Not to break flowers, leaves, or branches, or cut names or the like on trees.
3. Not to disturb the gardeners.
4. Not to offer them money or presents of any sort.
5. No shooting allowed. Horses and dogs are also prohibited.
6. The hours of admittance are from 6 o'clock A.M. to 6 P.M.[21]

CHAPTER 6

The 30-acre ornamental garden formed only a part of McIvor's realm: on the slopes above it were large areas of nursery ground, where no fewer than 451 species and varieties of trees and vast numbers of bedding plants were grown. The most popular trees with Nilgiri residents were acacias, willows (including a weeping variety taken from cuttings of a tree over Napoleon's grave on St Helena) and cypresses of which about 4000 each were grown, with much smaller numbers of rarer and choicer species such as araucarias and camellias. There were also two 'outstations' at lower altitudes in the Nilgiris, mainly for growing fruit trees and raising vegetables for seed – at Coonoor (10 acres) and Kalhati. McIvor also had responsibility for planting trees on roads to Jackatalla, and at least for a time in some remedial work on the restoration of the sholas close to Ooty. He also made an annual visit to the Nilambur teak plantations, though Cleghorn resisted the idea that he should take on greater forestry responsibilities, especially in view of the quinine demands that were about to involve a much greater part of his time.

Cleghorn was one of several Scottish medics who saw the introduction of quinine to India as a high priority, and in his November 1858 comments on McIvor's 1857 Report had written:

> The experiment of introducing the Quinine yielding Cinchona plants is well worthy of trial, especially in the Neilgherries. If possible, it should be conducted on a large scale, and with several species of the Genus to be planted in various parts of the Hills ... it is certainly most desirable that the British Government should make the attempt to obtain the Cinchona [cailsaya] either from Java, or from its native Habitat in the Andes.[22]

Although he would be on furlough during Markham's first quinine introductions, and in northern India for most of the period of McIvor's development of the plantations, Cleghorn's early support for the enterprise should not be forgotten.

The Lal Bagh, Bangalore

If Ooty is the most beautiful botanic garden in India, the equally well-cared-for Lal Bagh in Bangalore is certainly the grandest, at least given the present lamentable condition of Calcutta. Cleghorn had probably last seen Sir Mark Cubbon, Commissioner of Mysore, in September 1847 when, on leaving Shimoga, Cubbon had signed his sick note. In July 1856 Cubbon invited him back to Mysore to advise on the setting up of a public garden.[23] Andrew Jaffrey had been sent on ahead to do a recce and was waiting for Cleghorn who arrived on 10 July, having started out from Madras the previous day and taking with him his friend Alexander Hunter, the founder and superintendent of the Madras School of Art. Unlike Ooty where a garden had to be created from scratch, Bangalore had a long horticultural tradition, and Cleghorn considered its climate and soil 'much better suited for agricultural and horticultural experiments than either Utakamund or Madras'. The choice was not a difficult one, and the recommendation was a revival of the century-old old garden of Hyder Ali, an area of 'more than forty acres, well situated [apart from being a short journey south-east of Bangalore], and sloping gently towards the north ... [with a] tank at its upper extremity [that] admits of easy enlargement' to which Cleghorn applied the name of the 'Lal Bagh'. Cleghorn's wording has been given verbatim because the history of this garden, as variously related, is riddled with inaccuracies and confusion, not all of which is fully resolvable.[24] Undisputed facts from the various sources are as follows:

6. GARDENS OF SOUTH INDIA

1. Contemporary maps show that there were five garden plots in the area in the 1790s.[25]

2. Francis Buchanan, on a visit in May 1800, considered two of these to be significant, calling one (which was watered from a tank) Hyder's and the other Tipu's (watered from three wells); that both were in traditional 'Mohammedan' style, divided into compartments each of which was devoted to a different species, mainly fruits, but with recent additions from the Cape of Good Hope of pines and oak.[26]

3. Contemporary prints/drawings by R.H. Colebrooke,[27] Robert Hunter,[28] and Claude Martin,[29] show what is almost certainly a combination of these two plots dominated by lines of cypress trees, from which they were usually known as 'cypress gardens'. The one nearer the tank (and the Kempegowda Tower) was the larger and more open, conforming with Buchanan's description of Hyder's garden.

4. From the maps the larger of these, the southern one, was between 6.2 and 7.2 hectares, and the smaller, northern one of 5.5 hectares.

5. Benjamin Heyne, as part of Colin Mackenzie's survey, recommended obtaining the use of the 'Sultan's Garden' of about 40 acres for turning into a botanical garden, initially to house the plants from Roxburgh's abandoned gardens at Samalkot.[30]

6. At this period none of these gardens was ever known as the Lal Bagh (which referred to another of Tipu's gardens, at Seringapatam).[31]

In a fascinating recent article, using modern geo-referencing techniques to attempt to match the old drawings and maps with modern maps and topography, Meera Iyer and others usefully drew attention to the multiplicity of the gardens in this area in the 1790s.[32] However, they came to the controversial conclusion that the present Lal Bagh was not on any of them, and that Hyder's garden was the most north-westerly of the five, fed by a quite different tank, not by the one now known as the Lal Bagh Lake. The validity of this interpretation relies entirely on the accuracy of the old maps and drawings and, more particularly, the reliability of the geo-referencing techniques, which the authors admitted to have been 'challenging'. In fact a simple shift of the position of their two large southern gardens slightly to the south-west would place them firmly on the traditional site, with a combined area of about 31 acres, which would not conflict with the old drawings, and would have the virtue of agreeing with tradition, but also, significantly, with the evidence of Cleghorn's wording.

There is also uncertainty as to the history of the site following the time of Buchanan and Heyne, but it seems highly likely that it was this still nameless garden (initially leased by the Resident of Mysore from the Raja, but in 1802 made a 'fixed establishment', presumably meaning purchased from him), which, around 1814, came into the hands of Colonel Gilbert Waugh who introduced to it many exotic fruit trees. In 1819 Waugh offered to present his garden to the Government of India, which, on the advice of Nathaniel Wallich, was accepted and it was for a time treated as a 'branch' of the Calcutta Garden under Wallich's (remote) supervision, its expenses being paid by the Supreme (i.e., Calcutta) Government.[33] It was certainly Waugh's garden that was assigned by Cubbon in 1836 to the Mysore Agri-Horticultural Society, whose leading light was William Munro; the Society did not long survive Munro's departure in 1839, being wound up in 1842. After this the garden's rich and irrigable soil was used for raising the healthy crops of sugar cane observed by Jaffrey and Cleghorn when they visited fourteen years later, in 1856.

CHAPTER 6

Whatever its prehistory the garden was known henceforth as the Lal Bagh and Cubbon had the idea of restoring it as a public garden. Cleghorn saw it as part of 'a connected chain of Gardens at Madras and Ootacamund worked upon one general plan, interchanging and reciprocating with each other',[34] and in his report to Cubbon gave its potential role as follows:

> The great objects are the improvement of indigenous products, the introduction of exotics, the supply of these to the hills and plains when acclimatised, and the exhibition to the people of an improved system of cultivation in practical and successful operation. Seeds and plants should invariably be sold at a fair price to all applicants; but none should be given gratuitously, except for public purposes.[35]

In other words, exactly the same role as that of the Ooty garden. To achieve this he stressed that it was essential to appoint a European supervisor, though in the meanwhile a local *maistrie* called Heera Lal was employed to start the work of restoring the site, on which some of the buildings had clearly remained from the days of the disbanded Agri-Horticultural Society. The garden was not to be a subscription one (a method that had previously failed, both at Ooty and in Bangalore) but to come directly under the Commissioner, or, rather, his deputies: Major Gregory Haines (Superintendent of the Bangalore Division), Captain Francis Cunningham (Cubbon's secretary) and Dr James Kirkpatrick (an Edinburgh contemporary of Cleghorn). An initial idea that Jaffrey would become the superintendent fell through and the request for finding a suitable gardener, rather surprisingly, did not go to Edinburgh. Cubbon had received a letter from Sir William Hooker offering to find a 'skilled Overseer for the New Garden', and Cubbon and Cleghorn took up this offer. Cleghorn told Hooker that the chief requirement of the gardener was 'conciliatory manners' and that the greatest dangers that would face him were 'Intemperance of any kind' and 'Speculation in trade'.[36] Hooker (with the help of J.F. Royle) selected William New, who, doubtless duly warned of the dangers, arrived in April 1858. The terms were similar to those at Ooty – a monthly salary of 150 Rs, a house to be built for him for 2000 Rs, with the running costs for the garden slightly higher, at 300 Rs. Immediately prior to his appointment New had been in charge of parts of the gardens at Kew, having previously worked at the Belfast Botanic Garden, but biographical facts are in short supply.[37] Mr New set to work and two of his reports, with Cleghorn's comments thereon, are reproduced in *Forests & Gardens*,[38] representing the main source of information for this period. With him from Kew came Wardian cases of plants, and material was soon in flooding in from the botanic gardens of Saharunpur and Calcutta, Kew, Sydney and Melbourne. Private individuals (many of them military, as there was a large cantonment in Bangalore) seem to have been particularly generous – donating not only plants but garden furniture; the photographer Linnaeus Tripe contributed 'garden seats and select creeping plants'.

William New was clearly more assiduous with his record-keeping than McIvor and in 1861 produced a garden catalogue listing more than 1000 species, of which Cleghorn sent a copy to Balfour for publication in Edinburgh and also included in his own *Forests & Gardens*.[39] Cleghorn also had a selection of the garden's denizens painted, including the 'Moreton Bay Chesnut' (*Castanospermum australe*) and the European olive (see *Cleghorn Collection* nos 82, 83), both survivors from the Mysore Agri-Horticultural Society's garden, though further material of the *Castanospermum* was obtained from Australia – copies of both of these drawings were sent to the Agri-Horticultural Society in Calcutta.[40] In contrast to Ooty, with its emphasis on timber trees and vegetables, the Lal Bagh seems to have concentrated on ornamentals and fruit trees, though it was also used as a staging post for cinchona and tea. New held the post until 1863 when for unknown reasons he resigned and Allan Adamson Black was sent from Kew to replace him. Black (1832–1865), the son of the minister of an Independent Congregation in Dunkeld, had been appointed curator to Sir William Hooker's herbarium in 1853 on its move to Hunter House on Kew Green, effectively the first step towards its nationalisation. A single letter to his former employer survives from Black,[41] in which he gave some of the history of the Lal Bagh, and described the building of an aviary and 'new cages for wild beasts', but as usual with expatriate British gardeners he was more concerned with apples and strawberries. Black was consumptive and the Bangalore appointment had been with the hope that the warmer climate would restore his health. The hope was vain; he held the job for less than two years and took a sea voyage to visit his brother in Rangoon but died on the return journey off the Cocos Islands, after which William New resumed the job until his own death in 1873.[42] Cleghorn's involvement with the Lal Bagh was only ever advisory, and his last recorded visit there was in Black's time, in March 1865, on his way to what may also have been his last visit to the Ooty garden.

Private Gardens

In Ooty

One of the minor differences of opinion between Cleghorn and McIvor concerned the contribution of private individuals to increasing the horticultural diversity of southern India. McIvor was somewhat full of himself and denigrated their efforts, whereas Cleghorn noted that 'several private individuals are now in the habit of getting large consignments of valuable plants from England and the Colonies'.[43] Frederick Price, in writing the history of Ooty, was interested in the dates of first introduction of some of the significant species and regretted McIvor's lack of record-keeping; he was also keen to give credit to the efforts of some of these private individuals.[44] A previously unknown source of information on the introduction of exotics, as also of the almost entirely undocumented topic of private gardens in southern India at this period, is to be found in the annotations on Cleghorn's drawings, 19 of which were drawn from material grown in such gardens, mostly in Madras and Ooty, but one each in Bangalore, Chikmagalur and Yercaud. In addition to their botanical interest these annotations are also informative about the social circles in which Cleghorn was moving.

In Ooty, as discussed by Price, there were competing claims for the dubious distinction of exactly who first introduced Australian acacias and when. This may have been as early as 1825, but the main introduction was around 1830 from Tasmania, by either or both Colonel King and Captain Dunn. Cleghorn's acacia drawings were made mainly in the Government Garden, but one drawing of a then still unidentified 'wattle' (in fact *Acacia dealbata*) is inscribed 'new Acacia Mr Lascelles' (*Cleghorn Collection* no. 75). This refers to Arthur Rowley William Lascelles (b. 1830),[45] whom Price credited as the introducer of the walnut and of *Cupressus macrocarpa*, who had a house at Ooty variously given as Woodville or Woodlands. There is also a drawing of *Chrysophyllum roxburghii* drawn either in his garden or on an estate, as he was a young, 'irascible and combative' coffee planter,[46] Managing Director of the Moyar Coffee Company and agent for the Western Neilgherry Coffee, Tea and Cinchona Plantations Company. Lascelles evidently gave up this pioneering career after only a few years and returned to London to train as a lawyer, being called to the Bar in 1868, then emigrating to New Zealand. Credit for introduction of the first *Eucalyptus* goes to Cleghorn's friend Captain Frederick Cotton in the gardens of two Ooty houses, Gayton

CHAPTER 6

Park and Woodcot. Cotton and his wife were well known for their horticultural activities, as noted by Lady Canning in 1858;[47] but Cleghorn's single drawing of a gum tree (which he called *Eucalyptus perfoliata*, now reidentified as *E. cinerea* – *Cleghorn Collection* no. 79) was drawn (along with *Zizyphus mucronata* from the Cape of Good Hope) in the garden of Kempstow. This was one of several properties that in 1858 belonged to Mrs Brooke Cunliffe, wife of a senior Madras civil servant who also owned Kaity Lodge, another of the Ooty private gardens mentioned by Cleghorn. Though not relating to a garden plant, it was a friend of Cleghorn, John Rohde (who ran the Ooty jail), who introduced gorse (*Ulex europaeus*) to the Nilgiris, doubtless to make the 'downs' appear more English. A major introducer of trees was another friend of Cleghorn, Edward Brown Thomas, Collector of Coimbatore. Price credited Thomas with the introduction of the Spanish and the horse chestnuts, and he had a garden at Burliar (lower down and warmer than Ooty), where he grew tropical fruits and spices such as mangosteen, jak, allspice, cinnamon, rose-apple and cloves, which he generously supplied to the Ooty garden and the Lal Bagh. General Cullen from Travancore also had a house at Ooty, originally built by a Captain Macpherson and named after his Clan seat, Cluny. Cullen must have been receiving Himalayan seed, either directly from Saharunpur or via McIvor, as it was from here that Cleghorn had a specimen of *Pinus wallichiana* painted (*Cleghorn Collection* no. 74).

Private Madras gardens

Cleghorn's social-horticultural networks in Madras are reflected in some of the paintings made for him, mainly by Govindoo, showing rarities that were presumably not grown in the Agri-Horticultural garden, and reflecting independent lines of introduction through British nurseries or personal contacts. These individuals can be classified according to the area of the EIC service in which they were employed, with one striking exception. A drawing of a persimmon (probably the common one, *Diospyros kaki*) was painted from the garden of Dr Lobo at Saidapet. Michael Francis Lobo D.D. was a Roman Catholic priest who from 1853 to 1860 acted as Bishop of San Thomé, not the sort of cleric one would necessarily have expected the fervently Free-Church Cleghorn to have got on with, but showing the potential of plants as a means of overcoming sectarian divides. Much more expected are fellow Scots including horticulturally inclined medics. A drawing of the desert rose (*Adenium obesum*) was made from a specimen in the garden of Dunbar-born James Sanderson, Garrison Surgeon, and fellow-member with Cleghorn on many bodies and committees including ones connected with the Free Church, the Madras Railway and the Agri-Horticultural Society – the plant may well have been obtained while passing through the Red Sea on the way to or from Europe, for example at Aden. Also Scottish, from a distinguished Galloway family, was the merchant John Vans Agnew (1824–1874) from whose garden three interesting, and probably at that time rare, plants were drawn: the European climbing asclepiad *Periploca graeca*, the beautiful purple-flowered South American tree, the jacaranda, and another American tree *Triplaris americana*. In some cases, as in Agnew's, wives are named on the drawings as the garden owners, another example being Lady Montgomery, wife of a senior civil servant who provided the 'Tonghoo orchid', *Dendrobium albosanguineum*, which must have been sent to her by someone stationed in Burma – Taungoo in Tenasserim being a major army base after the 1852 Burmese war, as well as a centre for the teak industry. Other senior civil servants in the Madras Government whose garden treasures are represented among Cleghorn's drawings are Sir Christopher Rawlinson (*Caryota urens* – *Cleghorn Collection* no. 91), Walter Elliot (*Triplaris americana*), James Duncan Sim (*Cerbera odollam*), and the lawyers George Stanley Hooper (*Solanum wrightii* – *Cleghorn Collection* no. 90) and William Ambrose Serle (*Litsea quinqueflora*). A senior member of the Madras Army, General Alexander Tulloch, whose garden supplied the balsa-wood tree (*Ochroma pyramidale* – *Cleghorn Collection* no. 87), is also of interest in showing how closely connected these circles were. Tulloch's wife Emma was a member of the Wahab family from Northern Ireland, who in several generations married members of the Cowan family of Penicuik, so that one of the General's nephews, Charles William Wahab of the Bombay Army, married Catharine Cowan, an elder sister of Mabel who would shortly become Cleghorn's wife.

The People's Park, Madras

During his brief Governorship Sir Charles Trevelyan was responsible for bringing a new type of horticultural space to Madras. By this time public parks had become well established in Britain, where there were concerns about over-crowded, industrialised cities with unhealthy, and potentially rebellious, populations. Trevelyan's idea of a 'People's Park' was launched in 1859 and may have been one of the subjects on which he sought the opinion of his Conservator of Forests – Cleghorn was certainly on the Park's first committee,[48] of which G.S. Hooper was President, Dr Howard Montgomery the Honorary Secretary and Robert Brown the Superintendent. However, involvement in the realisation of the scheme on the part of the Governor certainly, and Cleghorn probably, was limited: Trevelyan was recalled and Cleghorn went on leave to be based in Madras only intermittently thereafter, though by the time of these later visits in the mid-1860s the Park had been fully established under the governorship of Sir William Denison.

The design for the Park was the subject of a public competition, the site being one of a little over 100 acres on the west bank of Cochrane's (later Buckingham) Canal, east of the suburb of Vepery and immediately to the west of the southern end of Black Town. Its purpose was to provide a place for 'out-door amusement and exercise for the great body of the middle classes of Madras, (who are almost exclusively Eurasians)' – the lower classes, it would seem, were deemed beyond redemption.[49] There was considerable interest in the competition: 16 plans were received within three months, and the prize was awarded to Robert Baldrey, doubtless the 'Mr R.J. Baldry' who had shown promise at Hunter's School of Art in 1850.[50] A chromolithograph of the prize-winning plan was published in the 1863 Madras Almanac, and it is probably not coincidental that a Robert John Baldrey later became Superintendent of the Government Lithographic Office (Fig. 28).[51] The scheme was ambitious – either side of a formal east-west axis, with a centrally placed bandstand, leading to a suspension bridge that crossed the canal to Black Town, was a series of serpentine walks and lawns, ponds and a lake for boating, a children's playground area, an archery ground and a cricket ground. Serious interest was paid to the trees, with groves of sacred banyan trees, breadfruit trees, palmyra, coconut and date palms and a nursery for raising them. In the key to the plan Baldrey gave a list of the trees and the reasons for their planting, including 'Margosa [neem, *Azadirachta indica*] – A graceful tree', 'Mugghadum [*Mimusops elengi*] – A most beautiful and ornamental tree with fine scented flowers', 'very graceful' bamboos, and the 'Hushogen or tree of liberty [perhaps *Polyalthia longifolia*] – A lofty noble ever green affording a very cool shade'.[52] The Park also housed a zoo, with tigers, leopards, deer, and even a giraffe, which was popular well into the twentieth century. Being at the centre of the city, immediately to the west of the Central Railway Station, the People's Park began to be encroached upon and its southern end was built on in the early twentieth century by public buildings such as the municipal Ripon Buildings, the Victoria Hall and Moore's Market. By the end of the century, most of the rest of the Park had been obliterated though two large sports stadia (one outdoor, and an indoor one on the site of the zoo) at least reflect part of Trevelyan's original ethos. The boating lake survived as the 'Lily Pond' until comparatively recently, but the only fragment of green-space that now remains is 'My Lady's Garden' with a reproduction Ashoka Column and, as guardian deity, a version of the Capitoline Venus, said to have been made by students of the Art College in the 1930s.[53]

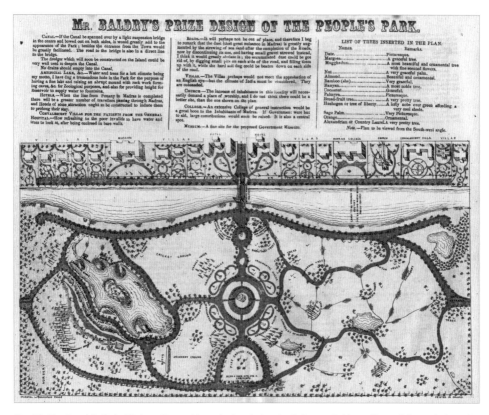

Fig. 28. The People's Park, Madras. Competition-winning design by Robert J. Baldrey. Chromolithograph from the *Illustrated Madras Almanac*, Asylum Press, 1863. (Sprot family collection).

7
Second Furlough and Marriage, 1860–1861

The year 1860 marked the end of a remarkable and productive period of Cleghorn's life as a teacher and horticulturist, and as originator and developer of a major government department. That it ended in illness, as had his Shimoga period, allows one to imagine his pilgrim's staff turning into a martyr's palm in the cause of applied science. In fact by the time he had wrapped everything up in Madras during July and August 1860 – both metaphorically for his reports and literally so in the case of the expensive extra baggage that housed some of his botanical collections – Cleghorn was feeling better. He had hoped to take the first steamer in September, but in fact took the second, the *Candia* via Galle, Aden and Suez, which left Madras on 27 September. Another reason for going home was connected with his 'prospects' and for the previous two years, in his letters home, he had been referring to the idea of 'trying for a wife'.[1] Promotion was entirely a matter of seniority and, despite his many achievements, he had been festering on the lowest rung of the EIC's medical service ladder, as an Assistant Surgeon, for no fewer than 18 years. On behalf of his un-ambitious son, his father Peter had been carefully monitoring the gazettes in this waiting game, but ascent could no longer be far off, and would finally come within the next six months, in February 1861.

Cleghorn and maritime transport seem not to have been a successful combination – this time there was no shipwreck, but the screw of the *Candia* became disconnected from its propeller in the Red Sea, and for ten days the ship lay disabled off the Nubian Coast.[2] He disembarked, crossed Egypt, and boarded the *Massilia* at Alexandria, but this time the ship got no further than the harbour mouth where its cylinder broke and Cleghorn had to transfer to the *Vectis*. The *Vectis* appears to have reached France without further adventure and at Marseille Cleghorn boarded a train to Paris. Here he visited the botanist Joseph Decaisne who since 1854 had been Professeur de Culture at the Muséum d'Histoire Naturelle and director of the Jardin des Plantes. They had a shared interest in Indian plants, as Decaisne had worked on the collections of Victor Jacquemont. Also in Paris, and rather more surprisingly, Cleghorn met General George St Patrick Lawrence to discuss the establishment of a branch of the Lawrence Asylum at Ooty, one of four such schools established in India as the brain-children of the General's brother, Sir Henry Lawrence, the Punjab administrator who had been killed in 1857 in the Indian Mutiny. The Ooty school was set up the following year, 1858, with major financial contributions from Walter Elliot.

Return to Britain

By early November 1860 Cleghorn had reached London and was staying in Kensington Square with Dr John Merriman, his companion of the 1849 Irish excursion.[3] There were discussions in the India Office about timber supply for the Indian Navy and for railways. He then headed north and on 1 December celebrated his father's 77th birthday at Stravithie. Since the time of his previous leave, communications had continued to improve and in 1852, thanks to Hugh Playfair, Dr John Adamson, Alexander Kyd Lindesay and other civic worthies, a spur from Leuchars now linked St Andrews with the North British Railway. The major semi-official task

CHAPTER 7

of this leave was the publication of his forest reports – but although this had been requested by the Madras Government before he had left India,[4] it was to be undertaken at his own expense, which he expected to cost £500.[5] The work was to be published in an edition of 500 by W.H. Allen of Leadenhall Street, formerly publishers to the EIC – although the Company was by this time defunct, they probably still seemed like the most suitable publisher for a work that would certainly not have appealed to publishers of travel literature such as John Murray.

Edinburgh Activities

The Botanical Society (Part 3)

As on his first furlough Cleghorn spent time in Edinburgh renewing contacts with the botanical community and visiting the Botanic Garden, where he doubtless made use of the herbarium and library for his work on *Forests & Gardens*. He also re-established links with the Botanical Society (BSE) of which he was elected a councillor for the year 1860–1 and attended all of its monthly meetings between December 1860 and July 1861.[6] At these he read papers with a strong emphasis on the economic botany of South India: the cultivation of tea and quinine, the varieties of yams and mangoes (illustrated with drawings of 40 varieties by Walter Elliot and Cleghorn, which sadly do not survive), the uses of the coconut; he also read William New's list of the plants of the Bangalore garden. Only one of the papers related to his forestry work – on the species of timber used for railway sleepers. Cleghorn as on many previous occasions either proposed or seconded the election of new members, of whom two are of particular interest – his assistant from Medical College and Forest Department, Francis Appavoo, the first Indian member of the Society, and William Coldstream, son of Dr John Coldstream of the Edinburgh Medical Missionary Society, with whom Cleghorn would later interact in the Punjab where the young Coldstream became a Civil Servant. On his first leave he had exhibited some of his Shimoga drawings, and at the meeting at RBGE on 13 June 1861 Cleghorn exhibited four from Madras. These were among the most spectacular (in terms of both size and subject) of Govindoo's drawings – of plants grown in the Agri-Horticultural Garden – the desert rose (*Adenium obesum*, received from Lambert Playfair in Aden), *Solanum arborescens* (from the judge G.S. Hooper, who had obtained the plant from Ceylon), the African tulip tree (*Spathodea campanulata* – Cleghorn Collection no. 49) and *Leea macrophylla*, a native vine-relative.

Walter Elliot (1803–1887)

The specimen of the *Leea* had been sent to the Madras Garden from the Northern Circars by Walter Elliot, with whom Cleghorn renewed his friendship in Edinburgh in December 1860. A year previously Elliot had retired to his family seat, Wolfelee in Roxburghshire, after a 40-year career of devoted service in the Madras Presidency.[7] From 1845 to 1854 Elliot had spent nine years as a revenue officer in the Northern Circars, and a very large number of his herbarium specimens (and a few drawings) from this period ended up in Cleghorn's collection. It was here that Elliot developed not only botanical interests (culminating in his authorship of *Flora Andhrica* in 1859), but his important zoological, numismatic and archaeological ones. In 1855 Elliot took a short home leave, during which he met Darwin at the British Association meeting in Glasgow. This led to a correspondence between the two men and Elliot's sending Darwin numerous specimens of Indian pigeon races and Burmese domestic fowl and, using his wide-ranging linguistic skills, translating for Darwin a Persian pigeon-fancier's manual.[8]

7. FURLOUGH, MARRIAGE & THE COWANS, 1860–1

On his return to Madras Elliot became Senior Member of the Governor of Madras's Council – with this position came the title 'the Honourable', which Cleghorn explained to Balfour was usable only in India adding ambiguously, if amusingly, that 'he is not Hon[oura]ble in Scotland'.[9] In this Madras period Cleghorn and Elliot had become friends, with numerous interactions over shared botanical and photographic interests, and both administrative and editorial work for the Literary and Agri-Horticultural Societies. Given Elliot's senior role in Government, his support for the new Forest Department was also significant and in 1856 Cleghorn described Elliot as 'a strong friend and a great help to me in all public matters'.[10] In the midst of this period, the years 1857 and 1858 were ones of high political tension in India, with the Mutiny raging in the north, though at least in part due to the buffering effect of Hyderabad, where mutiny was avoided, the 'contagion' did not reach Madras. Nonetheless Lord Harris's health broke under the strain; Elliot had to stand in for him and it fell to him to read the Royal Proclamation of the assumption of Queen Victoria's rule from the steps of the Banqueting Hall on 1 November 1858. When Elliot retired in December 1859, taking home 'many valuable drawings and specimens', Sir Charles Trevelyan chose a biblical-botanical quote to represent Elliot's polymathic interests:

> if ever there be anything that I ever wished to know connected with India, from the cedar tree that is in Lebanon even unto the hyssop that springeth out of the wall, I would go to Walter Elliot for the information.[11]

The Royal Society of Edinburgh & Evolution

On 3 December 1860 Cleghorn and Elliot, two exiled Madrasees, attended a meeting of the Royal Society of Edinburgh addressed by George Campbell, 8th Duke of Argyll, the Society's newly elected President (only the fifth since its foundation in 1783). The meetings were held in the western, ground-floor suite of rooms of W.H. Playfair's Royal Institution on the Mound, which had been built by the Board of Trustees (originally funded by money from the British Parliament as compensation to Scotland following the Union of the Parliaments in 1707). This complex edifice, a Scottish Somerset House, was built in two stages (1822, 1832) to house the Board's own offices and to provide premises (galleries, meeting rooms, libraries and museums) for three learned Societies – the Royal Institution for Fine Arts, the Society of Antiquaries, and the Royal Society, its internal complexity and two-storey structure disguised and unified externally within a severe, peristylar, Doric temple.[12] Cleghorn was not yet a Fellow of the Royal Society, though several of his friends (including Balfour, Douglas Maclagan and Christison) were active in it, so Cleghorn had probably been forewarned that the scientific duke's speech was likely to be one of unusual interest. At this point Balfour (who had just become General Secretary of the Society, a post he would hold for 19 years) had raised the possibility of Cleghorn's membership, but as he was having to fund the publication of his own book the hefty £30 subscription put him off (fellowship being strictly for gentry and upper ranks of the professional classes).

Argyll began his address with notices of the deaths of Fellows and a lengthy obituary of his predecessor as President Sir Thomas Makgill Brisbane,[13] though two of the others who were briefly mentioned had Indian connections. Mountstuart Elphinstone (1779–1859) was one of the greatest of Indian administrators who for a period was Governor of Bombay; he was also a relation of Elliot's, whose aunt Janet had married John, 12th Lord Elphinstone, Mountstuart's elder brother. This Lord Elphinstone's son John (the 13th Baron) served as

CHAPTER 7

Governor both of Madras (1837–42) and of Bombay (1853–60), and in the former Walter Elliot had acted for a time as his cousin's private secretary. Dr George Buist (1804–1860), from Angus, was, like Cleghorn, an alumnus of the universities of both St Andrews and Edinburgh, and had made a career in Bombay where he was known professionally as a controversial newspaper editor, but also for his scientific interests, particularly astronomical, geological and meteorological. For a time Buist served as Secretary of the Agri-Horticultural Society of Western India, and in 1850 (the same year as did Hunter in Madras) started an Industrial School for Natives under the patronage of Lord Elphinstone.

The main talk of the evening was devoted to current developments in science, and were revealing of Argyll's major personal interests in geology and evolution. The climax of the speech was a nuanced discussion of Darwin's *Origin of Species*, published just over a year earlier (24 November 1859), and which Argyll certainly recognised for what it was, even if his denomination of it as 'an event in the history of scientific speculation' was understated. Argyll praised Darwin for the range and detail of his evidence, particularly the chapters on the 'struggle for existence' and 'geographical distribution', as a major contribution to the progress of science; he also did not shirk from pointing out its relevance to the origins of man, on which he commented somewhat archly: 'the honour of this [fish/reptile] parentage ... will not be universally appreciated'. He could not, however, accept that natural selection could lead to novelty – a new habitat could not lead to the change of one species into a new one, only to its replacement by one already adapted. Like many critics of the time, he also believed that the theory could not account for the initial creation of species, as boldly claimed in the work's title, and that it was but a modification of 'the old theory of development' whereby 'the difficulty [of origin] is but postponed'. The Duke of Argyll has been called both a 'Christian Darwinist' and a 'Creative Darwinist'. Neither accurately describes his position – he accepted evolutionary change and deep geological time but for him, as for many others, there was still a need for creative intervention, and so he was at least a Deist – he even thought that species could have more than one origin. However, given his arguments with Darwin over the sufficiency of Natural Selection a better characterisation of Argyll's position is as a 'creative transformationist'.

Virtually nothing is known about Cleghorn's views on Darwin or evolution, though as an evangelical Christian, he is unlikely to have gone even as far as Argyll in deviating from biblical orthodoxy and creationism (a decade later they would have dealings over the much less controversial subject of the training of Indian foresters). Although Elliot had corresponded with Darwin and, as acknowledged in the *Origin*, had supplied him with Indian pigeon races, which formed a major part of Darwin's analogy between Artificial and Natural Selection, he too was a committed Christian and supporter of missionary societies in Madras. But one would love to know the conversation that passed between Cleghorn and Elliot after the meeting was over. Only two tiny pieces of evidence have survived that even hint at Cleghorn's own views on evolution. The first predates the appearance of the *Origin*, an annotation recently discovered in his copy of Hooker & Thomson's *Flora Indica*, on which he scribbled when writing a review of the work for the *Calcutta Review* in 1856. At this point Hooker was still being cautious and stated that mutability was not an innate character of species, a passage that Cleghorn marked in agreement, suggesting, unsurprisingly, his belief in fixed creations. The second was probably written very soon after hearing Argyll's lecture. And while it reveals nothing of Cleghorn's own thoughts it does show an awareness of

7. FURLOUGH, MARRIAGE & THE COWANS, 1860–1

current issues, when, in an introduction to an extract on how to collect good specimens of trees for inclusion in *Forests & Gardens*, he wrote 'the subject of geographical botany … is capable of throwing much light on the vexed questions of the nature and origin of species'.[14]

By contrast the views on evolution of others in Edinburgh evangelical circles are well known, and although there are dangers of making assumptions through association, it is at least worth outlining these. Closest to Cleghorn was John Hutton Balfour, author of works entitled *Phytotheology* and *The Plants of the Bible* (Balfour gave a copy of the latter to Cleghorn as a wedding present), who expressed his opinions in botanical textbooks that were widely used, including by Cleghorn himself in Madras (though prior to 1859). In the first edition of his *Class Book of Botany*, dedicated to the Duke of Argyll, in the notable year 1859, Balfour strongly opposed the idea of transmutation:

> there are not grounds for tracing back man to the monkey, or for supposing that, by a gradual series of changes, animals and plants have become more and more perfect and complete. All the facts of science lead to the conclusion, that there are distinct species, which vary within certain limits[15]

citing as support the works of the Cambridge scholars Whewell and Sedgwick and the more local ones of the eloquent Free-Church geologist Hugh Miller. That this statement was repeated unchanged in the 1871 edition of the same work should probably not be seen as overly significant, as for this only the section on 'organography' was revised. There are also complexities arising from the contemporary publication, in multiple editions, of no fewer than three different textbooks by Balfour. In the 'Introductory Remarks' of Balfour's first post-1859 edition of his *Manual of Botany*[16] is a passionate diatribe against those who 'shut out God from His works' though here, conspicuously, Darwin was not mentioned by name as one of this breed; Balfour, by contrast, considered God to be active in renewed creations and that it was wrong to look for 'a progressive development and eternal advance towards perfection … without further creative acts'.[17] The Remarks were repeated unchanged in the fifth edition,[18] but later on in this edition Balfour did acknowledge the work of his Edinburgh medical contemporary Charles Darwin, and offer an opinion on Natural Selection. He did acknowledge the possibility of transmutation and Natural Selection as a mechanism for change in species but denied that this process could be an 'inexorable law' independent of a deity. If it did occur in a way analogous to Artificial Selection then it must be under the control of a God who had planned it from the outset, that is 'under the guidance and direction of Him who works by means of instruments, and who carries out His mighty plans in an orderly and systematic manner'.[19] Hugh Miller, in a passionate attempt to reconcile new geological findings with Biblical faith, had opted for something similar: a series of *de novo* creations, and that impressions of 'advancement' were illusory.[20] One can only speculate on the likelihood of Cleghorn's sharing similar views.

The spring of 1861 found Cleghorn in London where, on 21 February, he received news of his long-awaited promotion to the rank of Surgeon on the Madras establishment. The previous night he had attended a banquet at Willis's Rooms in St James's, in honour of the recently knighted engineer Sir Arthur Cotton,[21] whose great Godavery irrigation scheme he had visited two years previously, and whose horticulturally minded brother Frederick was also a friend. On 23 March he had a meeting with the Earl de Grey (later the Marquess of Ripon) at the India Office, probably about the publication of his 'forest reports'. By April he was back in Edinburgh where on the 29th he read his Anamalai paper to the Royal Society of Edinburgh. This had doubtless been suggested by Balfour as part of a strategy to get him elected FRSE, which took place in November of the same year, after Cleghorn had returned to India.

CHAPTER 7

The Cowan Family

While the publishing of *Forests & Gardens* would be the major professional task of his second furlough, the second major achievement was personal – the 'trying for a wife'. Cleghorn's choice fell on Mabel Cowan whom he married on 8 August 1861 and with the marriage came the additional benefit of bringing Cleghorn into the fold of a remarkable clan. For information about this extended family (and the family tree is labyrinthine) I am indebted to the genealogical work of Roger Kelly, who lives in the ancestral Cowan home, Valleyfield House, Penicuik, and to a recent, richly detailed and long overdue, biography of Charles Cowan (Mabel's father) by Donald MacLeod that, while dealing fully with its subject's role in the 1843 Disruption, provides a rounded picture of the man and his family.[22] Charles Cowan's privately published *Reminiscences* is also a valuable source,[23] though almost as egocentric as those of his youngest son John James, which would win a prize for the most prosaically titled autobiography ever written: *From 1846 to 1932*.[24] One must, however, be grateful to John James for publishing the only known photograph of his sister Mabel.

The Cowan family was not dissimilar to Cleghorn's own, also having eighteenth-century roots in both Fife and the Edinburgh area. With the Cowans, however, the manufacturing aspect (paper-making rather than brewing) had taken precedence over the academic – though the latter were by no means neglected in terms of fervent education for self-improvement. This led to a much greater prosperity than the Cleghorns ever achieved, and a much more conspicuous public role – in both local and national politics. The family was also much more widely travelled.

Mabel's great great grandfather George Cowan (1695–1775) had been a farmer at Crail in the East Neuk of Fife only a few miles south-east of Stravithie. George's son Charles (1735–1805) moved to Leith and established a business that in 1779 led to the acquisition of a paper mill at Penicuik. Paper-making, at various mills on the River North Esk, was hugely developed by the first Charles's sons, Alexander (1775–1859) and Duncan (1771–1848) and the industry would dominate the town of Penicuik until 1965. Alexander's eldest son was Charles Cowan, Mabel's father, who was born in Edinburgh in 1801. The first Charles Cowan had a sister Lucy who remained in Fife and married George Hall an Anstruther merchant, resulting in two children, George and Elizabeth. Elizabeth was the mother of the Rev. Thomas Chalmers the theologian, and George's daughter Elizabeth Hall married Alexander Cowan and was Mabel's grandmother. Charles Cowan, Mabel's father, was therefore related to Thomas Chalmers both paternally and maternally, and evangelical Christianity is fundamental to an understanding of the Cowan family. Though extremely successful in business, their faith led to conspicuous practical philanthropy, undertaken from their base, the Valleyfield Mills in Penicuik. This philanthropy was based in a belief in progress through market forces and Adam Smith's 'invisible hand', but the worst excesses of exploitation and capitalism could be avoided if underpinned with Christian ethics. Alexander Cowan believed that his sons should be educated not only practically and in business but academically, from which the generation of wealth would spring as a secondary benefit. Having made money employers had a responsibility to promote happiness and good working conditions and he once said 'I would rather my descendants were Papermakers than Lords'. By the 1860s, when Cleghorn married into the family, the company had more than 600 employees and was making 2–3000 tons of paper a year. Markets were opening up in the USA and Australia, and the firm invested in railways and insurance. One of Charles's nephews, William Menzies, set up the Scottish American Investment Company, of which John Cowan was a director, so the male members of the family made large amounts of money, though Charles is said to have lost £100,000 with the collapse of the Glasgow Bank in 1878. The company ethos, however, was high-minded, employing no wives of male employees and no children under 13, while providing a Medical Club, good housing, a supply of clean drinking water and free schooling.

Father-in-Law: Charles Cowan (1801–1889), Paper-Maker

The patriarch of the firm, and successor to his redoubtable father Alexander, was Cleghorn's father-in-law Charles (Fig. 29). He had attended the High School of Edinburgh (where he was a contemporary of his future son-in-law's teacher James Syme, and, like Robert Wight, a pupil of James Pillans) and of the University of Edinburgh. After this, in 1817, he was sent for a year to the Auditoire in Geneva, a bastion of Protestantism, founded by Calvin in 1559, to which he was accompanied as tutor by the botanist Daniel Ellis. Here the sixteen-year old Cowan learned French and attended lectures in maths, moral philosophy, natural philosophy and chemistry. He also attended the botanical lectures of Auguste Pyramus de Candolle, then in the first year of his tenure of the chair of Natural History. This rich education gave Charles Cowan great intellectual curiosity and wide interests, which he developed through active membership of groups such as the Society for the Encouragement of Useful Arts in Scotland (founded by Brewster in 1821) and the National Association for the Promotion of Social Sciences. Cowan, like Cleghorn, was a friend of Brewster, which had led him to invite the British Association Edinburgh Meeting to visit Valleyfield in August 1850, though it is not known if Cleghorn participated in this.

In 1824 Charles married Catharine Menzies, daughter of a clergyman of the old Moderate Presbyterian school, the Rev. William Menzies, Minister of Lanark. Also deeply religious Catharine would instil her faith into her daughters. The Menzies family was another ecclesiastical dynasty; Catharine's sister Margaret married Dr John Coldstream founder of the Edinburgh Medical Missionary Society in which Mabel and Cleghorn would, in the Punjab, take a deep interest. One of Catharine's sisters, Jessie Ann was married to William Bonar three of whose brothers John, Andrew and Horatius were distinguished ministers of the Free Church. Of these Horatius was best known as a hymn-writer ('I heard the voice of Jesus say'). Horatius was married to Jane Catherine Lundie, sister of the Rev R.H. Lundie, who was married to Mabel's sister Elizabeth. And it was Lundie who officiated at Hugh and Mabel's wedding ceremony.

Fig. 29. Charles Cowan. Woodburytype, photographer unknown, from Cowan, 1878.

CHAPTER 7

The Disruption of the Church of Scotland

The most significant historical event in Scotland in the nineteenth century took place on 18 May 1843 when 474 ministers signed an 'Act of Separation and Deed of Demission' from the national church. They felt unable to accept what they saw as the intolerable power of the gentry who could impose ministers on unwilling congregations, a 'right' backed up by the Court of Session in Edinburgh and the House of Lords in London. It was a desperate act, and behind it lay a group of radical young clergy and a general upsurge of evangelical feeling – in part arising from contemporary challenges from increasing urbanisation and industrialisation, the rise in scientific knowledge and secularism, and those who saw a return to strict Protestant Christianity as a way of coping with these. There were many views at the time, and subsequently, from those such as Thomas Carlyle who saw the act as suicidal, to those who saw it as a re-establishment of the principles of the Reformation. The event resonates to this day, with an even greater potential schism on the political agenda, with not a few parallels relating to notions of fundamental Scottish identity and unsympathetic interference from Westminster (but also a warning for what might be the true cost, not only financially but spiritually, should such a sundering occur). As a close relation and disciple of Chalmers, and as a passionate evangelical Christian, it was inevitable that Cowan would support the Disruption (Fig. 30). He had been a Commissioner (representing the Northern Isles) to the previous six General Assemblies. However, he was excluded from the 1843 Assembly as a result of a pamphlet he had published, but he nonetheless followed at the end of the infamous procession from St Andrews Church in George Street (where the Assembly was held) to the hall at Tanfield, close to RBGE, where the Act was signed, witnessed by a crowd of 3000. As a prominent businessman Cowan's main role in the Disruption concerned its finances, in terms of how the new church was to pay its clergy and for its ambitious programme of church (and manse) building, education (schools and a new seminary) and foreign mission (notably in India, including Madras, which it took over in its entirety from the established church,). Cowan played a key role in the establishment and running of what was called the Sustentation Fund that made this ambitious programme possible. In its first year this raised an astonishing £223,028, ten times the Church of Scotland's annual budget. The Church mocked the Free Church for its emphasis on money, and Peter Cleghorn (who like Alexander Cowan and the Rev. William Menzies 'stayed out') warned his son of their being 'the greatest beggars'. For, as seen earlier, Hugh was also a committed Free Churchman, as was his wife Mabel.

The Free Church was, from its outset, a seriously intellectual body; that, at a time of great intellectual ferment, it would metamorphose, was almost inevitable. But within thirty years it had run out of steam and joined with the United Free Church. By this time the State had dropped its rules on patronage, which almost inevitably led to a general view, as stated by Cowan himself in 1878, that 'The Disruption is *now generally regarded* as having been not only a great crime, but ... a great national blunder'.[25] The emphasis was not Cowan's, as he excluded himself from this position, and remained a loyal member despite increasingly liberal views on ecclesiastical and religious matters. MacLeod (Cowan's biographer) as an historian from the Free-Church tradition saw this as a climb-down if not a betrayal, for example that Cowan's 'focus on "Nature" ... sometimes barely avoided pantheism'.[26] Certainly one of the running heads of Cowan's memoirs, 'the book of Nature not opened by ministers',[27] might support such a view.

7. FURLOUGH, MARRIAGE & THE COWANS, 1860–1

Fig. 30. The 1843 Disruption of the Church of Scotland.
Cartoon showing the Rev Thomas Chalmers (centre) dancing the 'Reel of Bogie'. (The rights of the Earl of Fife, patron of the parish of Marnoch, Banffshire, in the Presbytery of Strathbogie, had recently been upheld by the Court of Session (in the person of Charles Hope wielding the sword, left) to impose a new parish minister against the wishes of the people). Lithograph by W. Nichol after drawing by 'D.D.', c. 1840 (above).

Anonymous photograph of the painting of the signing of the Deed of Demission by David Octavius Hill. (Courtesy of the University of St Andrews Library, Govan Album, Alb-6-100.2) (below).

CHAPTER 7

Political career

If at all, Charles Cowan is remembered as the man who in 1847 unseated Thomas Babington Macaulay as one of the two MPs for the City of Edinburgh. Although by no means keen to push himself forward, Cowan, aged 46, found himself nominated as a candidate, standing against the somewhat remote and academic Macaulay, as a radical Whig representative of Free Church interests and a promoter of free-trade (including the removal of duties on products such as paper). That Cowan should be proposed was a sign of the times, with tempers still hot in the aftermath of the Disruption, and a feeling that the Westminster Government (especially the House of Lords) had shown a contempt for Scottish sensibilities in matters of religion. (There are parallels with the appointment two years earlier, in 1845, of Cowan's friend J.H. Balfour, rather than J.D. Hooker, to the Edinburgh chair of botany). Macaulay himself had supported the controversial, Tory-led, financial grant to the Irish Roman Catholic Seminary of Maynooth, an anathema to Scottish Protestants (though Cowan's views on Catholicism *per se* were not illiberal). To general surprise Cowan, with Sir William Gibson-Craig, took the seats, Cowan gaining 586 more votes than the great Macaulay – not that the latter was unduly perturbed: it allowed him more time to finish his magnum opus, the *History of England*. Cowan was also successful in the next two general elections (of 1852 and 1857), though in the first of these he was beaten to second place by a resurgent Macaulay. Cowan resigned from the Commons in May 1859, having gradually becoming less radical, though taking firm stands on several liberal educational debates including the removal of religious tests for Scottish university professors, and favouring non-sectarian school education. The abolition of tax on paper (introduced in 1711, though reduced in 1836), seen not only as unfair to paper-makers, but unacceptable ethically as a 'tax on knowledge', was, however, finally achieved only in 1861, after Cowan's resignation.

Old age

Cowan indulged in two sports – curling and shooting. For the sake of its grouse he, in 1852, purchased the 6000-acre estate of Loganhouse in the Pentland Hills, for £31,000 from Ferguson of Raith. It is for this reason that he was known as Charles Cowan of Loganhouse. Of his other properties he left Valleyfield in 1869, and in Edinburgh moved from Royal Terrace to Wester Lea, a large Jacobethan mansion in Ellersly Road, Murrayfield, newly built for him to a design by the architect Campbell Douglas. Cleghorn would stay here frequently after his retirement, though it must have held a tinge of sadness as Mabel's mother Catharine died there in 1872. Charles Cowan retired from the business in 1875 by which time his fervent evangelicism had been somewhat tempered. He came to believe that the new learning, both in science and biblical criticism, could be reconciled with his faith, and supported the controversial work of Robertson Smith published in the ninth edition of the *Encyclopaedia Britannica* (to be discussed in Chapter 14). Like his Bonar in-laws Cowan moved away from the narrow Calvinism of the Westminster Confession, and took to more liberal forms of worship including the singing of hymns. This led to his attendance at the Barclay Church in Bruntsfield (Frederick Pilkington's gothic masterpiece), with its charismatic minister, the Rev. James Hood Wilson, a friend of Cardinal Newman. It was Wilson who supported the American preacher Dwight Lyman Moody and harmonium-playing vocalist Ira David Sankey on their Edinburgh visit of 1873. Charles Cowan died at Wester Lea on 29 March 1889, his daughter Mabel having predeceased him two years earlier and Cleghorn himself outlived his redoubtable father-in-law by only six years.

7. FURLOUGH, MARRIAGE & THE COWANS, 1860–1

Cleghorn's Marriage

There had been so many points of potential contact in the Cowan and Cleghorn family biographies before the marriage that it has the feeling if not of inevitability, then certainly of great aptness – the Fife background; Charles Cowan's Continental education and travels, including the remarkable coincidence that he knew Professor Christian Brandis and his wife in Bonn long before their son Dietrich became Cleghorn's colleague in the Indian forest service. In London at the Great Exhibition in 1851, on which his future son-in-law worked, Charles Cowan had presented the Prince Consort with a pair of curling stones (taken by 'unfortunate Cockneys and Southrons' to be models for cheese).[28] After William Tennant had moved from Dunino to be schoolmaster at Lasswade he taught Sandie Cowan, Charles's younger brother. There were even links with eighteenth-century Scottish botany – while Hugh Cleghorn senior had attended the lectures of John Hope, the Cowan family was friendly with the Rev. Dr John Stuart of Luss, the Gaelic scholar and pioneer of the exploration of the Scottish alpine flora who had helped the Rev. John Lightfoot on his 1772 Scottish tour with Thomas Pennant. It is not known when the alliance of Mabel and Hugh was hatched, though Cleghorn may have met the Cowans during his winters in Torquay between 1848 and 1850, as at this time the family also wintered there for the health of their daughter Catherine, and it was in this south-coast resort that Mabel's youngest brother John James had been born in 1846.

The minister who took the wedding ceremony at Valleyfield on 8 August 1861 was one of Mabel's brothers-in-law, the Rev. Robert H. Lundie, husband of her elder sister Elizabeth (b. 1827). Despite having been in the first graduating class of the Free Church's New College, Edinburgh, Lundie spent his working life in the English Presbyterian church in Birkenhead and Liverpool. He also travelled in America and was involved in bringing Moody and Sankey to Scotland in 1873. The wedding took place at Valleyfield, the substantial 26-room house that overlooked the family mills at Penicuik. Originally built as two houses, Valleyfield had been designed by the King's Architect Robert Reid for the surgeon and the chaplain during the time that the mills had been used as a prison during the Napoleonic Wars. The Cowans themselves were notable architectural patrons and their architect-of-choice for building projects in Penicuik was Frederick Thomas Pilkington whose fantastical Gothic creations were lavishly adorned with carved naturalistic vegetation. Similarly Ruskinian, but in the guise of a Venetian palazzo, was the Cowans' 1864 Edinburgh warehouse/office that survives, if crumbling, beside General Register House, designed (and externally signed) by George Beattie of Glasgow.

Mabel Cowan

Regrettably little is known about Mabel as none of her correspondence survives and there are merely fleeting references in the letters of her husband (mostly relating to illness) and even briefer ones in the various Cowan sources. She was born at Valleyfield on 25 February 1838, the tenth of Charles and Catharine Cowan's 13 children (four died in childhood or early adulthood, nine were daughters). Three weeks later Mabel was baptised 'Marjorie Isabella' by her maternal grandfather the Rev. William Menzies of Lanark. The witnesses, the Rev. William Scott Moncrieff and Thomas Constable, might have augured for a literary career rather than marriage to an Indian forest officer, for the Minister of Penicuik was grandfather of Proust's translator, and Constable was son of the publisher whose bankruptcy almost led to Sir Walter Scott's downfall. The families were closely connected – Alexander Cowan had

CHAPTER 7

Fig. 31. Mabel, Margaret and Anna Cowan. Watercolour by Hope Stewart, c. 1848, from Cowan, 1933.

represented the Constable creditors and had personally paid for the education of Thomas; Charles Cowan (who as a child had heard *Waverley* read from proof sheets) had acted as agent in the melancholy task of selling Scott's manuscripts in London to pay off some of the debts,[29] and Thomas Constable married one of Charles's sisters, Lucy (known as 'Puss'), so was an uncle of Mabel. Despite the Cowans' belief in education, the girls, typically for the period, were kept at home under a governess, Miss Jane Newbigging, but Mabel's father took her on at least two of his frequent Continental tours – in 1860, the year before her marriage, to Russia in connection with the rag supply for the paper mills, returning via Finland, Sweden

7. FURLOUGH, MARRIAGE & THE COWANS, 1860–1

and Denmark. Three years earlier, in 1857, with her sisters Charlotte and Anna and brother Charles she had visited Germany and Switzerland (where the party climbed the Righi and met the Brandises), this time returning via Geneva, Lyons and Paris. Like Cleghorn, Charles Cowan was accident prone – a chapter of his autobiography is entitled 'Repeated Preservations from Danger'[30] – and was involved in no fewer than three railway accidents. Mabel was with him on one of these occasions, in 1855, when part of the express train in which they were travelling from Edinburgh to London was derailed (due to frost) between Cramlington and Killingworth in Northumberland.

Donald MacLeod, Charles Cowan's biographer, has described his impression of Mabel, and her familial relationships as 'a woman who was always fussing, protecting, and shielding her younger female siblings and who was very close to her father'.[31] These

Fig. 32. Mabel Cowan. Photographer unknown, from Cowan, 1933.

relationships and her character is suggested in a touching watercolour by Hope Stewart reproduced in her brother John James's autobiography, which shows the three youngest Cowan sisters, aged perhaps 10, 8 and 6, beside the steps of Valleyfield House, the youngest Margaret holding the family parrot, and Mabel's hand protectively placed on Anna's shoulder (Fig. 31).[32] The only one of Mabel's accomplishments mentioned by her husband is music. One of her musical instruments was stolen in the Nilgiris soon after she reached India, and her skill came in useful in the leading of hymn-singing at evangelical meetings in the Himalayas. Almost certainly Mabel was taught to draw and paint, but only two, very slight, botanical drawings in a somewhat eighteenth-century style that are probably hers survive – one of the Persian Lilac still with the Sprot family, and one done at Galle in 1861 on their journey to India in the Cleghorn Collection at Edinburgh University that her husband has annotated as possibly a *Flacourtia*.

While not blessed with children, the marriage appears to have been a happy one – Mabel did what she could to help Cleghorn's career, compiling an index for his first book, copying the official papers and correspondence that were part of the lot of Indian officials, and occasionally writing letters on his behalf. In the only description of Mabel from someone who knew the couple, her training for such a role was explained as follows:

> [the] amiable and accomplished Miss Cowan, whose gentle and winning ways, no less than her trained abilities, tact, and shrewdness proved, for a quarter of a century, a source of daily happiness and help [to Cleghorn]. As her father's amanuensis, while he was a Member of Parliament for Edinburgh, Miss Cowan gained broad views of life and varied experiences – both being extended by Continental travel and careful reading, so that she was not only a delightful companion, but one who entered keenly into all her husband's scientific and administrative duties.[33]

CHAPTER 7

Mabel's uncles and siblings

In this account of the Cowan family emphasis has been on Cleghorn's redoubtable father-in-law, but the three brothers of Charles who survived to adulthood, Mabel's uncles, were also significant. John (1814–1900) and James (1816–1895) became Charles's business partners in the Valleyfield Mills around 1841. James, after a rocky start, became Lord Provost of Edinburgh 1872–3 and MP in 1874. John would end up Sir John Cowan of Beeslack, rewarded by Gladstone with a baronetcy for his role as his Scottish fixer for the Midlothian election. Charles's next brother George had died aged only 20 and Alexander ('Sandie') lasted little longer, dying aged 25 in 1831. Sandie had considerable literary talent and helped Walter Scott with research for his biography of Napoleon. In a vain attempt to improve the tubercular health of himself and his wife, he went to the Continent. It was probably on a visit to him in 1831 that Charles first met the Brandises in Bonn, where Christian August Brandis (1790–1867) was Professor of Philosophy.

Mabel had twelve siblings, eight of whom reached maturity. As these became Cleghorn's brothers- and sisters-in-law, with whom they kept in touch after returning to Scotland in 1868, they also merit a brief mention. In his *Reminiscences* the patriarch Charles Cowan remarked on the fact that he had 'contributed three daughters to India, one to each of the three Presidencies'[34] – Mabel being his contribution to Madras. Her elder sister Kate (b. 1832) married Charles Wahab, an Ulsterman and engineer in the Bombay Army, who ended up a Major General and, after retiring to Scotland became useful in the Cowan paper mills. Mabel's younger sister Anna (b. 1840) married Edward Newnam, a lieutenant in the 17th Bengal Cavalry: they spent 21 years in India where he reached the rank of Lieutenant Colonel. Like Mabel, Kate had no children, but Anna had four who were brought up in Sheffield by their remarkable sister Charlotte (b. 1833). Charlotte was an early feminist, who fought for women's rights (including those of prostitutes) and was married to the pacifist, Quaker MP for Holmfirth, Henry Joseph Wilson; one of their daughters Helen Mary (1864–1951) was a suffragette and pioneering female doctor. Of the other sisters, Elizabeth has already been mentioned; Margaret (b. 1842), despite being an invalid, managed to live to the age of 92, and the eldest, Jean (b. 1825), was married to Thomas Chalmers (nephew of the theologian), a papermaker at Linlithgow. Jean died suddenly, aged 39 in her father's Edinburgh house, leaving five of her seven children under the age of ten, and a sense of shock and loss from which her mother never recovered. Mabel was also deeply upset by the news which she learned in Simla in June 1864, though her own feeble health thwarted her wish to return home to help look after the orphans.

Only two of Mabel's brothers reached adulthood, Charles William (b. 1835) and John James (b. 1846). Charles William took on his father's role in terms both of business and civic responsibilities – he was in charge of the paper-making company and inherited the territorial designation 'of Loganhouse'. Charles also had botanical connections, both direct and indirect – he was a keen breeder of daffodils, first at Valleyfield, then in retirement when he rented Dalhousie Castle. At the latter he inherited the garden of Christian, Lady Dalhousie (wife of the 9th Earl, mother of the Marquess), a competent botanist who had sent plants to the Dalhousie garden from both Canada and India, and as already noted her herbarium collection was among those worked on by Cleghorn as part of his curatorial activities in 1849/50. There were also connections with the botany of the Western Himalaya through Charles's wife, Margaret Craig, daughter of another paper-maker, one of whose sisters was married to the Edinburgh-trained Indian surgeon James E.T. Aitchison (1835–1898) who published on the plants of Afghanistan, Punjab and Sindh.

7. FURLOUGH, MARRIAGE & THE COWANS, 1860–1

John James Cowan, the youngest of the family (b. 1846), is best known as a distinguished collector of the paintings of his contemporaries, including one by Edouard Manet (*Le Pont d'un Bateau*, now in Melbourne), no fewer than 14 oils and 6 watercolours by James McNeill Whistler, and numerous works by the 'Glasgow Boys' and the Scottish colourists.[35] As a result of financial difficulties resulting from the General Strike, most of the collection was sold at Christies on 27 July 1927, but Whistler's portrait of his patron and friend *Arrangement in Grey and Green* (for which Cowan gave the artist 600 guineas, but, in spite of at least 60 sittings between 1893 and 1900, was never completed), was given in the sitter's memory to the National Galleries of Scotland in 1930. John James inherited Wester Lea from his father and continued to live there, at least for some years.

Honeymoon

Honeymoons were doubtless considered frivolous and following the wedding the newly-weds stayed with Mabel's parents in their grand Edinburgh town-house at 37 Royal Terrace where, on 2 September, there had been a party for the botanists Robert Kaye Greville, Robert Wight and George Walker-Arnott[36] – the conversation doubtless revolved around Indian botany to which each of them had contributed so much. Royal Terrace forms a piece of stupendously grand Neo-Classical architecture, designed by William Henry Playfair, which, with its somewhat humbler continuation, Carlton Terrace, had from the 1830s become something of a colony for both the Cowan and the Cleghorn families. High above the coastal plain, it looks out over Leith, and the Firth of Forth, to the Fife homeland of both the families, as is revealed in a fascinating study of the occupants of these houses by Ann Mitchell.[37] Of the older generation of Cowans, Duncan lived at No. 1 Royal Terrace (from where he took an interest in the laying out of the communal gardens of Calton Hill) and his brother Alexander at No 35, with three of Alexander's children nearby – Lucy ('Puss') Constable at No. 34, Charles at No. 37 and James at No. 38. Of a younger generation, Helen Bannerman (*née* Helen Brodie Cowan Watson) – author and illustrator of a now politically incorrect children's book of 1899 with charming pictures of tigers melting into a pot of ghee – was born at her grandfather Alexander's house though her mother Janet Boog Watson was from Alexander's second marriage, making her only a 'half-cousin' to Mabel. At 14 Carlton Terrace, until his death in 1866, lived Dr William Mackenzie and his family including his sister-in-law Helen Wyllie, Cleghorn's maternal aunt, though her sister Margaret Mackenzie had died in 1841. There were other Indian botanical connections in the terraces – Edward Madden lived at 26 Regent Terrace, and Thomas Anderson (senior) lived at 6 Royal Terrace. Anderson, a banker, was married to Jane Cleghorn (no relation, from a Wick family); their two naturalist sons both went to India. The eldest Thomas (1832–70, who married his cousin Elizabeth Cleghorn), was the botanist who helped Cleghorn to curate the Edinburgh University Indian herbarium; he became Superintendent of the Calcutta Botanic Garden (1861–8) from where he became involved in the quinine story, and in 1864 was the first Conservator of Forests for Bengal (effectively Sikkim and Darjeeling). His younger brother John (1833–1900), a zoologist, was professor of Natural History at the Free Church College in Edinburgh and then, from 1865 to 1886, Superintendent of the India Museum in Calcutta.

From Edinburgh the newly-weds headed south for London, where they stayed with William Bonar, husband of one of Mabel's Menzies aunts, Jessie Ann; Bonar worked in the City, where he was Secretary to the General Shipowners' Society. The major event while in London, in

CHAPTER 7

September 1861, was the publication of Cleghorn's first book, the *Forests & Gardens of South India*. It was published in London by W.H. Allen & Co. of 13 Waterloo Place (until recently publishers to the EIC) but printed by Patrick Neill & Co. in Edinburgh, where Cleghorn had done much of the work of compilation, and supervision of the making of plates and maps, while staying with Balfour at 27 Inverleith Row.

Author: *The Forests & Gardens of South India*
"Full of valuable information, and thoroughly reliable in all its statements" [38]

As requested by the Madras Government,[39] the book is a compilation of Cleghorn's first three annual forest reports, with numerous other associated official correspondence including extracts from works such as the jury reports of the 1855 and 1857 Madras Exhibitions. No attempt was made to make it readable by a general audience, nor to try to draw general principles or conclusions, but it was an act of generosity on the part of officialdom to make available a mass of information that would otherwise have been entombed in the archives of Madras, Calcutta or London. Cleghorn sent an early version of the book to Joseph Hooker for comment, and the accompanying letter gives informative details about the printing and publication – the edition was to be of 500 copies and the author was bearing the expenses, though the Madras Government had subscribed for 50 copies and the Mysore one for 25. The cost was estimated at about £200 (£120 for the paper, the rest included the cost of the woodcuts and binding).[40] The volume that emerged was modest in size, duodecimo in format (though advertised as 'post octavo' – Fig. 33), bound in green buckram and sold for 12 shillings.[41] Its 428 pages are dedicated to Lord Harris who had given Cleghorn his Conservator's job and the only adornments to the dry text are 17 woodcut figures, 13 plates and a map. The plates and map were made in Edinburgh: the map shows the teak, sal and sandalwood forests of South India (and the Tenasserim teak forests) and was engraved by the well-known cartographers William & Alexander Keith Johnston (who had produced Edward Forbes's biogeographical maps). Twelve lithographs were made by William Husband McFarlane (ten of these are based on drawings by Douglas Hamilton, six being

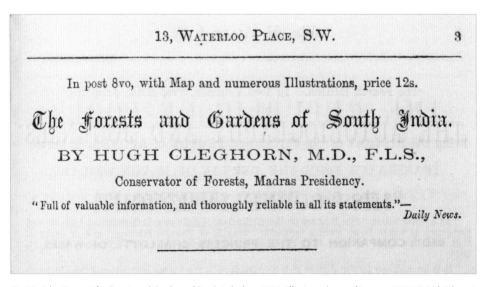

Fig. 33. Advertisement for *Forests and Gardens of South India* from W.H. Allen's catalogue of January, 1862. (British Library).

reprinted from the Royal Society Anamalai paper).[42] In addition to Cleghorn's three annual forest reports, there are reports on several distinct forest areas (Mudumulai, Godaveri, the Nilambur teak plantations, the Australian plantations at Wellington), on woods used for particular purposes (for burning, making furniture, carving or engraving, for naval purposes and making charcoal), and extensive papers on the subject of 'kumri' (shifting cultivation). Brandis's forest rules for Pegu in Burma are reproduced, as are those for Sind by N.A. Dalzell. Horticultural items concern avenue trees, soldiers' gardens, the public gardens of Bangalore and Ootacamund and hedge plants. The last, along with the report of the Anamalai expedition, display a characteristic Cleghornian trait – the republication of the same work in several different places and formats. As end-matter there is a bibliography and the work is completed with a thorough-going index, largely compiled by Mabel,[43] though some of the early work on it must have been done by John Sadler, Balfour's teaching assistant, as he was paid £2 for his contribution by Cleghorn.

The climate question

Given Richard Grove's claims of the negative environmental effects of deforestation as Cleghorn's fundamental motivation, it is worth seeking evidence for this in *Forests & Gardens*. In fact in the 412 pages there are only seven references (pp. x, 16/7, 20, 79, 171, 212). The statement in the Preface is certainly supportive of this view: 'supplies of water, and consequently of food and other produce, are in a great measure dependent on the existence of forests, especially in all the elevated parts of that vast country'. The next five refer exclusively to the Nilgiris, and had arisen in discussions with E.B. Thomas over the competing needs of land for plantations and the protection of forest on ridges to ensure continuing water supplies (see p. 91). The last reference is the quote in the Timber Report from the 1855 Madras Exhibition discussed on p. 73, and probably related as much to E.G. Balfour as to Cleghorn. The prevailing and overwhelming concern of the book is how to ensure a continued supply of timber for development projects, with strict preservation restricted to the minimum area for climatic needs.

Reviews

Five contemporary reviews of the work have been found, which appeared soon after publication – three in London publications, the *Gardeners' Chronicle*, the *Spectator* and the *Athenaeum*, one in the *Edinburgh New Philosophical Journal*, and one in an Indian medical journal.[44] The British ones all stressed the book's un-literary style, but commended it for its usefulness in terms of potential economic benefit to India in providing evidence of the best means for the wise management of its forests. The *Athenaeum* review,[45] (anonymous, but attributed to William Robertson) is overtly imperialistic in tone, fixing the blame for the destruction of forests firmly on 'native contractors and traders': by contrast 'the races destined to rule the other varieties of mankind hate waste'. The anonymous *Spectator* reviewer was acutely aware of worldwide threats to forests and familiar with the Indian situation: Cleghorn was praised as a 'most efficient and determined forester', and for making a detailed and useful book, but one that would have benefitted from 'striking out all surplasage'. The reviewer sympathised with Cleghorn's impossible position, operating under a Government 'divided between a passion for timber and a wish to restrict the denudation', although:

> his business is to preserve the forests, [he] fills his reports with plans for the swift cutting down of more and yet more timber, builds timber channels up precipices which would put the Swiss channels to shame and calls for "elephants! elephants!" as if he were collecting cows.[46]

CHAPTER 7

The reviewer pointed out that strict preservation was limited only to the tops of ridges where deforestation was starting to affect rainfall, and the only possible remedy for the depletion was the making of extensive plantations that were then only just beginning.

The Edinburgh review may be by J.H. Balfour,[47] the journal's botanical editor, though Cleghorn had suggested that, were he too busy, David Paterson (the Madras medical missionary then on leave in Edinburgh) might do it.[48] This too stressed the economic benefits to government of Cleghorn's management, but, interestingly, began with a statement on the benefits of forest conservancy to climate and soil. The most surprising of the four British reviews is that in the *Gardeners' Chronicle*.[49] This would have been taken to have been by its editor, John Lindley, had not Cleghorn thanked Joseph Hooker 'for so cordially noticing my little book in the Gard: Chron:', which he had seen at Saharunpur on his way to the Punjab.[50] Hooker was not known for generosity in such matters, so his approval must have been particularly pleasing to the author. Cleghorn's honesty as a reporter and his simplicity of language was praised and Hooker took the quotation of financial accounts and government reports to testify to its authority, claiming 'never [to have] read reports more explicit, perspicacious, and thoroughly to the purpose than these, though devoid of pretension to literary merit'. It is also remarkable for Hooker's pointing out that Britain was in no position to 'throw stones' on the topic of the destruction of Indian forests when it had treated its own New Forest and the Forest of Dean so badly. The fulsome conclusion was that the work was 'beyond comparison the best that has appeared on the vegetable products of any part of India'. This hyperbole calls for comment as, published as it was in one of the most widely read of horticultural journals, it cannot have been an off-the-cuff remark. It can surely not be intended as a slur on the floristic works of Roxburgh, Wight or Wallich, and the only remotely comparable works on *economic* plants are those of J.F. Royle (on cotton, on fibre plants, on drugs, and above all his regional study of the Himalayas). Hooker is known to have had a low opinion of the ('flatulent') Royle, so this invidious, if silent, comparison would seem to be aimed at him, though if so the target was beyond reach as Royle had died in 1858.

By far the longest review of Cleghorn's book was in the *Madras Quarterly Journal of Medical Science*,[51] written by an author with a taste for poetry. It is unfortunately anonymous, though there are internal hints (such as an expression of regret that Cleghorn had not yet produced his Manual of Botany) which suggest that the author was connected with the Medical College. The most likely candidate is Howard Benjamin Montgomery the journal's editor, and Cleghorn's successor as Professor of Botany, but the most surprising aspect is the source of the third of the poetic quotations, a parody of 'our English Ariosto', which turns out to be from William Tennant's *Anster Fair*. It seems unlikely for anyone but a Scot to have known of this work 40 years after its composition, but Montgomery was Irish, and surely Cleghorn didn't get someone in Scotland to write it? The author was clearly familiar with India as the article starts with a review of the Madras surgeons' contribution to the study of Indian botany, notes the pitiful inadequacy of the mere 53 peons assigned to the Forest Department, and, (in a discussion of the section on hedges), that cattle-trespass was the commonest source of village quarrels. The reviewer as well as commending the book to foresters, engineers and the 'botanical enquirer' was also generous and could 'cordially recommend it to the general reader as a book in which he will find much to interest [him]'. There does, of course, remain the intriguing possibility that Cleghorn wrote the bulk of it himself, and left the editor to add the recommendation that modesty, and his highly developed sense of rectitude, would surely have forbidden.

The author's devoted father, Peter, was not so enthusiastic, and described the work as one that 'would drive all other soporifics out of fashion',[52] hoping that Hugh would one day produce a 'work more generally interesting', presumably a naturalist's travelogue in the tradition of Hooker's *Himalayan Journals*.[53] A curious juxtaposition, and reminder of the popular literature of the time, is that the review that follows *Forests & Gardens* in the *Spectator* edition of 28 September 1861 was of Mrs Henry Wood's *East Lynne*.

The Global Plant Network

Before leaving London in the autumn of 1861 Cleghorn visited Kew on 17, 18 and 19 September to collect some plants to take back to India. These were packed in three Wardian cases, two closed boxes (of roots), and a small case with a glazed top with particularly precious contents.[54] While Cleghorn's role as a vector of exotic plants was tiny in the great scheme of the plant translocations that were going on at the time, and in which Kew played the crucial central role, the plants, their origins and destination, are characteristic of the whole enterprise. The first of the Wardian cases was destined for G.H.K. Thwaites, who since 1849 had been Superintendent of the Peradeniya Garden in Ceylon. The other pair was for Dr Thomas Anderson, Cleghorn's old friend whom he 'once looked upon … as a younger brother – but he has far outgrown me in his rapid Career',[55] referring to Anderson's recent appointment as Superintendent of the Calcutta Botanic Garden. The cases were designed to sit on the ship's deck and were substantial affairs (Thwaites's, which got damaged while crossing the Bay of Biscay, weighed 200 pounds). Each contained over 20 species, including herbaceous plants and small trees. The former, all tropical exotics, were of the sort featured in contemporary illustrated horticultural literature, and in circulation in the European horticultural trade, notably through the Belgian nurseries. There was a strong bias towards plants from the tropical Americas and West Indies, with the family Acanthaceae, and also Rubiaceae, being the most strongly represented. In the two Wardian cases for Bengal were palms, the Australian conifers *Araucaria cunninghamii* and *A. bidwillii*, and two New Zealand podocarps. Most of the herbaceous plants have long since disappeared from cultivation in the Subcontinent, but among the Bengal consignment *Justicia carnea* 'Superba' has survived, and – for better or for worse – the gaudy bougainvillea has become a fixture of Indian gardening and landscaping.

The roots in the closed boxes were for the public gardens of the Madras Presidency – for William New at the Lal Bagh, Bangalore and McIvor at Ootacamund. These were all of South American gesneriads, including five cultivars of *Gloxinia* and no fewer than 17 of *Achimenes*.

Return Voyage

On 20 September the Cleghorns sailed from Southampton on the ship *Indus*, seen off by Mabel's parents, her father having recently returned from a trip to Switzerland.[56] On the same ship was William Jameson, who had been home on a year's sick-leave, during which John Lindsay Stewart had acted for him as superintendent of the Saharunpur botanical garden. Through the Straits and into the Mediterranean where there were fine views of the Atlas Mountains and an already snowy Sierra Nevada. The ship put in at Gibraltar, where Cleghorn was impressed by the vegetable market; he collected seed of Spanish broom (*Spartium junceum*) with the dubious intention of introducing it to the Nilgiris (where it still occurs, though is not one of the worst of the potentially invasive aliens), and of Castille melons for the Madras Agri-Horticultural garden. They arrived at Alexandria on

CHAPTER 7

5 October where Mabel experienced her first taste of the 'riches of an Eastern Bazaar'. They went sightseeing, including Pompey's Pillar (a granite monolith erected in AD 279 to commemorate Diocletian's quashing of the Alexandrine revolt) and the much more ancient Cleopatra's Needle (one of a pair originally erected at Heliopolis by Thutmose III in 1450 BC – this one was relocated to London 17 years later, its brother to Central Park, New York).

The travellers arrived in Egypt to scenes of chaos and devastation (Fig. 34). The Nile was in spate, the Nilometer revealing that the river had risen more than 4½ feet, with resulting extensive loss of life both human and of livestock, and the destruction of crops and sections of the recently completed Cairo to Alexandria railway. Taking only minimal hand-luggage the couple had to travel by paddle-steamer – a nightmare journey against the powerful current. With no landmarks to aid navigation the boat hit a sandbank and the paddlewheel was damaged. Cleghorn's greatest concern was the small case with the glazed top entrusted to his care by Hooker. This contained eight precious plants of cinchona about which he was in a 'continual anxiety by night and day', though Mabel allowed the case a place in their cabin. At one point but for the 'strong hands of a couple of brawny-armed Glasgow Engine drivers it would have come to grief'. Despite the sights of Alexandria – historic and exotic – and the scenes of destruction, there were links with home as the swing bridge across the Nile at Kafr el-Zayyat (Fig. 34), on which they saw Sa'id Pasha walking, had recently been constructed by the Edinburgh engineer Robert Stephenson. This visit must have been a poignant one for the Pasha as in 1858 his heir, Ahmed Rifaat Pasha, had drowned when the train in which he was travelling had fallen from Stephenson's original car-float crossing. Cleghorn also noted with patriotic pride that the river steamer towing the Pasha's state barge was Glasgow-built. Cairo was eventually reached, where the Cleghorns stayed in the opulent Hotel des Anglais, which had been renamed 'Shepheard's' the previous year. Here the Cleghorns met William Muir, also returning to India where he was Senior Member of the Board of Revenue in the Calcutta Government. They shared strong evangelical beliefs, though at this point they had no idea that their paths would cross again in northern India and later in Edinburgh. The Cleghorns visited 'the Pasha's garden at Shubra – and [he] was desirous of seeing the Rhoda island collection of Ibrahim Pasha', but this had been neglected and 'parcelled out to Fellahs' since the death of James Traill.[57]

Fig. 34. The 1861 Inundation of the Nile, as witnessed by Hugh and Mabel Cleghorn. 'Below Kafr Zayat, showing the railway-bridge and train off the line'. Anonymous wood engraving, *Illustrated London News*, 2 November 1861.

7. FURLOUGH, MARRIAGE & THE COWANS, 1860–1

From Suez the couple sailed to Galle in Ceylon, which they reached on 26 October. Cleghorn had hoped for some time in Ceylon (the scene of his grandfather's exploits 65 years earlier) but this did not happen and it is not known if he even managed the trip to Peradeniya to deliver Thwaites's plants in person. A bombshell arrived which exploded the plans they had made to return to his old life based in Madras and Ooty, and a job largely concerned with the management of the broad-leaved tropical forests of the Western Ghats. This came in the form of a telegram from the Government of India that ordered Cleghorn to report to Simla to work on the Punjab forests, the most important of which were temperate and coniferous. The suggestion had come from Colonel Henry Yule,[58] Secretary to the Government in the Public Works Department, and the order was in the name of the Viceroy, Lord Canning.[59]

Cleghorn & Quinine

The most precious item of Cleghorn's out-bound luggage was the case of quinine plants that had been smuggled out of Peru and Ecuador only the previous year, and had meanwhile been convalescing at Kew. These formed part of the great quinine translocation venture – six were of the 'red bark' *Cinchona succirubra*, and two of the 'grey bark' *C. micrantha*, addressed to William New at Bangalore.

The year 1861 was one of great excitement in the introduction of cinchona to India as a new plantation crop – a major development in the fight against malaria, but also with the explicit aim of bringing the cost of treatment within the reach of the poor. In the very different world of contemporary bio-politics this has come to be regarded as a case of 'biopiracy', but it was the success of this great enterprise that Sir Joseph Hooker, in old age, regarded as his proudest achievement. 'Quinine' is a shorthand for a group of alkaloids extracted from various members of the genus *Cinchona* native to the eastern slopes of the Andes from Colombia southwards through Ecuador and Peru to Bolivia, whose miraculous properties had been known since the seventeenth century when, in the form of 'Jesuit's Bark', it was first brought to Europe. However, since 1751 the item of *materia medica* had been a monopoly of the Spanish Crown and access to its source was fraught with difficulties – geographical and political. The desire to cultivate the plants in India was in part a case of colonial rivalry, but there were also genuine worries about its non-sustainable harvesting in South America, and resultant increasing costs of the drug. As George King in 1876 somewhat jingoistically put it:

> the more active interference shewn in Bolivia has been quite as disastrous as the lazy *insouciance* of the more northern States [Colombia and Ecuador]. These Governments proved, however, one and all, exceedingly jealous of any attempt to procure for more civilized countries seeds or seedlings of the invaluable trees which they had done so little to conserve.[60]

For almost 40 years botanically minded medics in India, nearly all of them known to Cleghorn, had been pressing for this translocation. George Govan (by now retired and living in St Andrews), when stating the case for the establishment of the botanical garden at Saharunpur as early as 1816, had suggested that cinchona might be grown there. J.F. Royle, Govan's successor, repeatedly petitioned the EIC on the topic (suggesting that it might be grown in Khasia and the Nilgiris) – both from Saharunpur and after his return to London – as did Royle's successor Hugh Falconer (who suggested sending a collector to South America). Cleghorn's Calcutta-based medical friend Alexander Grant, and two Calcutta Garden Superintendents, Thomas Thomson and Thomas Anderson, also pressed the case. However, it was only after the demise of the EIC that the India Office (under Lord Stanley, Secretary of State for India) took action and in 1860 despatched Clements Markham to Peru to effect the transfer. This story has often been recounted so only the outlines need be summarised here.[61]

It had been decided that the Nilgiris would be the best place for the initial experiment and Cleghorn played an important role in choosing the sites. He was originally to have met Markham on his arrival at Calicut bearing the green gold from Peru, but illness intervened and he had to return to Britain on sick leave in September 1860. However, before this, Cleghorn had:

> made many excursions over the western ghats, comparing the soil and climate with the description of the Peruvian localities and Java plantations. It seemed to me that in the fine mould of the Nedivuttum Sholas, we have the porous, greasy, red subsoil alluded to by Mr Markham.[62]

It fell to William McIvor to undertake the initial plantings, for which he selected a 50-acre, wooded slope above the Botanic Garden at Ooty called Dodabetta, on which to grow the plants received, the Nedivuttum site being used only later.

Markham worked out a thorough-going strategy to collect, by means of help from others experienced in Andean collecting, a range of the best of the quinine-producing species, from three areas. He himself would concentrate on the highest-yielding 'yellow bark' *C. calisaya* in southern Peru and Bolivia; George James Pritchett would collect the grey barks (*C. micrantha* and two other species) from central Peru. The great South American collector Richard Spruce would target the red bark (*C. succirubra* now known as *C. pubescens*) in Ecuador, the transfer of which would be effected with the help of Robert McKenzie Cross, a Kew gardener sent out for the purpose (Cross would make several return trips for other species of *Cinchona*, and later for Para rubber). These fraught expeditions took place in 1860 and while they succeeded (against great odds) in their aim of obtaining seed and plants, the next stage of the process was ill thought out and effectively betrayed the immense efforts of the collectors. Instead of being taken directly by ship from Peru to Madras across the Pacific, the plants were taken on a complex eastward journey via Kew and the Red Sea where Markham's and Pritchett's plants all either died or were fatally injured. Markham reached Calicut in India in October 1860, but though his own and Pritchett's plants proved beyond saving, there were still a few plants in reserve at Kew, and Markham put his time to good use with a tour of South India, looking at potential additional nursery sites in the Nilgiris, Coorg, the Palnis, Wynad and Mahabaleshwar, which led to the establishment of a second Nilgiri plantation at Cleghorn's recommended site at Neddivuttum for the growing of the more tropical species – this was named after Sir William Dennison (Fig. 35), the new Governor of Madras (recently arrived from New South Wales) to whom Cleghorn would report on his return to India. (A third site at Pykara was added in 1862, and by 1869 there were 1200 acres under quinine in the Nilgiris).

The unsung heroes of this tale, however, are the Scottish gardeners. While Cleghorn was still in Edinburgh, Cross delivered his cargo of 463 succirubras and 6 calisayas to McIvor on 9 April 1861 in excellent condition, which McIvor proceeded to propagate with enormous success. Two months later, by June 1861, he had increased these to 2000 plants (mainly succirubra, but also a few calisayas), which by the following April had increased spectacularly to 14,450 plants of *C. succirubra*, 8106 of *C. officinalis* (newly raised from seed obtained by Cross on his second mission) and 8276 of *C. micrantha* (presumably from Pritchett seed). These impressive figures

7. FURLOUGH, MARRIAGE & THE COWANS, 1860–1

Fig. 35. The Quinine Experiment in the Nilgiri Hills.
Sir William Denison (seated), planting the first tree of the Peruvian Bark in the new Neilgherry plantations, with the help of W.G. McIvor (standing with spade). Anonymous wood engraving, *Illustrated London News*, 6 December 1862 (above).
Young plants raised by McIvor. Wood engraving from a photograph – frontispiece of C.R. Markham's. *Travels in Peru and India*, 1862. (RBGE: Cleghorn's copy) (below).

CHAPTER 7

themselves doubled over the following four months. Cleghorn's contribution (assuming it reached Ooty safely in early November 1861) can therefore be seen to have been a booster to what was already a thriving body of plants.

In this initial phase of establishment of cinchona in India there had been a helpful interchange with the Dutch who, from the start, had been far ahead of the British in matters quinonial. As early as December 1852, Justus Carl Hasskarl, a former director of the Buitenzorg Garden in Java, had been sent to Peru, resulting in the establishment of plantations in Java. Although at this stage most of its plants were the unproductive *C. pahudiana*, there were also some of *C. calisaya*. At the same time as Cleghorn and Mabel were on their way back to India, in September 1861, Thomas Anderson had been sent from his new post at the Calcutta Garden to Java. Plants of both *C. pahudiana* and *C. calisaya* (with a few of a third species, *C. lancifolia*) were generously supplied to Anderson, who took them to McIvor in December 1861. In return Anderson received some red and grey barks from McIvor, which would form the start of the plantations around Darjeeling (in 'British Sikkim') where it was the Scottish gardeners Andrew Jaffrey,[63] and later James Alexander Gammie, who once again saved the day. At the same time (1861) Thwaites in Ceylon was establishing plantations of *C. succirubra* and *C. officinalis* at Hakgalle in the highlands of Ceylon.

It was later realised that none of the forms grown in India was particularly rich in terms of quinine yield. The best proved to be a form of *C. calisaya* (called '*Ledgeriana*') that had been passed over by Kew when offered to them. Some seed did, however, reach McIvor though it did not succeed well in the Nilgiris, but did better in Sikkim. *C. ledgeriana* was largely taken up by the Dutch in Java, where it grows well, and it was therefore the Dutch who came to dominate the international quinine market from the 1870s, though India continued to be supplied by the Sikkim and Nilgiri plantations.

Cleghorn would have a second minor link with the quinine story four years later, during Clements Markham's second Indian visit of 1865/6.

8
Forests of North India, 1861–1864

The need for timber in a rapidly Westernising India, for railway sleepers, for construction projects, and (given the lack of coal) as fuel – both domestic, and for ships and trains – was pressing. Timber supply was in the hands of agents (who obtained leases either from local rulers or, in the case of British territory, from the Government), over which there had been some patchy attempts at control for the previous decade; but what was urgently required was a survey of standing resources and a co-ordinated plan as to how these should best be managed to ensure a continuing supply of timber. Brandis was already at work in the teak forests of Burma, and Cleghorn's immediate task was to survey the forests of the Punjab, which he did for a year after which he was asked to stay on, to 'institute a systematic plan of Conservancy and Management' for a Forest Department for the Punjab. Brandis was summoned from Burma and with Cleghorn the pair held the post of joint Commissioners of Forests for a year until February 1865, when Cleghorn returned to Madras and a six-month home leave.

To the Punjab

The Cleghorns crossed from Ceylon to Madras to make the necessary arrangements for the challenging new job and on 11 November 1861 Hugh sailed for Calcutta on the P & O steamer *Simla*.[1] Until plans became clearer and she could join her husband somewhere in the north Mabel was left at Coonoor in the Nilgiris, with a Madras medical colleague, Dr Walter Alexander Leslie and his wife Elizabeth. The Leslie house was burgled and Mabel lost some of her treasured personal items – trinkets, photographs and a musical instrument.[2] Though unidentified the instrument must have been portable – perhaps a guitar such as the one played by her Aunt Madge shown in Kenneth Macleay's conversation piece of her grandfather's large family.[3]

In Calcutta Cleghorn had meetings and received instructions from Henry Yule,[4] but also took the opportunity to press with the Legislative Council (including Sir Bartle Frere and Cecil Beadon, the Government Secretary whose ear he had already bent on the outward voyage) the case for support to allow his friend Joseph Hooker to continue his work on *Flora Indica*.[5] Although railway construction was by now making substantial progress, Cleghorn reached the 'Land of the Five Rivers' in the old-fashioned way by the Grand Trunk Road. Averaging 'six miles an hour including stoppages', he reached Cawnpur, via Benares and Lucknow, on 3 December and was in Simla on Christmas Eve. From here Cleghorn continued on to Lahore, the Punjab capital (today in Pakistan), stopping on the way to visit William Jameson at the Saharunpur botanic garden. From Lahore Cleghorn travelled to the most south-westerly point of his Punjab travels – to Multan (on the banks of the Chenab) and Muzzaffargarh (Fig. 36). This was in order to study the lowland jungles or 'rukhs', areas that had traditionally been preserved for trees, including babool (*Acacia arabica*), *Prosopis cineraria*, *Populus euphratica*, *Salvadora oleioides* and *Tamarix indica*, major sources of fuel for railway locomotives and the steam boats that plied up and down the Indus.[6] There were meetings with Colonel Hamilton, Commissioner of Multan, with the Senior Officer of the Indian Navy (its armed steam ships on the Indus had played an important role in the Sikh campaigns and the annexation of Sind), the Superintendent of the Indus Flotilla Company, and staff of

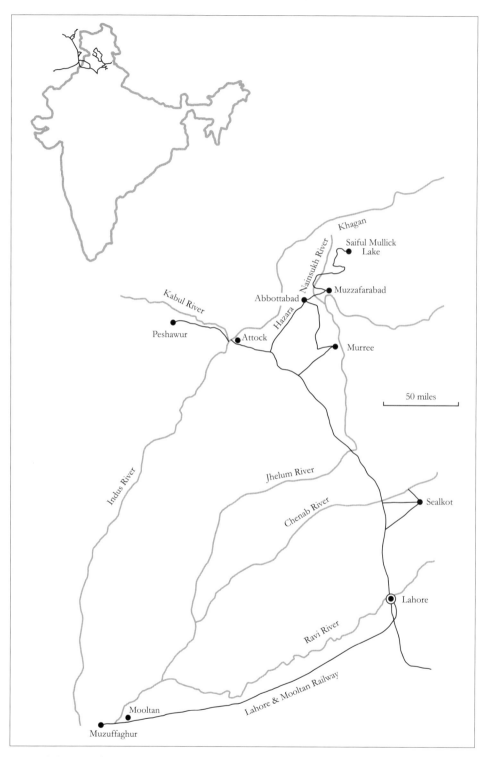

Fig. 36. Cleghorn's North Indian Travels, simplified from the map included in his *Punjab Forest Reports* (1864).
The western section: from Lahore (above).
The eastern section: from Simla (right).

8 FORESTS OF NORTH INDIA, 1861–4

the Punjab Railway. By this time Mabel had reached Lahore, but in March the couple went to Simla, the starting point for Cleghorn's exploration of the montane forests. In these the deodar (*Cedrus deodara* – known locally as *kelus* or *diar*, 'deodar', was also, confusingly, sometimes applied to *Cupressus torulosa*) was the major timber tree, being – as Brandis described it – to the West Himalaya what teak was to Burma and South India, and *sal* to Bengal and the 'Great Ganges Doab'.[7]

The aim was to investigate the timber resources of the valleys of four of the five great rivers of the Punjab – the Sutlej, the Beas, the Ravi and the Chenab, which all initially flow to the north and west before cutting through the outer ranges of the Himalaya to the plains and draining into the Indus. What followed over the next nine months counts as one of the most remarkable Himalayan journeys ever undertaken by a botanist, extending from the fringes of Garhwal in the east to the Kabul River in the west. Others, notably William Griffith and Thomas Thomson, had seen as wide a range of territory, but never in a single season.

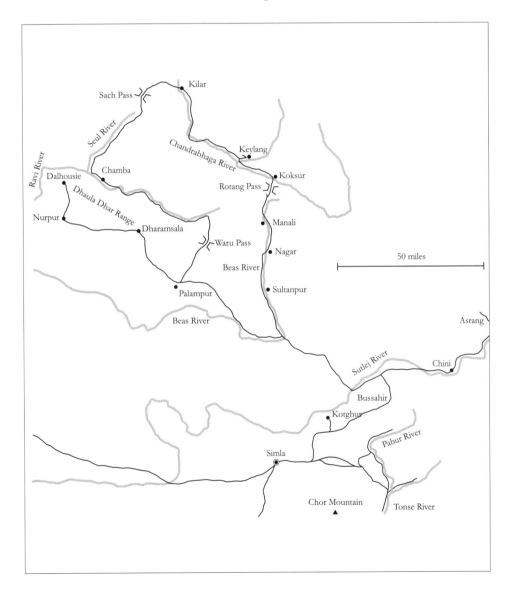

CHAPTER 8

Fig. 37. 'The Forest Mountain of Huttoo, near Nagkanda' (showing *Picea-Abies* forest on the Narkanda ridge east of Simla). Lithograph from Mrs W.L.L. Scott's *Views of the Himalayas*, 1852. (RBGE).

The explorer was aware of its significance, for which excursions to the Scottish Highlands two decades earlier, and a brief excursion to the Anamalais while in Madras, had been his only preparation. Even before embarking on the final leg Cleghorn described it as 'one of the most extensive and adventurous journeys ever made in the Himalaya'.[8] He started by travelling eastwards from Simla to the upper parts of the Sutlej Valley, but also across the watershed into the valleys of the rivers that drain southwards into the Jumna, and ultimately the Ganges (Fig. 37).[9]

This area was by now relatively well known, Simla having been settled since the 1820s and at this point having 980 houses. From the 1850s great effort had been put into building a road up the Sutlej Valley, a major trade route to Tibet and Central Asia, through the princely state of Bashahr ('Bussahir' to Cleghorn). The Simla Hill States were a mosaic of princely states and British territory, under a Superintendent in the person of Lord William Hay, son (and, indirectly, heir) of the 8th Marquess of Tweeddale who as Governor of Madras had given Wight such trouble. Cleghorn's initial travelling companion, on leaving Simla on 31 March, was Captain George Houchen, Superintendent of Hill Roads, their aim being to investigate the upper parts of the valleys of the Giri, Pabur and Tonse rivers not only for their deodar forests, but also for the (fuel-hungry) iron workings of the area. They got as far south as Banowli on the Tonse River, from where Cleghorn returned to Simla for the start of the second part of the explorations, on at least part of which he was accompanied by Mabel.

This second excursion was up the main Sutlej Valley,[10] an arid canyon on a vast scale and with a savage grandeur, starting at Kotgarh, a small settlement on the north-facing slopes of the Sutlej, the site of a Church Missionary Society station (Fig. 38). The journey is still a dramatic one and was considerably more so in 1862 – as may be seen in the 1860s

8 FORESTS OF NORTH INDIA, 1861–4

Fig. 38. Mission Houses in the Mountains.

'Mission Bungalow and School House at Khotghur' (the CMS station at Kotgarh). Lithograph from Mrs W.L.L. Scott's *Views of the Himalayas*, 1852. (RBGE) (above).

The Moravian Mission House, Keylong. Anonymous ink sketch, c. 1862. (Edinburgh University Library, Special Collections Department, E2015.23 9/2/2) (below).

photographs of Samuel Bourne, with wooden paths pinned to the side of sheer 5000-foot cliffs and a spectacular wooden cantilever bridge over the Sutlej at Wangtu.[11] Only a single stone pier of this ancient bridge survives, of glittering mica schist, close to a huge hydro-electric development that has long replaced timber as the major energy-generating resource

CHAPTER 8

of the valley. It was not a journey for the faint-hearted and it says a great deal for Mabel that she was willing to accompany her husband on such a dangerous track. Cleghorn's interest in economic plants was not limited to trees, and from Asrang (almost in Spiti) he reported on the native species of rhubarb with medicinally active roots.[12] These Himalayan species had been investigated by J.F. Royle, but were not as highly regarded as those obtained from western China. In the Himalayas Cleghorn took a much greater interest in the local names of economic plants than he had in South India, and recorded them in the plant lists that he published for the various valleys.[13] Mabel stayed at the village of Chini (now Kalpa) where, as Governor-General, Lord Dalhousie had stayed in 1850, and which Cleghorn thought 'one of the most attractive spots in the Indian Empire'. Although Mabel had seen the Swiss Alps she must have been astonished by the dramatic peaks of the Kinnaur Kailash range on the other side of the valley, which from Kalpa appear almost within touching distance. In addition to the main Sutlej Valley Cleghorn explored some of its tributaries, though due to bad weather he was unable to go up the well-forested Baspa valley. The most north-easterly point of all his Punjab excursions was reached at Kanam, only about twenty kilometres (as the raven flies) short of the Tibetan border.

Timber resources duly noted, it was time to turn westwards, returning down the Sutlej to Rampur, the capital of Bashahr, seat of a nine-gun-salute Raja, Shamsher Singh, then north over the Jalori La to the valley of the Beas. From Mandi the Cleghorns travelled up the Kangra Valley to Dharamsala, along the base of the spectacular outer wall of the Himalaya, the Dhaula Dhar range. By mid-June they had reached the hill station where Mabel was lodged with their friends Lieutenant Colonel and Mrs Edward Lake. Like Cleghorn, Lake (1823–1877), had been born in Madras, but in 1842 he had joined the Bengal Engineers. After a distinguished military career Lake had recently been made Commissioner of the Trans-Sutlej States, based at Jalandar with summers spent at Dharamsala; later he would become Financial Commissioner for the Punjab. Like Hugh and Mabel, Edward and Elizabeth Lake held strong evangelical beliefs; the Colonel was active in the Church Missionary Society, so Mabel was in safe hands as Cleghorn prepared for the most intrepid part of his Punjab travels –'the rugged ascents and descents of the Tibetan Alps ... [suited only to] goats and monkeys [and] men with hardened nerves'.[14] As in all aspects of his career, Cleghorn had done his homework thoroughly, in this case studying the collections, publications and maps of his predecessors, though for this area these were relatively few in number – the vet William Moorcroft and the Aberdonian Gerard brothers (Alexander, Gilbert and Patrick) in the earlier part of the century and, more recently, Alexander Cunningham, Godfrey Vigne, Werner Hoffmeister, Edward Madden, Michael Pakenham Edgeworth, Thomas Thomson and Victor Jacquemont, together with the plant collections of Royle and his *Illustrations of the Botany of the Himalayan Mountains*.

On 22 June Cleghorn started out from Holta, near Palampur in the fertile Kangra Valley, site of an early Government tea factory and plantation directed by Dr William Jameson from his base at Saharunpur. The route lay north over the Dhaula Dhar Range via the Waru Pass (3850 m),[15] which Cleghorn was probably only the second European to cross, where he collected *Rhododendron campanulatum*, then to Chamba down the beautiful valley of the Ravi, which though much larger in scale has more of the feeling of a Scottish glen than that of the deeper, drier Sutlej Valley. Chamba was a princely state ruled from 1844 to 1870 by Raja Sri Singh. Although the Raja, aged 27, was to some extent Westernised (he read the *Friend of India* and *Illustrated News*, and was interested in photography – he exhibited a photographic self-portrait at the 1864 Lahore Exhibition) there was patronising from Cleghorn and understandable suspicion on the part of the Raja. The former regarded the latter as an 'untutored Rajpoot ... [who] looks a sensual, thoughtless youth', though Cleghorn was pleased to learn that the Raja's young daughter had been spared from being 'destroyed in infancy' on account of a Viceregal proclamation. In turn, Sri Singh thought that Cleghorn must be on a secret mission from the Viceroy, by now Lord Elgin, to 'make a straight road to China'.[16] The town of Chamba was a sizeable settlement with a population of 6000 and 1000 single-storey, wooden houses, ancient temples and a polo ground, built on a series of terraces above the Ravi River. At the time of the visit the palace was being rebuilt and a new wooden bridge across the river constructed, the new developments being paid for with the money that the Raja was being paid by timber contractors.

From Chamba, in late July, when the rivers were swollen by monsoon rains, Cleghorn travelled northwards, passing up through the chir-pine zone and crossing the dramatic chasm of the Tisa Nullah with a bridge only twenty feet wide 162 feet above the river. The ascent of the Pir Panjal still passes through noble forests of spruce and fir up to an elevation of about 3000 metres where a dramatic tree line occurs: birches with snow-white trunks, then a black zone of dark oak dramatically edged with a yellow-green fringe of *Rhododendron campanulatum*. He was particularly interested by the rhododendron's growth form of which he drew a sketch, attributing its decumbent trunks to the weight of snow, but in reality probably due, at least in part, to soil creep. Above the trees are meadows, which in August were spangled with golden potentillas and geums, lapiz-blue borages, white *Anaphalis*, and clumps of the prickly *Meconopsis aculeata* with ice-blue flowers. Here the gaddis would still have been grazing their large flocks of sheep and shawl goats and the gujars their buffalo, though within a month they would be starting their annual journey back down to the plains. Cleghorn reached the Sach Pass at 4420 metres, and the bleak zone of ice and shattered rock, which, in May 1848, Thomas Thomson had been the first Westerner to cross. Then came the descent to the Chenab (or Chandrabagha) Valley into the district of Pangi, and back into forest – tall columnar spruce and fir, and then the magnificent, often multi-stemmed deodars, with patches of broad-leaved forest where the soil is richer and damper including wild cherries and the lovely hazel *Corylus jacquemontii* with its deeply lacerated, rococo cupules.

Cleghorn had already written up his report on the timber of the Sutlej and was here to observe the resources of the Chenab valley. On 1 August, at Killar on the north side of the Sach Pass Cleghorn met up with J.D. Smithe and J.A. Murray of the Chenab (or Pangi) Timber Agency to inspect the deodar forests of the Pangi Valley to the north,[17] where one of the most interesting natural products was a kind of bitumen called Drak-jun, Takjun or Salajit, which oozed from the rocks. This was:

> put in water ... [twice] boiled till it becomes homogeneous & then keeps many years of the consistency of opium – A good deal is taken to the Punjab – where it is used for healing wounds, & in a few grains internally – very bitter.[18]

The village of Killar was named for an interesting shrub of the family Hamamelidaceae then known as *Fothergilla involucrata* (now *Parrotiopsis jacquemontiana*) – used in Pangi to make the ropes of suspension or 'twig bridges'. From Killar the party headed north-east up the spectacular Chenab Valley, gouged between soaring mountain pinnacles. The path was

CHAPTER 8

terrifying, scratched along near-vertical slopes, close to the river, so narrow that ponies and mules could not use it: what was not carried by people in woven porter's baskets called *kiltas* had to be strapped to the backs of sheep. The valley gets rapidly drier, the deodar stops, to be replaced with bare slopes marked with slender black columns of pencil cedar (now *Juniperus macropoda* – Cleghorn eyed its flaky bark with a view to paper-making). Cultivation is possible with careful irrigation and small fields yielded crops of buckwheat and barley protected by 'Stone dykes with Rose briars to keep out Bears wch. are very numerous in [the] fruit season'. One of the most striking features, then as now, were the orchard-like willow plantations:

> striking objects in the landscape, often the only verdure near villages – the cuttings are planted with regularity – 3 in one hole – & are always pollarded at 8–10 ft. from the ground. The trees are in great demand for baskets [*kiltas*] in so treeless a country and the leaves are of great value to sheep during winter when there is no other food.[19]

The willow withies were also used, like the 'killar', for plaiting into ropes for 'twig bridges'. As ever Cleghorn was interested not only in the timber, but in any plant with potential for economic use – on the high pass there had been rhubarb and birch trees (*Betula bhojpattra* = *B. utilis*) the bark of which was used to write 'draft forest returns' (and even herbarium labels some of which are still attached to his Lahul collections at RBGE). In the Chenab Valley were *Datisca cannabina*, the root cortex of which produced a yellow dye, black cumin (*Bunium persicum*) and a local variety of *asafoetida*.[20] The last is a strong-smelling gum-resin used medicinally, and as a flavouring, produced by various members of the genus *Ferula*, the species occurring here being later described as *F. jaeschkiana*, though of only local commercial value. There was also human interest: the women's jewellery which included the *chirka*, an 'ornament worn in the hair … made of Talc [i.e. mica] fragments with Roses (red & yellow)'; in the villages were piles of mani stones 'prayer stones with letters engraved' and 'Scroll work in cut stone of a rather superior description', by which Cleghorn probably meant the beautifully inscribed fountain slabs characteristic of this area. He met 'a man from Chamba carrying a Kilta full of wooden vessels made from the knots of the Maple "Mandu"' being taken to Ladakh and in the village of Jaharma was a merchant called Bulleram who dealt in:

> Red Leather from Yarkund Bulgar …; Nirbisi from Lhasa (C[urcuma] zedoaria) a med[icina]l drug & aromatic; Yak tail – (black and white); Rhubarb – "Chuchi"; Munjit from Baltik. The Ladakhis bring goods (his principal dealings are in Amber, turquoise & Agate) to Tandi & return taking Opium, sugar &c from Koollu.[21]

By mid-August 1862 the party had reached Keylong, which lies on a terrace 300 feet above the Bagha River, just to the north of its junction with the Chandra at Tandi, at an altitude of 9400 feet.[22] Keylong is in Lahul (then a British territory), a dry region, ethnically Tibetan where the slopes are sparsely clad with pencil cedar and the air scented with *Artemisia*. It was then a village of 30 houses of interest to Cleghorn as the site of a Moravian Mission (Fig. 38). The Moravians, a central European Protestant church with a strong mission outreach, a belief in education, and in the spiritual value of industry, had long played a significant role in Indian botany from their outpost in the Danish colony of Tranquebar.[23] In 1856 August Wilhelm 'Papa' Heyde (1825–1907, son of a Silesian gardener) and Eduard Pagell (d. 1883) were sent to start a mission to Mongolia, choosing to approach it by way of Calcutta.

Having reached the CMS station at Kotgarh, they were not allowed into Tibet but the Indian authorities granted them land (190 acres for a farm) at Keylong and 50 trees of *kail* (*Pinus wallichiana*) for building a mission house, which Heyde and Pagell started in 1856 with the help of two Ladakhis. The following year the ascetic Heinrich August Jäshcke (1817–1883) arrived to take charge and mortified his brethren by ordering them to give up their horses, quoting Isaiah's 'How beautiful upon the mountains are the *feet* of him that bringeth glad tidings'. Jäshcke also arranged for three wives to be sent from Germany for himself and the others, Heyde's Maria having been born to Moravian parents in Surinam. Jäshcke was a brilliant linguist who kept his diary alternately in German, English, French, Latin, Greek, Danish, Polish and Russian. Cleghorn recorded that he 'knows 8 [languages] and speaks 15 dialects', the latter presumably of Tibetan, the language into which he had been sent to Lahul to translate the Bible. The missionary was also a competent botanist who made an extensive herbarium that seems not to have survived. Cleghorn owned a collection of 'Tibetan plants used in medicine or domestic economy collected by Rev. H.A. Jaeschke', which came to RBGE with the rest of Cleghorn's herbarium in 1896, but it would appear that these were kept in the Museum and destroyed in the 1960s. Curiously it was not to Cleghorn to whom Jäshcke gave his detailed notes on ethnobotany, but to J.E.T. Aitchison who submitted them to the Linnean Society for reading on 20 April 1865 and subsequent publication.[24] Life at Keylong was tough, especially during the long winters – the garden was covered in three to five feet of snow for the first three months of the year when the temperature could drop to minus 13°F, and the missionaries were reduced to making jam from the barely fleshy fruits of *Chenopodium foliosum*, though they later tried sea buckthorn berries. A native species of *Origanum* was, however, available to add flavour to their sausages.

At the point of Cleghorn's visit to Keylong in August 1862 (when Pagell was away on a visit to Spiti) a school, workshop and farm had been established. The school had 10 boys and 22 girls. Barley, buckwheat and lucerne were grown on the farm, which played a major role in the introduction of the potato to the region; in the garden were hops and apricots. A lithographic press (with three stones bought from the Rev. William Parker, the first Principal of the Lawrence Asylum at Sanawar) was supervised by the ecumenically-minded Tara Chunda ('a Hindoo in Kullu and strict Buddhist in Lahoul') whose son Hari joined Cleghorn's party. Cleghorn was shown four of the press's pioneering publications: a 'Primer with delineation' (presumably Tibetan), an Almanac, a volume of Christian Gottlob Barth's 'Bible Stories',[25] and the first part of the Tibetan translation of the New Testament ('Acts of Apostles with sketch map'). In the workshop coarse cloth was being woven and pasteboard made (presumably for use as book covers). Examples of this enterprise would later be sent to the 1864 Punjab Exhibition, including publications from the press, along with clothing, brick tea, gold, and samples of the local staple crops. Due to ill health Jäshcke had to return to Europe in 1868 long before completing the New Testament translation (which was not achieved until 1881) but the Heydes stayed on at Keylong until 1898. The locals proved resistant to conversion – the two loyal Ladhakis were baptised in 1865, but were the only converts for many years. However, the missionaries' medical and educational work was appreciated, as was that of their wives in teaching the locals to knit colourful socks, a cottage industry in the Kullu Valley to this day. The Keylong Mission limped on until 1940 and their substantial house until around 1980 when it was demolished and its stones used to build a large barn-like refuge for the Radha Soami Satsang Beas, a Hindu religious charity. The graveyard was ploughed up and now supports apple trees and healthy crops of glaucous cabbages. Some of the willow trees planted by the missionaries survive as venerable pollards, but the chief Moravian legacy is the potatoes in the fields on the terrace high above the river and the gaudy socks sold to tourists.

It was while staying at Keylong, that, by a process of deaths, retirements and seniority, Cleghorn achieved the next rung on the ladder of the medical establishment of the Madras Presidency, which from this Himalayan fastness must have seemed like another world. He was now a Surgeon-Major. From the Chandrabagha valley Cleghorn travelled with Smithe over the eastern end of the Pir Panjal range by way of the Rohtang Pass (3979 m). On the way down a semi-tame snake, who lived under a stone 'daily fed by Flour', played hard to get – he was 'torpid and gorged and refused to appear though repeatedly stirred up'. Growing here was *koot* (then known as *Aucklandia veracosta*, now *Saussurea lappa*) a member of the family Compositae with a valuable medicinal root, today grown as a lucrative cash crop in Lahul, but then collected from the wild. After the bareness of Lahul this was 'a return to a land of trees', and to cereal crops of rice and *ragi* (*Eleusine coracana*).[26] Passing through the British territory of the Kullu Valley, Cleghorn stopped to visit the hot springs at Vashist, where he measured the water temperature as 138°F. He also visited a tea garden at Kullu (then known as Sultanpore) that had been started eleven years earlier by Major Hay, Assistant Commissioner of Kullu (whose official residence was a former palace in Naggar, now run as the Naggar Castle Hotel). Sultanpore was a village of 300 houses with a dispensary where a Muslim *hakim* treated lepers and opium addicts. On 6 September bad news arrived from Fife: his father had suffered a stroke and was unable to sit up in bed, though it was a relief to know that his sister Isa was in attendance.[27] It was time to return, via Kangra, to Mabel who for three months had been patiently waiting in Dharamsala. He reached her in early September to discover that she had been undertaking horticultural activities, germinating some South African seeds sent by a Captain Lowther.

At Dharamsala Cleghorn took the opportunity to inspect the arboretum of the greatly revered, and horticulturally-minded, Financial Commissioner of the Punjab, Donald Friell McLeod whose name is still remembered in McLeod Ganj, since 1960 the seat of the Tibet Government in Exile. The arboretum contained 'many introduced Himalayan trees of great interest ... as well as many European fruit trees ... perhaps the only collection of indigenous Alpine trees in the Punjab'.[28] The only plane trees in Dharamsala today grow beside a building called Mortimer Hall; these may be descendants of some of McLeod's trees and be close to the location of his large house, which was almost certainly demolished in the 1905 Kangra earthquake. But McLeod was also importing exotic plants from the Exeter/Chelsea nursery of Veitch. Cleghorn also visited the jail, where prisoners were being taught useful skills including paper-making using local raw materials, an activity that must also have interested Mabel as a daughter of one of Britain's leading paper manufacturers. The husks of maize cobs were being tried as a novelty, but also the barks traditionally used for paper-making locally – birch, daphne (*jeku*) and a desmodium (*katti*). These were similar to Alexander Hunter's experiments in Madras and the link may not be coincidental, as in the Cleghorn collection are watercolours of *Daphne oleioides* (now *D. mucronata*) and *Desmodium elegans* made for him by Margaret Eleanor Prinsep (see *Cleghorn Collection* Fig. 46). She was Hunter's cousin, married to a member of the redoubtable Prinsep family, Edward Augustus Prinsep, a revenue official with whom Cleghorn worked closely. Leaving Mabel behind in Dharamsala it was now time to continue the survey of timber resources of the North-West Punjab, areas that are now in Pakistan, and on 18 September Cleghorn set off for Lahore via Dalhousie and Amritsar accompanied by the young Anglican missionary the Rev. John Barton, a recent graduate of Christ's College Cambridge – a keen botanist, geologist and mountaineer. At Nurpur ('City of Light'), a city of 18,000 inhabitants, Cleghorn visited a leather factory, where goat skins were tanned with the bark of *Cassia fistula* and dyed red with lac. The next stop was Dalhousie, a new hill station planned in 1854 on land bought from the Raja of Chamba and named for the Marquess, as yet with only 26 houses inhabited by 70 Europeans and 70 children. From here the deodar forest at Kalatope at the western end of the Dhaula Dhar ridge was surveyed in the company of Edward Prinsep and Thomas D. Forsyth, Commissioner of Punjab. This forest, of magnificent deodar and *Abies pindrow*, is still maintained as a nature reserve.

From the hill station Cleghorn returned to the plains, to Amritsar, looking at tree-planting along the canals of the Doab – native species were mainly used but Australian eucalypts and acacias were also being tried. He then continued to Lahore where he visited the garden of the thriving Agri-Horticultural Society that, like the Madras society, had an important role in the distribution of seeds of useful trees and plants. In Lahore plans were made for a six-week trip to the north, to the valleys of the Jhelum and its westerly tributary the Nyusuk, via the new hill station of Murree (then with 112 houses), to survey the deodar forests of Hazara and Khagan.[29] At Murree, which he reached by mail cart, Cleghorn stayed with his old schoolfriend the engineer Colonel Robert Maclagan, secretary to the Punjab Government in the Public Works Department, a keen botanist (whose herbarium is now at RBGE) whose brother John had been Cleghorn's assistant in Madras. Also at the hill station was the missionary the Rev. Robert Clark with whom Cleghorn had travelled out to India in 1851, now on the way to Kashmir with his wife as medical missionaries. A professional development also occurred in Murree, as it was here that Sir Robert Montgomery, Lieutenant Governor of the Punjab, asked Cleghorn to stay on for another year 'on special duty' to advise on the newly organised Forest Department of the vast province.[30]

In Abbbotabad Cleghorn met another missionary, the Rev. Isidor Lowenthal, who promised to provide him with a list of Pashto and Kaghan plant names.[31] Lowenthal was a converted Polish Jew who had been sent to Afghanistan by the American Presbyterian Board in 1856, where he translated the New Testament into Pashto, though he met his end before he could translate the Old Testament. Accompanying Cleghorn was Dr Leonard Horner Lees, whose father George had been assistant to Cleghorn's old maths lecturer Thomas Duncan in St Andrews. Lees had recently served in the Opium War in China but his regiment was now back in Peshawar to which he was summoned back to deal with a cholera outbreak. After this Cleghorn pressed on via Mansehra and Gari-Habibullah up the Kaghan Valley through the district of Hazara. An unpleasant incident occurred when a dog stole a loaf of bread and his trigger-happy companion Lieutenant Blair, an engineer, to the delight of the locals, shot the poor creature. On 2 November the party reached the beautiful alpine lake of Saiful Maluk, at an altitude of 3200 metres, where Cleghorn was pleased to find an abundance of wild rhubarb. The visit, however, had to be cut short as snow was starting to fall on the hills – some of the party were suffering from sore throats and 'my faithful Madrassee being threatened with Frost bitten fingers, we beat a timely retreat'.[32] This is the only reference found anywhere in Cleghorn's archive about his loyal staff, so, as with his 'Marathi artist', it is particularly unfortunate that he neglected to mention his name, and one can only wonder as to the nature of the man's duties, or how long he had worked for Cleghorn.

CHAPTER 8

From these lofty regions Cleghorn returned to the Doab, which he crossed until reaching the most north-westerly point of his Punjab travels, Peshawar (the 'doorway to Kabul') on 16 November. The route was via Haripur, and Attock with its substantial fort built by Akbar. The Indus crossing here was by means of a bridge of boats, its carriageway strewn with stems of the asclepiad *Periploca aphylla*, a switch plant. On approaching the bridge the wheel of the truck in which Cleghorn was travelling hit a wooden railing and the driver was precipitated into the icy waters of the Indus – the man managed to cling to the bridge and was pulled out by Cleghorn with the help of a chowkidar. 40 miles beyond the Indus was Peshawar 'in a long fertile valley – almost entirely surrounded by hills covered with snow [for] 4 months'; a city of great strategic importance with a cantonment four miles in length, then containing 10,000 troops, situated between the city and the Khyber Pass. Here Cleghorn admired the inhabitants – 'athletic – manly – chiefly Mahomedan', and visited the museum, where he saw Bactrian coins, a series of wood samples arranged by J.L. Stewart and sculptures excavated from stupas at Eusofzie. He heard the Rev. Robert Clark deliver a sermon in Pashto and a native doctor (*hakim*) called Mohamed Kassim provided him with a list of medicines from Afghanistan and Bokhara. From Peshawar Cleghorn explored the valley of the Kabul River,[33] the Kophes of antiquity, that rises in the Hindu Kush to the west of Kabul and formed the route through which Alexander had invaded India in 327 BC. In early December Cleghorn returned to Lahore where he established himself, and where Mabel was able to join him;[34] here he would be largely based for the next two years with the summers spent in Simla, away from the city's searing heat. Some plant specimens from this period survive in the Edinburgh herbarium.

Medical Mission

It was not, however, all forestry and much of the content of Cleghorn's letters to J.H. Balfour from this period is devoted to discussions of missionary activities (Fig. 39).[35] The end of the year 1862 was marked by a great Missionary Conference in Lahore, which took place from 26 December to 2 January, and was attended by many of the senior administrators of the Punjab, all of them friends of Cleghorn who shared his evangelical convictions – Sir Herbert Edwardes, Edward Lake, Donald McLeod, Thomas Forsyth, Edward Prinsep and Robert Maclagan.[36] Such men saw no incompatibility between evangelical faith and active soldiering and John Ruskin regarded Edwardes, of whom he wrote a memoir, as the embodiment of the 'Christian knight'. In a discussion at the Conference on lay co-operation Cleghorn stated his belief that medical missionary work was one of the best ways of persuading the 'mild Hindoos and irascible Pathans' to overcome their aversion to Christianity. Believing that 'the human heart is the same everywhere', he said:

> the object of Medical Missions is to win the affection and confidence of the people in imitation of the example of the *Great Physician*, "who went about healing all manner of diseases".[37]

He was also able to report on his recent visit to the Moravians at Keylong,[38] who were unable to be there themselves as the Rohtang Pass was already closed by snow. On the afternoon of Hogmanay he chaired the ninth session, in which papers on 'Hill Tribes', 'the Sikhs' and 'Vernacular Christian Literature' were read. Cleghorn got himself elected to two committees – one (under McLeod) for advising missionary bodies on secular matters, the other (with Lake and Maclagan) to consider the use of medical missionaries; the latter becoming the 'Peshawur and Lahore Committee of the Punjab Auxiliary of the Medical Missionary Society'.

8 FORESTS OF NORTH INDIA, 1861–4

Fig. 39. Mission in the Punjab.

Sir Donald McLeod (Lt Governor of the Punjab, centre, front) and senior civil servants, Murree 1865. Showing several people known to Cleghorn: Capt. Charles Mercer (curator of the Punjab Exhibition, standing in white); Edward Prinsep (civil servant, standing far right); Col. Edward Lake (military administrator, seated far right); the Rt. Rev. George Cotton (Bishop of Calcutta, (seated second from left). Anonymous photograph in Brandreth Collection. (© The British Library Board, APAC, P 211/1(61)) (above).

The Rev. John Barton, a missionary with whom Cleghorn travelled in 1862, expounding the Gospel to Punjabis male and female. Stained glass window, c. 1910, Holy Trinity Church, Cambridge (below).

CHAPTER 8

For Cleghorn this topic was a major passion. The Edinburgh Medical Missionary Society (EMMS) had been formed in 1841[39] – its two vice-presidents (Thomas Chalmers and William P. Alison) were both known to Cleghorn and James Syme was a board member; its first secretary was Dr John Coldstream a close Edinburgh friend of J.H. Balfour, whose fervent evangelical beliefs made him feel duty-bound to put missionary work above his natural inclinations that tended to academic zoology.[40] The Society sponsored medical training for young, committed Christian men who would evangelise in a lay capacity – the ideal, in Cleghorn's view, being 'an educated English [sic!] gentleman – as well as a Christian surgeon'.[41] The EMMS had already despatched Dr John Evans to Bengal in 1852 and in 1856 (in conjunction with the Free Church of Scotland) had sent David Paterson to Madras, where, as already noted, Cleghorn had been closely involved with his work and the Dispensary opened in Black Town in 1857. The great interest in the Punjab was to provide a medical missionary to work in Kashmir, where medical skills in lithotomy and treating goitres (with iodine), ophthalmic diseases and hydrocele (an accumulation of fluid in the scrotum) were especially needed.[42] The suggestion for a medical missionary for Kashmir had been voiced at the Christmas conference, but there was already interest from the Rev. Robert Clark (1825–1900) the CMS missionary who since 1854 had been on a mission to the Afghans based in Peshawar. Further stimulus came from Dr Coldstream's son William (1841–1929), who arrived in the Punjab as a Civil Servant in 1862.[43] Missionary work in Kashmir was problematic,[44] facing strong opposition from the Muslim population, though the Hindu Maharajah at least partly supported the medical work. Robert Clark was allowed to spend the summer of 1864 in the beautiful Vale, where he taught and his wife, Elizabeth (*née* Browne), daughter of a Calcutta medic who spoke several Indian languages and had trained as a nurse in London, dispensed medicines and advice, hoping to pave the way for further medical missionary work. It was, however, a dangerous time for missionaries, and in this same year (1864) two of the Princeton-trained American Presbyterians were murdered – the Rev. Isidore Lowenthal who had provided Cleghorn with Pashto plant names, at Peshawar, and the Rev. Dr Levi Janvier near Hoshiarpur. Both were killed by Sikhs, the former possibly by mistake (shot for an intruder by his chowkidar), the latter battered to death for anti-British rather than specifically religious motives. There was also suspicion of missionaries within EIC circles and when asking Balfour's help in finding a surgeon with 'missionary spirit' for the Lawrence Asylum at Sanawar, Cleghorn told him not to mention what might be called the 'M-word': 'the name of Missionary is a bugbear to many in high places (Sir John [Lawrence] excepted)'.[45] In these concerns Cleghorn was wholeheartedly supported by Mabel, with her own strong beliefs and family tradition. Her grandfather Alexander Cowan was a keen supporter of the EMMS, and Dr John Coldstream was her uncle by marriage, his wife Margaret (*née* Menzies) being her mother's sister.

Lengthy negotiations took place, which included fund-raising in the Punjab (£400 had been raised by 1864), and approaches to the CMS, through its secretary the Rev. Henry Venn, to get them to act as joint sponsors. A suitable young man was sought in Scotland, which resulted in 1863 in the selection of Dr William Jackson Elmslie (1832–1872) who graduated, 'with éclat',[46] from Edinburgh the following summer. Elmslie arrived in Calcutta in October, in time for the Cleghorns to meet him the following month on their return from Simla. With Thomas Farquhar they worked out a programme for the young Aberdonian: to find a 'moonshee' from whom to learn Urdu, to study over the winter in the Lahore Medical

College, but on the way there to visit the mission stations at Benares, Allahabad, Agra, Delhi, Ambala, Ludhiana, Jalandhar and Amritsar. Cleghorn and Farquhar put Elmslie on the train on 18 October,[47] and for the next six years, despite strenuous opposition, he was based in Kashmir. The Maharajah did not permit him to stay there over the winter, for which he had to return to the Punjab. Elmslie went home on furlough in 1869, returned to India in 1872, the year in which the CMS published his Kashmiri-English dictionary in London, but he died in Kashmir the same year, aged 40.

Inherits Stravithie & First Work with Brandis in Simla

In the early part of 1863 Cleghorn's forestry work was mainly in the plains, the area known as the Doab, crossed by a series of major irrigation canals. Here there were timber depots to be inspected, where timber brought down from the mountains by river was stored (at Ropar, Madhopur, Sialkot, Jhelum and Hashtnagar), and a continuation of his investigation of the 'brushwood tracts' or 'Rukhs' overseen by *rukhwallahs*. Cleghorn must also have spent considerable time collating the materials acquired on the previous year's extensive travels, and condensing them into reports that a year later would be collected to form his second book.

A major 'life event', if a not unexpected one, took place on 9 June 1863, when his father Peter died at Stravithie at the age of 80. It was a time of emotional stock-taking, as two of the aunts who had played such an important role in his motherless upbringing also died this same year: Jessie Cleghorn in Fife and Jane Campbell in Edinburgh. The huge separations, both temporal and geographical, gave added poignancy to such events for those working in India, and in Cleghorn's case brought with them additional responsibilities of property ownership, and deeply conflicted loyalties between personal duties in Fife and the 'public duty that at present forbids the idea of leaving India'.[48] Of course 'public duty' was not entirely untainted with the financial rewards that would be required to support Stravithie, but Cleghorn stuck to it and would not return on home leave for another two years. Cleghorn clearly had absolute trust in Charles Stuart Grace, whose long-established firm of solicitors in St Andrews had for decades run his grandfather's and father's affairs and acted as factor for the estate. Grace must have acted as executor to Peter Cleghorn's will, wound up his affairs, and overseen the smooth running of the estate until Cleghorn could find time to return.

The summer migration away from the heat of the plains, by now an established Raj custom, took the Cleghorns first to Dharamsala in June 1863 and in August to Simla, which the following year would be formally declared as the official summer capital of India. In December 1862 Cleghorn had written to Joseph Hooker 'Dr Brandis has been summoned from Burma to Calcutta & there is a prospect of something like a Forest Code for India being compiled ere long'.[49] It now fell to Cleghorn to take a part that he had not previously imagined – of working with Dietrich Brandis on what amounted to the organisation of forestry on a national basis for the whole of British India.

Setting Up an Indian Forest Department

Lord Canning, the Viceroy who had sent Cleghorn to the Himalayas, had left India in March 1862, broken by the death of his wife. He was replaced by James Bruce, 8th Earl of Elgin (1811–1863) who came with a lengthy record of overseas service, notably in Canada and, more recently and much more controversially, in China where he found himself personally at odds

CHAPTER 8

with implementing the aggressive foreign policies of Lord Palmerston. What turned out to be his final posting followed his witnessing, and at least partial responsibility for, one of the most terrible acts of cultural vandalism ever condoned by the British state – the looting and destruction of the Imperial Summer Palace near Beijing. Having earlier deflated the claims made by E.P. Stebbing on Lord Dalhousie's part as the originator of a national forest policy for India, it is now time to give the credit where it is really due. In fact Stebbing did at one point in his undigested book admit that it was an exchange between Elgin as Viceroy, and the Secretary of State for India, Sir Charles Wood, in the latter part of 1862, that 'inaugurated the birth of the Forest Department in India'.[50]

Elgin had already made up his mind that what was needed was a new, national department, that should be organised by Brandis, brought from Burma 'on special duty'. The new body was to be placed under the Public Works Department, which 'practically was [the one] most deeply interested in successful forest management', and that what was required was 'the concentration of authority and responsibility'. Sir Charles Trevelyan, Finance Member of his Council, recorded his disapproval; he considered that control should remain with the Revenue Department, but Elgin got his way. The reasoning behind the Viceroy's belief that control of forests must be a matter for the State was, not unreasonably, that 'no individual can have a personal interest in doing more than realising the highest possible present amount from any forest tract of which he may get possession' and that 'the idea of giving a proprietary right in forest to any individual should be abandoned'. But it was necessary to do more than this – Government forests must be demarcated and 'made inalienable'; existing rights to the forests on the part of the people should be recognised, though extinguished (with compensation) wherever possible. The Government forests must be distinguished from the 'waste' lands that were allowed to be sold to private owners. Elgin also realised that in order to effect this change in policy 'the object might best be attained by an Act of the Legislature', which would, in due course, take place in 1865. Another important issue concerned what is now termed 'sustainability', and Elgin stated that the role of the government forests was to deliver 'the largest possible quantity of produce ... consistent with their permanent usefulness'. This then is a genuine Forest Charter, not Beadon's 1855 letter to Captain Phayre in the name of Joseph Dorin, under Dalhousie's government.

None of this was enough on its own: Elgin's timing was astute, as his despatch gained the unqualified support of Sir Charles Wood, Secretary of State for India in Palmerston's Liberal administration of 1859–65. In his reply Wood agreed that 'it requires the stability of a settled administration to prevent the present destruction of forests' in order to 'leave a due supply for future generations'. Wood also noted that it was 'very satisfactory to me to learn that you have come to the same conclusion as Her Majesty's Government, that individuals cannot be relied upon for due care in the management of forests'. Furthermore, Wood accepted that short-term financial gain was not the only consideration: 'I firmly believe that a considerable profit will be derived from the forests ... still, profit is not the only object to be kept in view' and there would be expenditure ('outlay'), that would be paid back only in future years.

This is the background to Cleghorn's work in Simla, where in the summer of 1863 he caught up with Colonel Richard Strachey, the engineer who, a decade earlier, had (at least nominally) been a joint author of the British Association report. Strachey had in the meanwhile been making a name for himself with irrigation projects and was now Secretary to the Government of India in the Public Works Department, which gave him a place on the Viceroy's Council, and made him Cleghorn's boss. Strachey 'chalked out' work for Cleghorn and Brandis for the next year,[51] which was formalised as their joint position as 'Commissioner of Forests' for India for the period of a year from 1 January 1864,[52] at which point John Lindsay Stewart took over from Cleghorn as Officiating Forest Conservator to the Punjab Government.

In August 1863 Brandis and Cleghorn wrote two reports, one relating to the Punjab, the other to the Madras Presidency, showing the ambitious scope for the new national Forest Department, though, in fact, Bombay and Madras were to be left largely to their own devices. The Punjab report recommended the taking over of a limited number of the *rukhs* by the Forest Department permanently, in order that they could be managed to provide both fuel for the railways and revenue, as the lower hill forests seemed unlikely to be able to fulfil either of these functions.[53] The paper on Madras is a much more important one,[54] but one whose significance has been almost entirely overlooked. It was elicited as Beddome (who was holding the fort for Cleghorn in Madras) had sent them a set of draft forest rules for comment. What resulted was the first explicit statement of a three-fold classification of forests into Government Reserved Forests, Village Forests, and a third, intermediate, type that was not the property of Government or strictly 'reserved' but managed according to rules – what later became known as Protected forests. It is ironic that these designations should have first been used in connection with Madras, as the Madras Government refused to endorse them, and for many years resisted the attempt to recognise such distinctions. Despite this the rules, and Cleghorn's and Brandis's comments upon them, were circulated to other regional governments, where they doubtless had considerable influence at a time when those bodies were drawing up similar guidelines.

It is rather remarkable that the two colleagues were able to concentrate on such work as in this same month, August 1863, Cleghorn's health was giving trouble in the form of 'hepatic congestion', which he self-medicated with an alarming-sounding mixture of 'Nitric Acid and Taraxacum',[55] and on the 26th Brandis's wife Rachel died. She was a daughter of the great Serampore Baptist missionary, Joshua Marshman, and it was at least in part through this connection that Brandis had been talent-spotted and brought to India in 1856 – one of her sisters was married to Henry Havelock who had been asked by Dalhousie to find a suitable candidate. But there also was a botanical aspect to this network as Rachel's first husband was Joachim Voigt, the Fredriksnagore physician and botanist, and so, rather surprisingly, it was actually in Copenhagen (where Brandis had studied, and where his paternal grandfather lived) that Brandis had met her.

Cleghorn must have regained his strength as in October he was back in Lahul.[56] Once again Mabel accompanied him but she was left with the Heydes at Keylong from 8 to 17 October while he travelled in the neighbourhood. Mabel and Maria Heyde became friends and they continued to correspond for several years – Mabel would send porters' baskets (*kiltas*) of goodies from Simla and in return Maria sent Mabel wool – using the wool of the long-haired goats of the *gaddis'* flocks, one can imagine her knitting cashmere waistcoats to keep her husband warm on chilly Simla evenings. Maria Heyde's diary survives and in an 1866 entry is to be found a rather surprising reference to a sewing machine – the details are unclear, but it seems that Mabel was being used as an intermediary to send one from Simla to Keylong.

The two Fife land owners, if of very different degree, both spent the summer of 1863 in Simla. Elgin lived in the house called Peterhof (now, following a fire, hideously rebuilt in a style that has aptly been described by a local historian as a hybrid between a 'flying saucer

CHAPTER 8

and a bus station'),[57] but it is not known where the Cleghorns stayed at this point. Cleghorn asked J.H. Balfour that any spare seedlings raised from seeds that he sent to RBGE should be sent to Elgin's Broomhall estate near Dunfermline,[58] which might well be the origin of some deodars that can still be seen there. After the summer both the Viceroy and his Commissioner of Forests went separately on tour, Elgin's aim being to reach as far as Peshawar. The Cleghorns, as already noted, had gone to Lahul when Jäschke received a note that Elgin wanted to meet him, so with Cleghorn he did a double march to Koksur for a meeting on 13 October. Elgin's health was not strong, and having crossed the Rohtang Pass, he suffered a heart attack after struggling across the ninety-foot Koksur Bridge over the Chandra River. This was what the British knew as a 'twig bridge', a type of suspension bridge that, given the shortage of wood in the region, was made of three ropes of plaited branches which in Lahul were made of birch. These fragile structures, which often lasted only a single season, swayed alarmingly as the pedestrian walked only a few feet (in this case 12) above the raging waters. Though at this very time a more substantial bridge on this major route to Lahul, Ladakh and Spiti was in the course of construction, by the engineer Richard Elwes, this was not in time to spare the Viceroys' frayed nerves. Elgin returned back over the Rohtang and had to be carried to Dharamsala where Cleghorn would also end up in November. Though Cleghorn was not Elgin's personal physician (a post held by W.B. Beatson), three weeks before the end came he had to break the news that the earl's days for:

> Durbars and public displays had gone by – & that little time was left for meditation & retirement ... It is a solemn duty to speak of eternity to any patient but to a great statesman (Viceroy) it seemed more difficult. He thanked me & said "<u>you have given me a great deal to think about tonight</u>".[59]

Elgin died on 20 November and was buried in the graveyard of the church of St John of the Wilderness, now (though not then) surrounded by gloomy deodars. Lady Elgin sent out a window of Belgian stained glass, and an ugly monument was erected in the churchyard – doubtless designed by a military engineer with only a rudimentary grasp of gothic, loosely based on an Eleanor Cross but with an unhealthy multiplication of apical crockets (Fig. 40). His monument in Calcutta Cathedral (by George Gilbert Scott) is aesthetically much more satisfactory.

Fig. 40. The death of the 8th Earl of Elgin. Anonymous wood engravings, *Illustrated London News* 23 January 1864.
The bungalow at Dharamsala where Cleghorn probably attended him shortly before his death on 20 November 1863 (above).
The church of St John in the Wilderness (looking south over Kangra). The Viceroy was buried in the railed enclosure to east (left) of the church, on which a Gothic monument was later raised (below).

From Dharamsala Hugh and Mabel returned to Lahore, where the year 1864 started with a prayer week, with meetings held in the American Presbyterian Mission Church. The imminent arrival from Britain of Sir John Lawrence as the new Viceroy (Sir William Denison, Cleghorn's old boss from Madras, had been acting following Elgin's death) gave cause for hope over the plans for the Medical Missionary scheme for the Punjab and Kashmir.[60] Lawrence, despite his military fame as leader of the relief of Delhi, was highly regarded for his earlier administrative work in the Punjab and was appointed Viceroy for this first-hand knowledge, in order to strengthen Indian infrastructure. In fact Cleghorn's hopes were not to be fulfilled as Lawrence, while a committed Christian, tactfully did not go out of his way to support missionaries publicly.[61] Lawrence was one of three remarkable Irish brothers; Sir Henry had also been a notable administrator in the Punjab but had been killed in the Mutiny. Henry Lawrence was one of very few Britons to leave money for charitable purposes in the country that had made him rich and, as noted earlier, Cleghorn had been involved in advising the third, military, brother George on the setting up of the Ooty 'branch' of the Lawrence Asylums (which were memorials to Henry).

The Punjab Exhibition, Lahore, 1864

For Cleghorn the most spectacular event of 1864 was the Punjab Exhibition of Arts and Industry, which, after several years of planning, was opened by the Lieutenant Governor, Sir Robert Montgomery, on 20 January.[62] The motives behind the exhibition, which followed in the wake of the great London exhibitions of 1851 and 1862 and, closer at hand, the Madras ones of 1855 and 1857, were various. Its overt purpose was to improve prosperity through stimulating arts, manufacture and agriculture, and it certainly represented a large investment on the part of the Government – the exhibition building alone cost Rs 60,000 (£6000 – around £300,000 in today's values).[63] Behind this it is also possible to see it as playing a role in the Anglicisation of territory that had come into British possession after the Second Sikh War only 15 years earlier, and also in healing the more recent and still raw scars of what Cleghorn and his fellows saw as the great betrayal of the 'Sepoy revolt' and the savagery that followed. Though personally spared from witnessing this violence in Madras, and thanks to the efforts of John Lawrence it had barely spread to the Punjab, many of the individuals with whom Cleghorn was in contact had personally been involved in the fighting. As medics both John Lindsay Stewart and Thomas Anderson were at the siege of Delhi and received medals (Anderson with clasp) and Colonel Edward Lake had helped to 'put down' the mutiny at Jalandar.

The organisation for the exhibition was similar to that employed in Madras – a Central Committee with four Vice-Presidents (one of whom was Cleghorn's friend Robert Maclagan) under a President (Donald McLeod). Cleghorn was one of 33 committee members, eight of whom were Indian; there were also Local Committees in towns such as Jehlum, Peshawar, Amballa and Rawlpindi, each of which was co-ordinated by a Secretary; Cleghorn's friend William Coldstream was Secretary to the General Committee. The exhibits received, which included many from private individuals both European (soldiers, medics, missionaries) and Indian (Hindu, Jain, Sikh, Armenian and Muslim, of all degrees – from the Maharajas of Patiala and Kashmir, numerous Rajas and Ranas, to farmers, merchants and craftsmen) as well as the Local Committees and from several Jails. The exhibits were all listed in an *Official Handbook*, arranged in four major Sections (Raw Materials, Manufactures, Machinery, Fine Arts), divided hierarchically into 34 Classes (three of which were split further into Divisions). The

CHAPTER 8

list of exhibits reads like that 'certain Chinese encyclopaedia' of Jorge Luis Borges, starting with a 'Ferrugineous Rock' and ending with 'A metallic bird which sings when wound up'. It must have been the Curator (Edinburgh-born Captain Charles McWhirter Mercer) and the Assistant Curator (D.C.M. Gordon) who had the challenging task of organising this miscellany into coherent displays, their curatorial job titles pre-echoing the use of the building after the exhibition closed, when it was taken over by the Lahore Museum to form the 'Wonder House' (Ajaib-Gher) described in Kipling's *Kim*.

The substantial building was designed by Edwin E. Baines, District Engineer of the Punjab Railway, in a style claimed by the architect to be 'Belgian Gothic' (Fig. 41).[64] This seems a little wide of the mark and the exterior, as shown in contemporary photographs and woodcuts, has something of the appearance of an oriental railway station – the pair of spired and louvred ventilators on the cast-iron crested roof ridge with something of a Burmese look.

Fig. 41. The Punjab Exhibition, 1864.
Anonymous wood engravings, *Illustrated London News*, 14 May 1864.
Opening of the exhibition and exterior (above).
Interior view of exhibits (below).

The long, narrow structure, a central hall with a handsome wooden king-post roof (that must have consumed a good deal of deodar), and a gothic arcade either side opening onto lateral aisles, was, however, functional, allowing large numbers of visitors (nearly a thousand a day) to file past the exhibits, which, in addition to glazed wall-cases, were largely displayed on a long, two-tiered, central table draped in green cloth. Lighting came from triple lancets in the gable walls, and a clerestory of dormered *oeils de boeufs* that pierced both sides of the main hall, between which were hung trophies of arms. The whole building was erected in six months, not by the use of modern materials (as in the case of the Crystal Palace) – it was made of Lahore bricks and timber, the roof tiled – but by sheer force of numbers, using an average of a thousand workmen per day.

As an exhibitor, Cleghorn's own contribution formed but a minuscule proportion of the 11,100 exhibits – thse were all in Section A, Class 4, Division 1G, 'Woods', in which his friend Brandis exhibited samples from the Central Provinces. Cleghorn provided wood samples of 33 different species from the districts of Simla, Kumaon and Jaunsar Bawar and with these, rather oddly, were nine plant products from the Simla bazar including 'Dried violets, Spikenard, Aconite root, Bark of Myrica [*edulis*], Root of Kurrooa, [*Picrorhiza kurroa* and] Sliced root of kafur kuchri' [*Hedychium spicatum*], which would surely have been more appropriately placed in Class 2 (Chemical and Pharmaceutical Substances). After the event Cleghorn wrote a Catalogue of Woods, and the jury report on Woods and Fibres.

8 FORESTS OF NORTH INDIA, 1861–4

Photography in the Punjab

By the time of Cleghorn's fourth Indian period, the 1860s, photography had continued to grow and develop in the Subcontinent, with professional studios springing up in the Presidency cities and larger towns, of which none was more famous than Bourne (and Shepherd) of Calcutta. Staying with him for the Punjab Exhibition were three notable Scottish surgeon-naturalists, William Jameson, Superintendent of the Saharanpur botanic garden, who exhibited seven sorts of tea from the Government factory at Holta; Thomas Caverhill Jerdon, India's leading zoologist; and Cleghorn's young colleague in the Forest Department, John Lindsay Stewart. Aware of the significance of such a gathering (when naturalists tended to work in isolation, and widely scattered) Cleghorn had the event commemorated by commissioning a photographer to record the group (Fig. 42). This photograph has been discussed elsewhere,[65] but it is worth noting that Cleghorn was pleased enough with it to send copies back to a discriminating audience in Scotland: for Balfour, Walter Elliot, Mabel's mother, and Dr Oswald Home Bell in St Andrews – Bell was the stepson of Isobel Adamson, whose brother was Cleghorn's old friend Dr John Adamson with whom Bell had been in partnership as a physician before becoming Chandos Professor of Medicine.

The photographer of the naturalists' group is uncertain but one possibility is Samuel Bourne who was in Lahore for the Punjab Exhibition, at which he showed some of his photographs. Hugh Rayner,[66] however, has pointed out that there were other professional

Fig. 42. Group of botanists at the Punjab Exhibition of Arts and Industry, 1864.
Left to right: John Lindsay Stewart, Thomas Caverhill Jerdon, William Jameson (Superintendent of Tea Plantations, standing), and Hugh Cleghorn. Albumen silver photograph (191 x 235 mm), attributed to Samuel Bourne. (National Gallery of Australia, Canberra, 2007.81.58).

CHAPTER 8

photographers working in Lahore at the time including John Burke, William Baker and James Craddock, and it could even have been taken by a good amateur photographer such as Charles Mercer. Bourne had arrived in India only the previous year and the photographs he exhibited must have been his first ones; they were of Simla and a journey he had taken up the Sutlej between July and September 1863, showing scenes that would have been familiar to Hugh and Mabel.[67]

Although Cleghorn never seems to have taken photographs himself, as in Madras in the 1850s, he continued to collect the work of others; these were related to his professional interests, but unfortunately none of them has survived. For example, in the notebook with details of his Lahul excursion is a list of five photographs of Lahul taken by the bridge-builder Richard Gervase Elwes,[68] and in Simla in June 1864, despite their substantial cost '8 to 15 shillings each' he was making a large collection of photographs including 'some Photographs of trees & one Panoramic View of the Govt. Tea Plantations – wch. is most instructive'.[69] The photograph of the Kangra tea plantation was sent to Balfour who showed it at a Botanical Society meeting on 8 December 1864 to accompany a paper that had been sent unillustrated by William Jameson.[70] The tree photographs must have included those by Colonel Charles Waterloo Hutchinson RE, showing 'characteristic vegetation of the Deodar and other Himalayan trees', which Cleghorn later exhibited at the 1865 Birmingham British Association meeting.[71] It may have been these photographs of Himalayan forests, or similar ones by Bourne, that the author was shown at Stravithie in 1998 and that have since disappeared without trace. Cleghorn would continue to show photographs at BSE meetings in retirement: of a fine specimen of a peepal (*Ficus religiosa*),[72] and at one of the last he attended on 9 April 1891 he showed photographs of a date palm 'sent to him from Port Said by a relative'.[73]

The tragedy of John Lindsay Stewart (1831–1872)

In the 1864 photograph of the Naturalists' group appears, in profile, the sensitive face of John Lindsay Stewart, shortly to take over as Conservator of Forests for the Punjab after Cleghorn's transfer to the Government of India. Another equally rare group photograph has recently been shown by Hugh Rayner also to include Stewart, taken later the same year it shows a rather different figure (Fig. 43). In the entourage of the Raja of Chamba (Cleghorn's 'sensual youth'), the forester can be seen to have 'gone native', his head turbaned and face covered with wild whiskers. Stewart had always intrigued me – there was something patronising in Cleghorn's attitude to his younger colleague of whom he had written: 'I trust he may find a partner in life to soften the tones of his voice & manner a little'.[74] And just what might lie behind the 'naturally of a highly nervous temperament' in the obituary that Cleghorn had to write of him? In Scotland I had found the family tombstone in Fettercairn kirkyard and one of the aims of my own Punjab expedition in August 2014 was to attempt to find his grave in Dalhousie. In the charming Neo-Norman church of St John, the pastor showed me the remarkably intact church registers (though forbade me to photograph them), in which Stewart's cause of death on 5 July 1873 was given as 'softening of the brain',[75] and I was then pointed in the direction of the Christian cemetery.

Like Alexander Gibson 31 years earlier, Stewart was born on a Mearns farm, and Dalladies lies less than two miles north-east of Auchenreoch where Gibson ended his days, though on the Kincardineshire side of the River North Esk. The farm was clearly an unhealthy

8 FORESTS OF NORTH INDIA, 1861–4

Fig. 43. The Raja of Chamba and his Retainers, from the album 'Indian Photographs'. Sri Singh (seated second from right) and John Lindsay Stewart (seated far right). Albumen print (221 x 283 mm) by Samuel Bourne, c. 1864, photographer's number 553a. (The Alkazi Collection of Photography, 2002.06.0001(39)).

spot, a hotbed of tuberculosis: the Fettercairn gravestone revealed that John was the youngest child of nine, but his father died at 46 and eight of his siblings (including two Janes and two James Guthries) had predeceased him. The baptismal register records plain John, no 'Lindsay', another mystery being why he studied medicine not in Aberdeen or Edinburgh (like Gibson) but in distant Glasgow? Perhaps the two were connected, the name possibly adopted from a land-owning local sponsor with links to Glasgow, or else it could have been a territorial designation adopted informally to prevent confusion with other John Stewarts (the nearby village of Edzell is at the heart of the traditional 'Land of the Lindsays'). At Glasgow he studied botany under Walker-Arnott, and chemistry under Thomas Anderson, who a decade earlier had been Cleghorn's fellow Secretary of the Royal Medical Society, graduating MD in 1853.[76] In 1856 he went to Bengal as an assistant-surgeon, but his career was spent entirely in the Punjab and NW India where he was active botanically over a very wide geographical range on which he published extensively. But, given the times, Stewart also saw active military service – he was present at the siege of Delhi and took part on a military expedition to the Yusufzai territory on the North-West Frontier. In 1860–1 Stewart acted for William Jameson at Saharunpur and the Kangra tea plantations, he was then civil surgeon at Bijnour before succeeding Cleghorn as Punjab Conservator in 1864. As Conservator he continued to travel widely and to accumulate botanical specimens and knowledge – showing a particular interest in local uses and names of this richly polyglot region (at RBGE his own annotated copy of Jameson's Saharunpur garden catalogue is almost obsessively annotated with such names). This work culminated

CHAPTER 8

in his Lahore-published *Punjab Plants* of 1869, full of interesting details both personally gathered and taken from literature. This same year he returned home on a two-year furlough with a commission to write a 'Forest Flora', for which additional specimens for the Central Provinces and Oudh were sent to allow a wider geographical coverage. In the early part of his leave, in July 1869, Stewart visited the Strathspey forests with Cleghorn. The writing-up of the book, undertaken at Kew, clearly got bogged down, later 'explained' as due to incipient illness. The project, however, ended in tragedy – the work was not completed on Stewart's furlough, and the hyper-critical Joseph Hooker was dissatisfied and considered it unpublishable – doubtless his customary beef with colonial field botanists who had the temerity to stray into the territory of the Metropolitan taxonomist[77] – disapproval of the use of local names and field characters, which potentially led to narrow species concepts rather than magisterial herbarium syntheses. On the expiry of his leave, the book in limbo, Stewart returned to Lahore, where it was believed that:

> the fact of Dr Hooker having rejected his work on the Flora of India has so affected Dr J.L. Stewart … that on Friday last a medical board at Láhor not only pronounced that his life was in danger, but that his reason had left him.[78]

This was written in June 1873 and, as recorded in the church register at the hill station of Dalhousie, the poor man died on 5 July at the age of only 41½ and was buried two days later.

I eventually found his grave in the cemetery, a depressing place overgrown with gloomy deodars, every tomb shorn of its cross, the lush undergrowth riddled with sinister pinkish-brown slugs like slimy chipolatas. I scanned the terraces systematically from the top down, finding graves from the same period near the gate, and had almost given up when I came across it, on the very lowest terrace, far from his contemporaries. Cleghorn recorded that he was buried under a particular oak tree, doubtless deemed appropriate for a forester, but one can't help but wonder if his departed wits might have caused him to take his own life, and that this is what lies behind his distance, in death, though later 'a few of his old Friends in the Punjab' imported a tomb for him of extremely high quality, fine-grained sandstone, from the firm of Hund Ram & Son of Delhi. The headstone is beautifully inscribed with biographical details and a sarcophagus bears the note 'Distinguished by his contributions towards the study of Himalayan Botany' – even if it was this that had ultimately cost him such heartache and, perhaps, his life.

Face was saved by Brandis completing Stewart's *Forest Flora of North-West and Central India*, but even with phenomenal Teutonic energy it took him two years to do so. In the book's preface is a gracious acknowledgement of Stewart's work – the use of his notes and specimens, but what is curious is why Brandis felt he had to write all the descriptions anew, and he stated that his species circumscriptions in certain of the genera were very different to Stewart's. In the RBGE herbarium Stewart's legacy lives on in the form of a large number of specimens that he gave to Edinburgh University during his leave, but unfortunately this was not his top set, and there is no copy of the field notes that would allow interpretation of the specimen numbers. He intended to leave his own herbarium to Glasgow, but what happened to it is unknown, perhaps lost in the distress that must have followed his death.

8 FORESTS OF NORTH INDIA, 1861–4

Brandis & Cleghorn

Dietrich Brandis was appointed Inspector-General of Forests for India on 1 April 1864. As they worked closely together and remained mutually supportive friends for the rest of Cleghorn's life (Brandis not dying until 1907), it is necessary to say something about the background and remarkable talents of Brandis that go a long way to explaining why it was he, rather than Cleghorn, four years his senior and with much wider Indian experience, who obtained the senior post.[79]

Fig. 44. Brandis & Cleghorn.
Dietrich Brandis. Anonymous photograph, from the *Indian Forester* vol. 10(8), August 1884, in an article on 'the founder of forestry in India' on the occasion of his retirement. (RBGE: Cleghorn's copy) (left).
Hugh Cleghorn, 1865. Photograph by Dr John Adamson. (Courtesy of the University St Andrews Library, Alb-8-80) (right).

It has already been noted both that it was in Burma, in January 1857, that the pair met for the first time, and that, altogether more curiously, Mabel's father Charles Cowan had been friendly with Brandis's parents for 30 years. The German and wider European intellectual background is the key to understanding Brandis and goes back a generation – his paternal grandfather Joachim Dietrich Brandis was court physician to Queen Marie of Hesse-Kassel – wife (and cousin) of the Danish Fredrik VI in Copenhagen. Furthermore, the grandfather for whom he was named was married to Lucie a sister of H.F. Link, director of the Berlin Botanic Garden. The background was not, however, only north-European, because from the age of 13 the young Dietrich had spent three years in Greece, where his father was one of the advisors to the recently installed young king, Otho (son of Ludwig I of Bavaria), and where he was introduced to botany by Karl Nikolas Fraas whose ideas on historical changes in flora and climate caused him to be dubbed by Karl Marx 'a Darwinist before Darwin'. Brandis then studied for a year at the University of Copenhagen (with the physicist Hans Christian Ørsted, the zoologist Johan Reinhardt and the pioneering plant geographer Joakim Fredrik Schouw) before returning to Germany to the University of Bonn (the Prussian University of the Rhineland). It has been claimed that Brandis was a 'scientifically trained forester',[80] and even that he was brought up in the Black Forest, but both are completely untrue – he did not attend one of the technical forest schools in which German foresters were trained, rather, he received the most advanced academic education possible in no fewer than three European universities. The list of lecturers that Brandis gave in the 'Vita' in his PhD dissertation is a hugely impressive one, covering the arts, and both natural and physical sciences – six of the 13 are mathematicians, chemists, astronomers or physicists.[81] Of biologists are the botanist Ludolph Christian Treviranus, the zoologist G.A. Goldfuss and the physiologist L.J. Budge. Of those in the arts faculty there is a poet and an historian, both politically active (these were the days leading up to 1848), and a philologist. Nor was the spiritual side of his life neglected, as he was also taught by the theologian K.I. Nitzsch. From Bonn Brandis went to the Hanoverian University of Göttingen, where his main interest was, once again, in chemistry (including studies with his maternal uncle the mineralogist J.F.L. Haussmann), though he also studied botany with August Grisebach and the bizarrely named Scato Lantzius-Beninga. Returning to Bonn, Brandis's PhD thesis of 1848 was on the chemistry of alkaloids, only minimally botanical in that some of those discussed are plant derivatives: brucine from *Strychnos nux-vomica* (brucinum); cinchonium and chininum from species of *Cinchona*. However, of the titles of theses that he was prepared to discuss with opponents when he came to defend the thesis, several relate to biology, one of which, on plant geography, must reflect his time in Copenhagen under Schouw.

Nonetheless, although he had not studied them himself, Brandis would have been aware of the techniques of 'scientific forestry' that had developed in Germany as one of the 'cameral sciences' over the previous century. With his broad scientific (including mathematical) training, he would easily have been able to pick these up for himself, and to apply them as he did in Burma – using techniques such as his 'linear valuation survey', and the working out of yields and felling rotations that were the basis of the German system. Brandis believed that 'although climate and vegetation in India are different [from Europe], yet the fundamental principles of forest management are the same everywhere'.[82] This ability must have been abundantly clear to the Calcutta administration when compared with Cleghorn's more empirical approach. Although Cleghorn's love of acquiring, reading and absorbing relevant literature, first seen in his British Association report, and the books that he acquired after his Madras forest job in 1856, any background that he had in forestry was that of the Scottish tradition of arboriculture as represented by Brown's *The Forester*, which he recommended to his own assistants in Madras (along with works by Monteath, Loudon, Pontey and Mathew[83]). There is, however, little to show that he tried to apply this to an Indian setting, and his methods appear to have been largely based on observation and intuition as to what was both practical and appropriate.[84]

It was not only this question of background, ability and methodology; Cleghorn in the early 1860s had other priorities and interests. He seems always to have maintained a loyalty to Madras, but he also knew that sooner or later he would have to return to his lairdly

8 FORESTS OF NORTH INDIA, 1861–4

duties in Fife. Added to these came worries concerning Mabel's health, which would also demand a return to temperate climes and medical supervision. He may well have made it clear to his superiors in the PWD that he had no interest in the Inspector-General's post, and in March 1864, when he was thinking of leaving India permanently, he wrote in a letter to Balfour: 'If the wheels of the Forest Coach run smoothly, and my industrious German confrère returns – there is no hindrance to my leaving younger men to take up my Cantling [i.e., Division]'.[85] Another consideration is that Cleghorn found the responsibilities of the job stressful, and in January 1864 even before he had started the joint appointment with Brandis he complained of 'feeling a good deal the tension on the brain, which Departmental responsibility involves'.[86]

Brandis's connections with the Indian missionary world through his first wife Rachel Marshman must have also formed a part of the bond between the two men. In George Smith's biography of Rachel's father's friend, the Rev. William Carey, is a tantalising glimpse of a continuation of these connections: 'Sir D. Brandis and Dr Cleghorn at various times visited this arboretum [Carey's garden at Serampore] and have referred to the trees whose dates of planting is known, for the purposes of recording the rate of growth'.[87] No published record of these results, nor the dates of the visits, has been traced, though they must have been in the early 1860s (showing the long persistence of the Serampore garden). The establishment of quantified growth rates for Indian trees was a pressing need if scientific forestry techniques were to be applied, and Carey's garden with trees of known age represented a rare opportunity for the acquisition of such data. Much later Sir William Muir would reminisce about socialising with the Cleghorns at Barrackpore, the site of the Governor-General's country house and a green lung for Calcutta, that lies just opposite Serampore, and this must also date from the same period.

This is not the place to go into Brandis's post-1868 career in Indian forestry, which initiated a major new era and one with an influence that spread far beyond the Subcontinent, though it is worth noting that Brandis's own views on the rights of local users were notably more liberal than those of the upcoming generation of Indian foresters. It is, however, worth remarking on the taxonomic botanical works that Brandis wrote in later life; these have tended to be overlooked by environmental historians, but he had been prepared for them by his early training under Schouw, Treviranus and Grisebach. His *Indian Trees* was researched at Kew and published in 1906 by which time he was 77 years old. But before this, also prepared at Kew, on a two-year sabbatical, had come the *Forest Flora of North-West and Central India* (1874), the completion of Stewart's work already described. In the preface of this work Brandis paid a handsome tribute to Cleghorn – that 'without his devoted assistance', the 'advice and counsel of a true and faithful' friend, he could not have 'ventured to publish this book in a language which is not my own'. This is probably unduly self-deprecatory coming from a man whose two wives were both English, and whose dealing with officialdom had been entirely Anglophone for two decades, but it suggests one way in which Cleghorn contributed to the partnership. Brandis's last public expression of support of his friend came with a generous article on 'Dr Cleghorn's services to Indian forestry' first published in the *Transactions of the Scottish Arboricultural Society* in 1888, and in the same year also in the *Indian Forester*.

CHAPTER 8

Simla, Summer of 1864

During the summer of 1864 Cleghorn and Brandis attended to forestry matters, working towards the first Indian Forest Act, but Cleghorn also devoted much of his unofficial time to religious interests. In June George Cotton, Bishop of Calcutta, had, in the unecumenical spirit of the times, declined to give permission for Presbyterian worship in the Anglican church of Christ Church, so Dissenters had to meet in the Masonic Lodge (which happened to be next to the PWD office).[88] Mabel led the singing and the company was distinguished, including the Viceroy (Lawrence) and Sir Charles and Lady Trevelyan – Trevelyan having returned to India in 1862 for a three-year stint as Finance Member of the Viceroy's Council.[89] That there was nothing personal about Cotton's refusal is shown by the fact that in September he addressed an audience of about 50 in the Cleghorns' drawing room on the subject of a tour he had recently made in South India;[90] and in October 1866 Cleghorn would be shocked to learn of the bishop's untimely death in a drowning accident (an occupational hazard of Indian bishops – Heber had died, or was possibly murdered, while bathing at Trichinopoly). Brandis, who must still have been distraught from the death of his wife, left India in July on a six-month sick leave to England and Germany, so Cleghorn carried on the activities of the Forest Department solo, though was not, on this occasion, formally designated as officiating Inspector-General. At this point 'the office' was within the Public Works Department, then based in Herbert House, on the slopes below the army headquarters. For at least some of this summer the Cleghorns, showing their closeness to the seat of power, were staying with one of Sir John Lawrence's Private Secretaries, Dr Charles Hathaway.

From dated and localised herbarium specimens, and from suggestive but inconclusive evidence, a trip by Cleghorn to Lahul in August 1864 is suggested. The specimens could, of course, have been collected by others on his behalf, but another piece of evidence relates to a revolutionary book that was published in May of this year, in both New York and London: George Perkins Marsh's *Man and Nature*. Four years later Cleghorn told its author that he had read it on the 'slopes of the Northern Himalayas, and into Kashmir and Tibet',[91] two areas that Cleghorn never visited, but with a bit of geographical licence might cover Lahul with excursions down the Chenab or into Zanskar. The book was reviewed in the *Edinburgh Review* in October 1864, so if this was when Cleghorn read it, his London bookseller must have sent it to him very promptly. In the same introductory letter to Marsh Cleghorn claimed that:

> in the course of my duty I have endeavoured to direct the attention of the Local Authorities to the general changes and prospective consequences of Railway works now in progress in Upper India. The result of my observations [i.e. on deforestation, especially in montane regions] has been strongly corroborative of the views which you have so usefully promulgated.

It is worth saying more about Marsh's book – an awakening call for what became the environmental movement, and a work of the most astonishingly wide-ranging scholarship. *Man and Nature*, which is still far too little known in Britain, despite a superb biography of its author,[92] was an epoch-making work that pointed out the devastating environmental effects that could be caused by man, especially through deforestation, and which had been perpetrated throughout the course of history of which Marsh was such a well-informed scholar. But as pointed out by Lowenthal the book is a 'diatribe not a jeremiad'. Marsh believed that man's work had to be taken into account as part of Nature, but that his possession of a soul and reasoning powers placed him in a distinct category that gave him the possibility of altering his own destructive course. That the 'improvement' of Nature, which was a God-given right and duty, was to be done with intelligence and sensitivity, and that there was hope even for restoration of what would today be called damaged ecosystems.

In October the Indian Government returned to Calcutta, and by now the revolutionary possibility had come about whereby it was possible to do a large part of the 1200-mile journey by train. The Cleghorns left Simla on 20 October and reached Calcutta in 16 days, making several stops *en route* – on the first part of the journey, to Delhi, they stopped at Saharunpur (to visit the Botanic Garden), Roorkee and Meerut.[93] However, the only railway in the Punjab at this point was between Amritsar, Lahore and Multan, so the part of their journey to Delhi must have been either by *dak* or horse-power.

Author Again:
Report upon the Forests of the Punjab & the Western Himalaya

The reason for stopping at Roorkee was that it was the location of the Thomason College of Civil Engineering, the training school for the PWD. This had been founded in 1847 and its first principal (until 1860) had been Cleghorn's friend the engineer Robert Maclagan – its main role being the training of the men then undertaking the great irrigation projects in the Ganges plain and the Punjab. The College also had a press, supervised by James Johnston, where the printing of Cleghorn's *Report upon the Forests of the Punjab & the Western Himalaya* was reaching its final stages, its preface being dated 18 October 1864.[94] Had he lived to see it this would have been the despair of his father[95] – far from being a 'book more generally interesting than your last' (and for which Hugh had plenty of experience on which to base one) it was of exactly the same sort as *Forests & Gardens* – a compilation of reports with no linking narrative and not even an introduction to provide a context for the individual items. While primarily of practical use to foresters, the dedicated reader would, however, have been able to form from it an impression of the challenges involved in establishing a system of sustainable timber supply from the Punjab forests required for railways and fuel. It also contains more of botanical interest than the earlier work. Other than nine text figures the work is unillustrated, though bound into it are sketch maps of four river valleys, one by Lieutenant John Chalmers, a cousin of Mabel's who was acting as superintendent of the Ravi and Chenab timber agency.[96] In a pocket at the end are inserted two maps of the Western Himalaya, one a Survey of India topographical one, the other a large and handsome one lithographed by the Thomason College Press, based on the map by John Walker in Alexander Cunningham's *Ladak*, on which Cleghorn has marked his own travels (Fig. 36) and the distribution of the deodar forests from Kaghan in the north-west to Garhwal in the south-east.

The subjects of the reports can be placed into five categories: topographical (on individual river valleys or districts); specific topics (fuel supplies, production of charcoal); Forest Rules for particular areas; reports and memos by other individuals (administrators, timber agents, and one local ruler – Shamsher Singh, Raja of Bashahr) and, of interest to the botanist, notes on the vegetation of particular areas with plant-lists recording local names and uses. It is noteworthy that there is not a single reference to the effects of deforestation on climate in the entire book.

CHAPTER 8

The book was probably printed in a tiny edition and copies today are even rarer than *Forests & Gardens*. Cleghorn's own heavily annotated copy at RBGE is annotated with a list of those to whom he sent the book – in India two went to the 'Office', presumably the PWD in Simla, one to the *Friend of India*, two to Calcutta libraries and ten to friends and officials from the Viceroy to Thwaites in Ceylon. Six went to individuals in Britain or Europe and seven to British libraries. Assuming that the personal Indian copies probably returned with their owners this still makes only 23 that might have ended up in Britain. Given this rarity it is hardly surprising that the book appears not to have been reviewed in Britain. Only two reviews have been found, both in Indian journals, and both are more by way of summaries of contents and assessments of Cleghorn's forest work than critical literary evaluations. The first, in the *Friend of India*, though anonymous is known from a cutting stuck into Cleghorn's copy, to be by Brandis.[97] It praises Cleghorn's first steps in documenting the available forest resources of the Punjab and for obtaining Government control by means of leases from some of the 'forests under native chiefs' (i.e., Chamba and Bashahr), but is notably critical of the author's (and Cleghorn's) own paymaster, over its failure to make the *rukhs* that were so important for fuel supplies for the Punjab over to the Forest Department: 'as in other questions relating to forest management, we must wait for better times and a more far-sighted policy on the part of the Government of India'. Also notable are what were probably attempts to make the review of interest to a general reader – that the deodar was botanically probably conspecific with the cedar of Lebanon and therefore the same as that from which Solomon's Temple was constructed, and something that Cleghorn himself had noted (from Strabo) that Alexander's fleet on the Hydapses (= Jhelum) was made largely of deodar from Kashmir.[98] The second notice, of unknown authorship, is from the *Calcutta Review*, a joint review of three reports by J.L. Stewart, of Cunningham's *Ladak*, Hoffmeister's *Travels* and an article by Edward Madden on Himalayan conifers, in addition to Cleghorn's book. It is by way of a general analysis of the whole question of Punjab forests and their management, both political and practical, though the statement that the various reports:

> present a mass of matter which is frequently intensely statistical or technical, and which contains nothing very lively or sensational, but which has been put together with intelligence and is calculated to give full information on the various subjects treated[99]

describes Cleghorn's second book to a tee.

Fifty years later E.P. Stebbing was much more impressed with this work, of which he reprinted substantial chunks either verbatim or in summary. He described it as 'one of the most valuable pieces of investigation on the forests extant for this period, and probably could have been written by no other man serving in India at the time'.[100] It was also lauded by Stebbing as an example of that:

> devotion and self-sacrifice [with which] the British officials were throwing their whole hearts into ameliorating the condition of the people and bringing so far as possible order and the *pax Britannica* into regions which had known no orderly regime through the centuries. Those who ask what the British have done in India and for India can find their answer in Cleghorn's remarkable Report of what was being accomplished on the Punjab Frontier sixty years ago.[101]

In a post-colonial light it is easy to mock such bombast (and its encouragement to the author to dispense with punctuation), but, just perhaps, it contains a grain of truth. It is certainly a telling example of the belief commonly held by the 1920s that the Forest Department was regarded, along with the more frequently cited railways, as one of the supreme achievements of the British Raj.

8 FORESTS OF NORTH INDIA, 1861–4

Delhi to Calcutta – by Rail

With the recent filling of gaps in the line between Delhi and Agra, and Allahabad to Benares, the railway from the east bank of the Jumna at Delhi to Calcutta had finally, only six weeks earlier, been completed. There were stops on the way to visit schools and mission stations, the major break being at Allahabad where the Cleghorns met the CMS missionary the Rev. Brocklesby Davis and his wife Ellen (*née* Sherwood).[102] Ellen was also a missionary and one with a strong Edinburgh link, as she was the adopted daughter of the Episcopalian minister the Rev. D.T.K. Drummond, one of J.H. Balfour's closest friends. Also in Allahabad they visited William and Elizabeth Muir. A Bengal civil servant, Muir was at this point a member of the Provincial Board of Revenue, who was later (1868) appointed Lieutenant Governor of the North-West Provinces. He is best known for his interests in education (which in 1872 led to his founding of the Muir Central College, affiliated with Calcutta University) and as a pioneering Islamic scholar, which he incongruously combined with the promotion of evangelical Christianity – as demonstrated in his controversial, but scholarly, *Life of Mahomet* (1858–61) based on Arabic sources. Many years later Cleghorn would encounter Muir as Principal of Edinburgh University, a post he held from 1885 to 1902: at the launch of the Cleghorn Memorial Library, and over the establishment of a forest lectureship for the university.

On reaching Calcutta the Cleghorns were met by Dr Thomas Farquhar, a fellow evangelical and medical colleague, who for a time was personal physician to Sir John Lawrence. With Farquhar was Dr William Elmslie who the previous month had arrived in India as the new medical missionary for Kashmir. Hugh continued forestry work until Brandis's return from Europe in January, with which ended his first northern period – initially in the Punjab and thereafter the first of his spells with the Government of India. This was the occasion of a remarkable tribute from the Viceroy, Sir John Lawrence, in a formal Government of India Resolution dated 10 January 1865, the more remarkable because Lawrence was known as 'a chief who gave little praise to his juniors':[103]

> His Excellency the Governor General in Council ... avails himself of this opportunity to express his sense of the great service rendered to the State by Dr Cleghorn in the cause of forest conservancy. **He may be said to be the founder of Forest Conservancy in India**. His long services, from the first organisation of forest management in Madras, have, without question greatly conduced to the public good in this branch of the administration. In the Punjaub also, though there has not yet been time for important practical results to become apparent, his Excellency feels confident that Dr Cleghorn's labours have prepared the way for the establishment of an efficient system of conservancy and working the forests of that province.[104]

This statement (to which the emphasis has been added) calls for comment, not least because it is so extremely unfair to Alexander Gibson, whose pioneering efforts in Bombay had started informally as early as 1840; and in 1847 it was Gibson who was the first official in India to be formally 'gazetted' as Conservator of Forests.[105] What might lie behind such partisanship – was it simply that individuals were quickly forgotten once they left India – and Gibson had left five years earlier in 1860? Or was it deemed that Gibson's contribution was less important than Cleghorn's? Or did it merely reflect Cleghorn's (frequently noted) good nature and ability to get on with people of all ranks, and perhaps also the values and beliefs, not least in evangelical Christianity, that he shared with Lawrence? The return of Brandis to his post of Inspector-General freed Cleghorn to go 'home' to Madras, but before this came the passing of the milestone of the first Indian Forest Act.

CHAPTER 8

The Indian Forest Act VII of 1865

It has sometimes been said that the first Indian Forest Act was drafted by Cleghorn and Brandis,[106] but this is not strictly true. Baden Henry Baden-Powell (half-brother of the founder of the Scouting movement) explained that it was drafted (and in a hurry) by a lawyer, based on a report provided by Brandis.[107] However, in Simla in the summer of 1864 Cleghorn must certainly have been closely involved in the discussions that enabled Brandis to produce his briefing document and the Act received the Governor-General's assent on 24 February 1865.

Milestone it may have been, the first legislation to try to ensure some degree of forest protection on a national basis, but it was of a preliminary nature and extremely brief – running to only 4½ pages and 19 clauses. Its main purpose was to 'give effect to Rules for management and preservation of Government forests', defining these in the broadest of terms as 'land with trees declared subject to Government rules' that had to be notified in an official gazette. Most importantly, however, there was a safeguard that any such rules must 'not abridge or affect any existing rights of individuals or communities'. No rules were laid down in the Act itself, as these were to be drafted by local governments, but an outline of suggested topics was given under three headings. The first related to plant life – the protection of all, or named, species of trees, prevention of fires, control of the collection of what came to be called minor forest products and control of villagers' activities such as cultivation, grazing and burning of lime or charcoal. The second heading covered rules for streams passing through these Government forests and their use for floating timber. The third heading related to the custody of Government timber: controls on felling, and the regulation of ownership marks. Clause 4 regulated the behaviour of the officers put in place to administer the forests. Clauses 5–17 all related to law enforcement and penalties for breaches. The Act was to cover Bengal, the North-West Provinces and the Punjab, but neither Bombay nor Madras, though these 'minor Presidencies' could use it if they notified its adoption in a gazette. The Act was to operate from 1 May 1865 and to be known as 'The Government Forests, Act, [no VII of] 1865'.

Although this Act was in place for the rest of Cleghorn's Indian career, it is worth summarising some of its limitations because he did, after retirement, continue to take a major interest in the subject. The Act was almost certainly seen as only a preliminary step and in 1868 Brandis prepared an amended Bill and Memorandum, which was widely circulated, and discussed at the second Indian forestry conference held at Allahabad in 1873/4.[108] At this conference the civil servant Baden Henry Baden-Powell presented what could be called the 'case against' the 1865 Act, which he regarded as deeply flawed and drafted by someone with no real understanding of the issues.[109] Historians with political agendas, such as Mahadhav Gadgil and Ramachandra Guha,[110] have presented Baden-Powell as a malign influence, typical of the hard-nosed administrators who wanted rigid state control of the forests and to deprive Indians of access to hereditary resources. These were resources which Gadgil and Guha idealistically consider to have been wisely managed in the past by virtue of a combination of a polytheistic belief system and, especially, by the caste system which they justify as a form of ecological niche specialisation (what has been termed the 'Merrie India' school of Indian historiography). They accused Baden-Powell of making a specious distinction between 'rights' and 'privileges' of locals, with the aim of severely limiting the number of provable 'rights' by disallowing most of them as the *non-legally-binding* privileges that had accrued without any official approval. The problems were those experienced both by civil servants (including legal issues), and by practical foresters at the sharp end of trying to apply the Act. There can be no doubt that especially the latter group had a genuine commitment to preserving forests, both for what became known as their 'direct' benefits (producing timber and other products) and for their 'indirect' benefits (the protection of soils and climate). Baden-Powell's paper is somewhat verbose and impenetrable but in 1878 Theodore Hope, a member of the Viceroy's Legislative Council, noted the main deficiencies of the 1865 Act,[111] and these can be summarised as follows:

1. No distinction was made between forests that it was essential to 'reserve' (i.e., strict government control) and those under more general control.
2. There was no procedure for enquiring and 'settling' (legally accepting) villagers' rights or regulating them (for example by buying them out or transferring them).
3. It contained nothing about the control of private forests for the general interest of the state or the people.
4. It gave no authority for the levying of duties on timber.
5. It contained nothing about the protection of Government forests.

Regarding the first of these points, it should be noted that it did not go nearly as far as the memo on the Madras Forest Rules written by Cleghorn and Brandis in 1863, with its recognition of three categories of protection (Reserved, what came to be called 'Protected', and Village forests). After discussion with foresters and administrators these deficiencies, with many other refinements, would be embodied in the much more extensive and far-reaching Indian Forest Act of 1878.

Mabel's Health

Mabel was only 24 when she went, newly married, to the mountains of northern India. She accompanied Hugh on his expeditions to the Sutlej in March 1862 and to Lahul in 1863, but as early as October 1862 came the first mention that she had been 'ailing a little'.[112] For most of the next three years she was based in relatively secure surroundings either with Cleghorn or with loving friends, at Lahore, Dharamsala, Simla or Calcutta, but her constitution must have been frail. Though her strong faith always allowed her to remain 'bright and cheerful' and 'in spirit ... joyous', by June 1864 she was 'very spare, Fish Oil, cream, salep misri seem powerless in enabling her to gain flesh',[113] and by September she was 'very emaciated attenuated' and required a 'soft clime for the chest'.[114] The nature of this illness is unknown, her husband diagnosed it as 'simple debility with relaxation of nervous and muscular system'.[115] This seems unlikely to have been related to the gynaecological troubles of five years later, but it meant that it was vital for her to return to Britain, and that she would never be able to return to India.

Fig. 45. Cleghorn's Herbarium: a typical sheet of the material mounted in the Madras Museum in 1865.
The dye-plant *Ventilago madraspatana*, collected by Cleghorn in 1857; red dye has leached from fragments of root bark placed in the paper capsule top centre. Herbarium presented by Alexander Sprot in 1896. (RBGE).

9
Third Furlough, Officiating Inspector-General & Farewell to India

On 25 February 1865, the day after the Forest Act received its assent from Sir John Lawrence, the Cleghorns returned to Madras on the steamer *Nemesis* to prepare for taking the ailing Mabel back to Scotland, and the long delayed assumption of Hugh's role as Laird of Stravithie.

Madras & Cleghorn's Herbarium

The couple spent less than two months in Madras during which period Cleghorn made visits to the Lal Bagh at Bangalore and McIvor in the Nilgiris. The visit to Ooty was to see for himself the progress of the cinchona plantations and in the Government Garden he was pleased to see some young cherry trees (possibly *Prunus cornuta*) grown from seed he had sent from Pangi in the Chenab Valley. The main task was the sorting out of books and herbarium collections as at this point Cleghorn thought that he might well not return to India, though the decision on this would depend on Mabel's health and the state of the Stravithie finances. Were the debts on the estate to prove substantial he would have to return to India, as if he retired now his pension currently stood at £220 per annum.[1] His precious books and a set of mounted ('glued') herbarium specimens were therefore packed up and sent by ship round the Cape, some by the ships *Indemnity* and *Hotspur*, and two boxes by a jute ship directly to Dundee.[2] The herbarium would remain at Stravithie until after his death, when it was presented to RBGE by his nephew Alexander Sprot. It included not only his own specimens, but a large number made by Walter Elliot in the Northern Circars and some made by his young colleague Dr Charles Drew who had died in 1857. There were also many collected in the Agri-Horticultural Society garden including, some by Andrew Jaffrey, of which some represent vouchers for paintings made in the garden by Govindoo. The sheets are of a rather bright blue paper and their hand-written labels are headed 'Madras Museum' – which suggests that the mounting was probably undertaken by staff in the Government Central Museum, where Cleghorn (unlike Robert Wight) left a set of duplicates that must have been greatly appreciated by the Superintendent, Edward Balfour. Though this was Cleghorn's private herbarium, it was probably a selection of material from the one he had started in the Forest Department, consisting of his own collections and those of Elliot and specimens sent from much further afield – including Hooker & Thomson ones from North India, ones from Thwaites from Ceylon, and from the Calcutta Botanic Garden (presumably through Thomson and Anderson). In August of the following year (1866) this Forest Department herbarium was itself transferred to the Madras Museum (Fig. 45).[3]

Although botany was included in Balfour's original scope for the Museum in 1851, nothing appears to have happened until Cleghorn's involvement and there appear to have been no botanical collections in the Museum in 1859.[4] Cleghorn has been credited with a foundational role,[5] but it is almost certainly incorrect to have taken 1853 – the earliest date on many of the specimens (including some of the Elliot and Drew ones) – as the foundation of the collection, which as noted above was not given to the Museum until 1865. In 1909 the Madras Herbarium was moved to the Agricultural College in the drier air of Coimbatore and transferred to the Botanical Survey of India in 1957, where Cleghorn's specimens still form an important part of the extensive herbarium of the Botanical Survey of India's Southern Circle.

Third Furlough

In the middle of April 1865 Cleghorn was granted a six-month leave on 'private affairs without pay',[6] and the couple set sail in early May. On the way to Egypt they stopped at Aden, where Cleghorn shopped for drugs in the bazar, and they were both uplifted with the idea of having ship-board services and singing hymns in the 'land of the Bible'. On 15 May the boat was in the Red Sea opposite Mount Sinai (2285 m), where they doubtless reflected on Moses and his receiving of the Ten Commandments.[7] This time the voyages to Egypt and through the Mediterranean went without a hitch and the couple reached Southampton around 3 June. From Hampshire the couple took a train to London where they stayed at the boarding house of Mrs Cooper at 23 Princes Square, Bayswater though Mabel, who must have been desperate to see her family and bereaved sister, would almost certainly have gone straight back to Edinburgh. Cleghorn had not yet retired, so there were official visits to be made both to the India Office and to the Office of Woods and Forests (the department responsible for Crown lands, including the two Royal forests – the New Forest and the Forest of Dean). Rather surprisingly, for it was only about six months since he had last been in the city, Dietrich Brandis was also in London and must have taken part in these meetings.[8] There was also a meeting with Edward Waring, a Madras surgeon who had sent Cleghorn drawings and specimens from Travancore and Burma. Waring had retired to Britain and since March had been busy compiling a *Pharmacopoeia of India*, with advice from a team mainly of retired Company surgeons: Sir Ranald Martin, Sir William Brooke O'Shaughnessy, Alexander Gibson, Daniel Hanbury, Thomas Thomson, J. Forbes Watson and Robert Wight – the book was published in 1868. This was also a time for catching up with botanical colleagues, and Cleghorn attended meetings of the Linnean Society and visited Kew for discussions with Thomas Thomson. At Kew he must also have seen the Hookers, though this was to be his last meeting with Sir William who died later in the summer.

On 12 July the two Indian colleagues took the train to Edinburgh as Brandis wanted to inspect the best-managed of the Scottish forests, and was therefore probably already considering their possible role in training young recruits for the Indian forest service. Three days later the pair attended a meeting of the Botanical Society of Edinburgh, an event that was something of a galaxy gathering, not only from the presence of Cleghorn and Brandis, but because two other distinguished Edinburgh medical alumni were also in attendance. John Kirk (1832–1922) had made a name for himself for his medical work during the Crimean War and his success as naturalist on Livingstone's second Zambesi expedition of 1858–63. The fourth notable present at the meeting in the Histology Classroom at RBGE was Robert Wight, Cleghorn's old friend, by now at 69 the grand old man of Indian botany. Cleghorn read a paper entitled 'Supplementary notes upon the vegetation of the Sutlej valley' based on his excursion of May 1862,[9] the main account of which had been read in his absence earlier in the year.[10] Brandis contributed a paper on Indian forests, comparing those of Burma with the Sal forests of the Terai and those of the Central Provinces. Also read that evening, and of interest to Cleghorn and Brandis, was the translation of a paper by Antoine César Becquerel on the influence of forests on climate.

9. FURLOUGH & LAST INDIAN DAYS, 1865–7

Stravithie Rebuilt

The rest of the summer was divided by the Cleghorns between Edinburgh and Stravithie. At the latter his grandfather's mansion-house was found to be in poor shape, and it was at this point that the Edinburgh architect John Chesser was commissioned to rebuild it.[11] Chesser was not particularly distinguished, but he was cheap, and his final bill came to only £137/7/6;[12] he was not known for country houses,[13] rather for his extensive terrace developments for the Heriot Trust in the West End of Edinburgh (such as Belgrave Crescent of 1874, where Alexander Hunter ended his days). In 1860–4 Chesser had been building similar, if smaller-scaled and rather more elegant, terraces for Robert Hope-Scott on his Hope Park estate in St Andrews, which is doubtless what lies behind Cleghorn's commission.

The present Stravithie House (Fig. 46 right) looks very different to the modest structure that Cleghorn, with a strict eye to economy, reconstructed out of his grandfather's dilapidated mansion – this is due to a Baronial wing that Alexander Sprot added in 1896/7 to designs by James Gillespie and James Scott of St Andrews. The newer wing is not only handsome in itself, but greatly assists the picturesque effect of the whole, for it must be admitted that Chesser's house was not particularly attractive, largely a result of the decision to rebuild the old house rather than to start again from scratch. This probably proved a false economy as, despite initial optimistic quotes of £1600, the final lowest quote was for £2176/10/- and at the end of the day it almost certainly cost more than that.

The old Wakefield House, known only from a single early nineteenth-century drawing (Fig. 46 left) had a main, three-bayed, three-storeyed central block from which a wing at the rear projected at right-angles (forming the limb of a T-plan); this rear wing was round-ended and probably contained the staircase. To either side of the main block, on the garden front, was a pedimented, two-bayed, two-storeyed pavilion, clumsily appressed to the main one. It is possible that this could, at least in part, have been determined by an earlier eighteenth-century house, as the result is decidedly old-fashioned for 1806. It is known that the elder Hugh gradually increased the size of the house from 1806 up to 1827, possibly by raising the height of the central block and adding the pavilions (providing rooms necessary for his numerous spinster daughters);[14] such piecemeal addition to an old fabric might account for the poor condition that Cleghorn found himself landed with by the mid-1860s.

Chesser decided to demolish and entirely rebuild the central block and rear wing on the foundation of the original cellars, intending to keep only the lateral wings, though in the end only the eastern one was retained. These works reduced the overall size and accommodation in keeping with Cleghorn's requirements – no children and a live-in domestic staff of only three or four. The result was far from elegant – all traces of the charming Georgian symmetry were removed, especially in the fenestration, replaced by a nod in the direction of Scottish vernacular (if falling short of full Baronial) in the multiplication and crowstepping of the gables on what became the main facade, the carriage-approach from the north. But this was at the expense of the private, garden facade, which is decidedly suburban with the addition of a full-height stack of bay windows to what on the first and second floor are the drawing room and principal bedroom respectively. The site slopes steeply to the south and the ground floor was raised on this side to provide a kitchen and servants' quarters, so that the main (first) floor was approached by stairs from the more picturesque entrance facade, mostly enclosed

CHAPTER 9

Fig. 46. The transformation of Wakefield House
Anonymous drawing, c. 1820, showing the house from the south-east, as rebuilt by Hugh Cleghorn from 1806 and the lodge on the St Andrews-Anstruther road that was completed by 1818. (Photograph in possession of David Chalmers) (above).

The house as rebuilt by John Chesser in the 1860s. Anonymous photograph, c. 1870. (Collection of David Chalmers) (below).

9. FURLOUGH & LAST INDIAN DAYS, 1865–7

The north front from the north-east: from the tower to the left is the Gillespie & Scott addition for Sir Alexander Sprot, 1897; to its right Chesser's house for Cleghorn. (Author's photograph, 1998) (above).

The south front from the south-west: to the right of the central bush is the Gillespie & Scott addition. (Author's photograph, 1998) (below).

CHAPTER 9

within an entrance lobby, to either side of which lay a breakfast room to the east and the all-important library to the west. The former was on the site of the original rear wing, which led to the main staircase being moved to the surviving east pavilion.

The result of this, in terms of accommodation, was extremely simple, and little money was lavished on ornamentation. The plastering by D. & A. Craigie of Edinburgh cost £180,[15] with cornices that might be found in any Edinburgh villa of the period; likewise the marble fireplaces, which were estimated at £60; the iron balusters of the main stair and other metal-work cost £25/18/6 from John Kelly of Edinburgh, and George Potts (also of Edinburgh) was paid £12/18/8 for the encaustic tile pavement of the lobby and £55/4/3 for painting the interior of the house. The main floor has only four rooms. Off the entrance lobby was the Breakfast Room, which doubtless doubled as Cleghorn's estate office, where he probably received tenants and paid his outdoor servants (who thus did not need to enter the main house). At the top of the lobby stairs is a wooden screen with frosted glazing, which, with a stained glass window for the main staircase, cost £25/0/7 from David Small of Edinburgh. The lobby stairs led into a corridor that ran the length of the house, with the staircase at its eastern end, and at the west an awkward termination consisting of a fireplace beneath a window. Off this axial corridor, on the south-facing side, were a handsome drawing room to the west and a dining room to the east, and adjacent to the entrance lobby lay the inner sanctum of Cleghorn's library (Fig. 47). Above this on the upper (second) floor were only five bedrooms with two dressing rooms and a bathroom, accounting for the surprising fact that, despite the rebuilding, the number of rooms did not increase between the Censuses of 1861 and 1871. By this time all his aunts had died and it must have been clear that Mabel was not going to have children so the bedrooms were required only for themselves, and for the guests – including his Sprot nephew and his family and the Cowan in-laws – whom they doubtless intended to entertain. After Cleghorn returned to India supervision of the building works lay with Mabel and her father Charles Cowan, and it was he who suggested that the joinery and carpentry work was given to his local firm, John Ewart of Penicuik (their work came to £810).[16] The clerk of works for the project was Robert Dalrymple on a weekly salary of £2 and the project seems to have run smoothly except that at the end of it the stone masons, James Swan of Cupar, had to sue Cleghorn for non-payment of £114/15/4 for work above their £880 estimate.

Fig. 47. Cleghorn's library at Stravithie today, showing his original bookcases. (Author's photograph, 2015, with permission of David Chalmers).

The Cleghorns appear never to have owned property in Edinburgh, as there were always plenty of friends and relations to stay with. Hugh must frequently have visited Balfour and the Botanic Garden, and also his friends connected with the Edinburgh Medical Missionary Society. Many of Mabel's extended family lived in Edinburgh, but she also visited the city for medical consultations with Sir James Young Simpson, suggesting that by now some of her medical problems were gynaecological. This furlough allowed Cleghorn to attend his third annual meeting of the British Association, which, from 6 to 13 September, was held in the sooty surroundings of Birmingham. This year the Vice-Presidents of the botanical section ('Section D') included Balfour, Babington and Bentham. The section's President was Thomas Thomson and Indian botany was strongly represented,[17] with other participants including Walter Elliot, Michael Pakenham Edgeworth, William Munro and Robert Wight. Cleghorn read a paper on the deodar forests of the Western Himalaya that he had surveyed in 1862 and 1863, illustrated with his map and with photographs taken by Colonel Charles Waterloo Hutchinson of the Bengal Engineers.[18]

Stravithie finances must have proved every bit as dire as he feared. This, despite 23 years of service, meant a return to India for at least another 14 months, if Cleghorn was to qualify for a larger pension. He must have agreed to this with a heavy heart, as there was no way that Mabel could return with him; in the end he would stay in India for another two years and would not see her again until November 1867. Mabel was left in the hands of one of her husband's medical contemporaries, Andrew Halliday Douglas (1819–1908). Douglas was also keen on Medical Missionary work, but the names on his gravestone in the Dean Cemetery lead one to question whether he was the safest pair of hands in which to have entrusted his dear wife. Douglas's first wife Susan died aged 29, his second Jessie aged 32, his third (another Jessie) aged 42, and even Marjory, the last of four and clearly made of sterner stuff, still predeceased him though at the more respectable age of 72.

Return to India & Excursion with Clements Markham

Late September found Cleghorn back in Southampton. He sailed through a Mediterranean in the grips of the fourth of the cholera pandemics, which had started in India in 1863 and reached Europe by way of Aden and Egypt.[19] On this occasion the threat to Cleghorn's well-being in Egypt came not from inundation but pestilence: 60,000 people had died in the country in June and July of this year (1865). This did not put him off and he took the opportunity to meet an unusual missionary. Mary Louisa Whately (1824–1889) was a daughter of the Archbishop of Dublin, who through her books (*Ragged Life in Egypt*, 1858, and *More About Ragged Life in Egypt*, 1863) was already well known for her pioneering work in establishing schools for the teaching of Muslim children, both boys and girls. On the Red Sea coast Cleghorn boarded the Bombay-bound ship *Carnatic*. Fellow passengers included a number of 'gay dragoons' and Clements Markham on his second trip to India, to observe the progress of the *Cinchona* experiment. Cleghorn appears to have done some botanising *en route*, as when he returned to India he sent seed of the scammony (*Convolvulus scammonia* – its root the source of a powerfully cathartic resin) to the Agri-Horticultural Society in Lahore, which he had collected in September in the 'Levant', which presumably means Sinai as the plant does not occur west of the Red Sea.

Carnatic reached Bombay on 30 October, whereupon Cleghorn resumed his government service, but was refused the permission he sought from Madras to be allowed to accompany Markham to Ceylon.[20] This was a disappointment, having missed out on a commission to explore the

CHAPTER 9

island on his previous return-journey to India, due to his summons to the Punjab. So it was back to Madras, but in early December Cleghorn was permitted to travel to the west coast to re-join Markham. By this time Markham had returned from inspecting the cinchona plantations that Thwaites had established in 1861 at Hakgalle in Ceylon. He now wanted to investigate the country between Cape Comorin and Palghat – the hills of Travancore and Madurai (i.e., the Palnis), then north to the Nilgiris to observe progress there under McIvor. Cleghorn met up with Markham at Trivandrum where they had discussions with the Raja and his Dewan. Since Cleghorn's previous visit with Lord Harris in 1858 both the Resident and the Raja had changed – old General Cullen had died in 1860 to be replaced by Francis Maltby then, in 1862, by Henry Newill. The Raja was now Ayilyam Thirunal, though his Dewan was still Madava Rao. Markham and Cleghorn headed inland from Kottyam (where they stopped at the CMS station), in order to visit the Travancore Government's quinine plantation. This was in the Ghats at Peermade, at an altitude of about a thousand metres, from where there were extensive views down to the Backwaters. This plantation or 'Government Garden' had been started in 1861 by Maltby who had a nearby house called Maryville, and was now under the superintendence of a Mr Hannay who came to the job with experience in Australia. Cleghorn wrote a report on this garden in which he recommended the trial of other crops, including tea, vanilla and potatoes.[21] The botanical explorers received a telegram from Vere Henry Levinge, Collector of Madurai, inviting them to Kodaikanal, and they crossed the Ghats by a little-known route crossing the Periyar River and emerging in the Kambam Valley. Cleghorn was pleased at the spread of coffee plantations, a reminder that his primary interest was not in conserving forests for their own sake if they could be turned to useful agricultural account. From here they skirted the edge of the Ghats northwards to Periyakulam then on to Kodaikanal in the Palni Hills. This hill station had been started in 1845 by American Missionaries from Madurai, in whom Cleghorn must certainly have taken an interest; Markham inspected cinchona plants in Levinge's garden, and they must both have admired the picturesque lake that had been constructed two years earlier at Levinge's suggestion. From the Palnis, in the first days of 1866, Markham and Cleghorn went to the Nilgiris, where Markham was delighted with what McIvor had managed to achieve with the quinine plantations:

> When I selected the sites for plantations in 1860–61, they were covered with dense jungle. Now the scene is so altered as to be beyond recognition: acres of ground are covered with rows of Chinchona trees, with their lustrous crimson-veined leaves and fragrant flowers; and in one place they have grown to such a size that the foliage of their branches mingle and entirely conceals the ground. They have become so important and distinctive a feature of the fine range which overhangs the Moyaar Valley, that it should be known henceforth as the Chinchona mountains.[22]

While Markham continued his Indian sojourn, Cleghorn returned to Madras, though almost nothing is known about what he did there in the first four months of 1866. No information has emerged about any forestry activities, and it could be that Beddome continued to act as Conservator. But if Cleghorn was thinking that his final period in India was going to be a gradual wind-down to retirement he was in for a shock. A dramatic change of circumstance occurred when Cleghorn was summoned back to Calcutta, this time to act for Brandis during another period of European leave. This was Brandis's third trip to Europe and although it was initially designated 'sick leave', he turned it to productive account. On this occasion the deputising for Brandis was officially recognised by Cleghorn's designation as 'Officiating Inspector-General', which was slated to run from 7 May 1866 to 14 March 1867, though it was later extended to a full calendar year.

9. FURLOUGH & LAST INDIAN DAYS, 1865–7

Second Government of India Period: Officiating Inspector-General of Forests

Cleghorn arrived in Calcutta on 1 May 1866 on the ship *Nemesis*, accompanied by a 'native male servant'.[23] The Government of India would already have left for the hills so Cleghorn and his anonymous companion probably immediately headed for Simla. Here he would be based in the PWD office until early November, but under a new boss as Strachey had, the previous year, been replaced as Government Secretary for the PWD by Lieutenant Colonel Craven Hildesley Dickens of the Royal Artillery.

Soon after starting the new job Cleghorn explained to Balfour that 'much urgent business requiring constant attention has obliged me to curtail private and Demi Official correspondence to a minium',[24] so the only source of information about what Cleghorn did as Officiating Inspector-General is in the form of the memos and reports that he wrote, many of which were published in a Blue Book,[25] and many of them summarised by Stebbing.[26] Eleven such documents are to be found in these sources, but Cleghorn must have written many other unofficial reports and minutes dealing with the daily activities of the Forest Department; though by now fieldwork seems to have been a thing of the past. Although Bombay or Madras still did not come within the ambit of the Department, most of the other local governments had by this time set up forest establishments and drawn up a set of operating rules along the lines suggested in the 1865 Forest Act. These rules covered topics like the designation of Government-owned Reserved Forests, listing of reserved timber species, rules for felling, and rates charged for wood by timber merchants and by locals for specific rights (including grazing, access to minor products and to non-reserved trees). The annual reports of the regional Conservators covered these subjects, and gave statistics on numbers of trees felled, sleepers produced and financial statements. The majority of Cleghorn's memoranda from this period are responses to these annual regional reports.[27] In 1866, relating to this area of work, Cleghorn wrote a set of Rules for the Native States of the Central Provinces, which are more or less identical to Beddome's ones for Madras.[28]

Large parts of Cleghorn's memoranda are merely summaries of the report commented upon, but up to a point they reveal his own interests – not least his concern for obtaining revenue for the government though not at the cost of over-exploitation. They also show his hatred of wasteful felling and shifting cultivation, the value he placed on non-timber products, and a concern that the 'natives' should not be unduly oppressed. The most revealing is his long memo on the 1865/6 report for the Central Provinces,[29] where a Forest Department was prospering under Major George Pearson (who had started off in the Madras Army) and Lieutenant James Forsyth, with 25 Reserved Forests demarcated and five species of timber protected (sal, teak, shishum [*Dalbergia sissoo*], saj [*Terminalia alata*] and beejasal [*Pterocarpus marsupium*]). Rules were in place to control felling and locals' rights both in the Reserved Forests and to resources from the Unreserved Forests, from which revenue was being claimed for the first time. Pearson had also been making pioneering measurements of trees to attempt to establish growth rates, an essential technique in scientific forestry to allow sustainable cropping regimes to be worked out. Cleghorn's underlying philosophy is shown in the comment that the basis for the principle of strict conservancy in Reserved Forests was 'looking both to the prospective wants of the Railway and the general welfare of the people'.

CHAPTER 9

He also restated one of his deepest convictions, dating from Mysore and Madras days, on the evils of shifting cultivation (known in the Central Provinces as 'dhya'), over which he stressed that 'the rules cannot be too strictly enforced'. On the use of the Unreserved Forests Cleghorn wrote:

> The duty of the Conservator and the District [Revenue] Officers is to supervise methodically the local supply, and to secure its permanence, unless more important industries render it desirable that the forest lease be converted into a clearing lease. It is also their duty to see that the collection of dues is not oppressive to uncivilised tribes dwelling within the forest.

Another of Cleghorn's great interests was expressed in this memo – the use of 'minor forest products': fruits and materials used as dyes and drugs. But, as with timber, what most concerned him was that there should be 'a safe-guard against waste', which he thought justified charging locals 'Forest Dues' or 'seignorage' that would provide another source of government revenue. It is also worth noting that the 'canny Scot' makes an appearance in this memo, Cleghorn spotting inconsistencies in the finances both within the report, and between it and that of the previous year.

The Mysore Forest Department had been established in January 1864, but no Reserved Forests had yet been mapped in the kingdom. Cleghorn's memo was on a report by Godlieb James van Someren (who shared Cleghorn's evangelical persuasions) and drew attention to the huge wastage caused by felling for railway construction. One interesting point made by the Inspector-General, in stressing the need to avoid premature felling, was that the maturity of a tree was not necessarily indicated by girth alone, and he recommended the use of local people called 'Koorambers' to do the felling, being the best judges in such matters.[30] The Bengal Forest Department had also been established in 1864 under Cleghorn's Edinburgh friend Dr Thomas Anderson, who ran it in tandem with superintendence of the Calcutta Botanic Garden and the Darjeeling Cinchona Plantations. At this point the Bengal forests that were being exploited were the sal forests of the Terai and the temperate hill ones around Darjeeling ('Sikkim' and 'Bootan' [i.e. around Kalimpong]). The reserved species of the latter were oak, horse-chestnut, walnut and cherry and two species of magnolia (*M. excelsa* [= *M. doltsopa*], *M. campbellii*), though to today's sensibilities it is horrifying to think of such beautiful species being felled for turning into railway sleepers. Plantations were also being established and show that the planting of the exotic conifer *Cryptomeria japonica* and of *Eucalyptus* in this area dates back to this period. The terrible toll of the terai jungles (or rather its mosquitoes) on the lives of Forest Department employees is also revealed in this report: no fewer than 32 Indians and one European overseer had died in this year alone.

In addition to memos on regional forest reports, Cleghorn wrote three more general reports during his year as Inspector-General. The first was on the fuel supply for the Punjab and Delhi Railways;[31] the second on a sylvicultural matter – the practice of 'girdling' (i.e., ring-barking) of trees prior to their felling;[32] the third, related to a subject of long-term interest to him – the environmental effects of deforestation.[33] The fuel report is a thorough document, and one of which he was clearly proud, as he sent a copy to Balfour, which although marked 'for private information', Balfour read to the Botanical Society of Edinburgh and published in summary. It drew on Cleghorn's great knowledge of the issues concerned – both financial and botanical – and related to the question of how best

9. FURLOUGH & LAST INDIAN DAYS, 1865-7

to supply the necessary fuel for locomotives at a time when it was estimated that existing resources of the *rukhs* would last only seven to ten years. Clearly large amounts of new planting were required, but who should do this (local zamindars or the Government?), where (on canal banks?, on good soil, but not by taking land away from agriculture, though babool could grow on what was already becoming a problem – salinified or 'reh' soils), and should the species grown be natives or exotics? Another point discussed was the use of coal, which would remove, or at least reduce, the need for burning timber. The East Indian Railway from Calcutta to Delhi was already using coal, of which 370,000 tons per year were by now being produced at Ranigunj in West Bengal.[34] The Geological Survey of India, established in 1851, was run by Thomas Oldham who was actively surveying for new coal fields, but there was no way in the foreseeable future that the Punjab or Bombay railways could be fuelled by anything other than wood. One section of the report recommended that accurate quantification of timber in each area was necessary in order 'to calculate stock and prospective out-turn'. These figures should be provided under the headings 'age of trees', 'cubic content', 'annual growth', 'amount disposable annually' and 'accumulated quantity'. Such statistical information lay at the heart of German scientific forestry methods and shows that Cleghorn had been learning from his colleague Brandis. Another point of interest is what Cleghorn termed the 'secondary advantages' of plantations: they would provide employment, set a good example of tree-planting, and 'preserve some moisture and prevent the drifting of sand, whereby a considerable area of waste may be converted into arable land'. In the end, as 'no profits would be secured for a long period', which explained why private enterprise had not come forward to supply the wood required, the making and development of plantations was deemed a legitimate concern of the Forest Department, and that the Government should put substantial resources into developing them.

When trying to discover what Cleghorn did, one frequently despairs at the sheer volume of documentation that he and his contemporaries in government service (from the Viceroy downwards) had both to generate and plough through. The question of girdling trees prior to felling them – a form of seasoning of timber while the tree was still standing – was one that was much discussed, and information had been sought from a wide range of interested parties over the whole of India. The volume of responses is daunting, as is the diversity of opinion elicited. But Cleghorn had to wade through it all and come up with yet another report. Bearing in mind the perceived advantages of girdling (the timber was easier to float; it was easier to count how many trees were to be cut, rather than counting the easily obscured stumps after felling) and disadvantages (the timber was brittle), his conclusions were cautious. In areas where the practice had been given up (the majority) 'no facts have been adduced which would warrant the re-introduction or perpetuation of the system generally', but that in the (largely wet) areas where it was still practised, notably Burma, 'its entire and immediate abandonment under present circumstances does not appear to be advisable, but this ultimate result might be kept in view'. This report is typical of his careful reading of large numbers of documents used to reach cautious conclusions, habits demonstrated as far back as his British Association report of 1851. He looked not only at the reports received from India, but broadened them – in this report, for instance, he quoted a European periodical, *The Artizan*, of May 1866, showing that he kept up to date with literature from a wide geographical range, a characteristic evident from the contents of his library.

CHAPTER 9

The climate question again

In the third of his Simla memos Cleghorn returned to an issue he had first considered for the British Association report – the question of adverse physical effects arising from deforestation. In 1865 a report to the Madras Government by the engineers Colonel Lawford and Major Sankey had drawn attention to the damage to roads in the Western Ghats from landslides caused by local deforestation, and more distant deleterious effects on the rivers flowing from them – damage to bridges due to increased height of floods and drought between monsoons in the lower Caveri River. They engineers claimed (as reworded by Cleghorn) 'that Coorg and Wynaad have become gardens, but apparently in some degree at the expense of Trichinopoly and Tanjore' – a comment that has resonances to this day.[35] These effects were attributed, both by the locals and by Lawford & Sankey, to extensive forest clearance in Coorg and Wynad, especially for creating coffee plantations. The engineers recommended remedial steps – no forest clearance 50 yards either side of roads, no planting of coffee on steep slopes, and that engineers should be consulted before granting any land for clearing. In London a copy of this report had been noticed by the Secretary of State for India, the Earl de Grey, who pointed out 'the danger of imprudently disturbing the general arrangements in the economy of nature' and asked the Madras Government to investigate. Cleghorn's memo on Lawford's and Sankey's report, written in Simla on 12 September 1866,[36] while cautious (as usual) and calling for more information on the river flows, was firm in its recommendations but took the engineers' ones to be remedial and not curing the real cause. He stated that 'the influence of thickly wooded ranges in equalizing and regulating the flow of water precipitated from the sky ... cannot be denied', and he had seen the problems for himself only recently when travelling in Travancore and Madurai with Markham earlier in the same year (January 1866). Markham reckoned that in the Ghats an astonishing 180,000 acres had been cleared for the cultivation of tea, coffee and quinine, and considered that 'physical changes and climatic influences followed the destruction of the forest on the eastern slopes, as any natural effect follows its cause'.[37] Cleghorn's recommendation was therefore that:

> The protection of the wooded crest of the ghauts and of a certain proportion of the forest below, particularly in the eastern slopes and along the banks of streams, and also, in certain situations, the restoration or replanting of the hill sides would be more likely, than any other means, to prevent or diminish the evils alluded to in these papers ... [and that] ... the observations already made are sufficient to justify the action in the direction directed, without delay.[38]

This quotation has been given here at length, because it is so *unusual* in Cleghorn's official writings, previously evident only in his recommendations on the Nilgiri Sholas in 1857: an explicit statement on the climatic effects of deforestation and the recommendation of taking remedial action. But as always this was tempered with economic considerations, as would be seen more strongly in a follow-up to this memo.

After Cleghorn's return to Madras on Brandis's resumption of the Inspector-General's post, Cleghorn was sent by the Madras Government to investigate the matter in person.[39] In August 1867, 'when the south-west monsoon was in full force', he visited Bhavani at the junction of the Bhavani and Caveri rivers but found no 'recent erosion of banks or watermarks worthy of notice', and although he heard that the locals complained of decreased rainfall and more precipitate drainage, which they attributed to deforestation in Coorg, he was sceptical as he could find no way of 'reducing the statements to a numerical basis'. A study of the twenty-year records of the flow of the Caveri at Tanjore (much further downstream) showed no 'material alteration of volume in the Cauvery and Coleroon rivers'. But by this time the Government of India had already instructed the Commissioner of Mysore and Coorg to form 'forest reserves, which are not to be alienated, and to the strict conservancy of wooded crests and mountain ridges' of which Cleghorn did approve, even if he questioned some of the evidence.

Eighteen months later, at a Royal Geographical Society meeting in London, Cleghorn would hear more on this topic. George Bidie (successor to E.G. Balfour at the Madras Museum) read a paper on deforestation in Coorg in which, in the most tactful manner, he implied that on his field trip Cleghorn had been looking for evidence in the wrong place.[40] The effects of deforestation at this stage were still primarily local to Coorg, where they were all too apparent, and conditions in Tanjore were irrelevant as so much water had been taken off the Caveri for irrigation by time it reached there that any effects from Coorg would have been evened out. Bidie stressed the need for forest protection in Coorg to stop the damaging effects of the increased runoff following deforestation. The *first* comment in Cleghorn's response to Bidie's paper was approval of the increased capital that the coffee planters were bringing to the Western Ghats, 60 of them a year were going out, each cultivating about 150 acres, and so bringing £1500 of capital to the area. Only after this did he admit that forest conservancy was necessary in particular areas (such as along river banks), but that so far the deleterious effects were not 'so marked as to warrant any interference with the clearings to any great extent'. Cleghorn did, however, end by admitting that the 'changes from month to month' should be monitored by cordial co-operation between Forest officers and Canal officers (the latter being responsible for irrigation in the lower reaches).

Retirement Plans

With quicker maritime transport, but especially since the arrival of the telegraph, the world had become a much smaller place and Cleghorn learned of the vacancy of the botanical chair at Trinity College Dublin due to the death of W.H. Harvey, the distinguished algologist and author of *Flora Capensis*.[41] Cleghorn would soon have to return to Britain to be with his invalid wife and he had to think of the future. He was still only 46 years old and if he reflected on the lifespans of his father and grandfather, could look forward to a long and healthy life, a period that, given his high principles and sense of duty, he felt should be usefully filled. Mabel told Balfour: 'I know he dreads the want of work when he comes home. The position he would occupy at Dublin would give him a much more extended sphere of usefulness than he could have in Fife'.[42]

Given Cleghorn's extensive experience in the field of applied botany, and several years of teaching experience in Madras, the idea of a botanical chair was eminently reasonable, not least because such a post tended to require residence at the relevant university for only a few months of the year (in this case April, May and June). Cleghorn thought very seriously about the Dublin job and wrote out an application that he sent to J.H. Balfour, while deferring to the older man's opinion as to his suitability, and also taking into consideration the views of Mabel's medical advisor Halliday Douglas, giving them collectively the power of veto over submitting it. In the end Balfour withheld it, at least partly on the grounds that the Glasgow chair held by an ailing George Walker-Arnott was likely to fall vacant sooner rather than later, and Glasgow was much closer to Fife than Dublin. There would come to be an irony in this, as Alexander Dickson who got the Dublin chair, would, only two years later, compete with Cleghorn over the Glasgow one.

CHAPTER 9

A Last Himalayan Excursion

Knowing that he would almost certainly never see the Himalayas again, Cleghorn made one final natural history excursion. At 3647 metres the Chor Mountain (now Churdar) in Sirmour District is the highest peak of the outer range of Siwalik Hills, a three-day journey south-east of Simla. The locality was well known – George Everest had used it in 1834 from which to make measurements of Himalayan peaks, but of greater interest to Cleghorn was that botanical collectors had been sent there by Govan and Royle from Saharunpur. Cleghorn spent a week there at the end of October 1866;[43] he collected plants and sent seeds of about 25 species of trees (including deodar, *Abies pindrow* and *Cotoneaster bacillaris*), shrubs and herbs (including *Meconopsis aculeata*[44]) to Hooker at Kew.[45] He must also have taken his gun, as it is recorded that no fewer than 64 species of bird from the expedition were identified by Ferdinand Stoliczka, a young Moravian palaeontologist and zoologist who was one of Oldham's staff on the Geological Survey.

It was also at this time that Cleghorn made a rare foray into taxonomy – with an interesting, and well-illustrated paper on the conifers of the NW Himalaya, published in the Calcutta-based *Journal of the Agricultural and Horticultural Society of India*. This provided detailed notes, written jointly with J.L. Stewart, on the distinguishing characteristics (including the barks) of these important timber species that they both knew so well. Also given was a table showing the multiplicity and confusion of local names with the conclusion that: 'The bewildering confusion of vernacular names, which has existed since the time of Pliny, may reconcile the traveller to the comparative unity of Botanical nomenclature'.[46] In fact, this has proved to have been something of a vain hope: the botanical names of the conifers have proved anything but stable and are now as follows: *Cedrus deodara*, *Pinus wallichana* (then *P. excelsa*), *P. roxburghii* (then *P. longifolia*), *P. gerardiana*, *Picea smithiana* (then *Abies smithiana*), *Abies pindrow* (then *Picea webbiana*), *Cupressus torulosa*, *Taxus wallichiana* (then *T. baccata*) and *Juniperus macropoda* (then *J. excelsa*).

While in the Himalayas Cleghorn no longer had access to any Indian artists, so there is an almost complete dearth of botanical drawings from this 'northern' period of his Indian life. He did, however, have access to a few amateur Western artists, two of whom provided drawings of cones for this paper – William Coldstream drew two, and the other four are by Colonel Michael Anthony Shrapnel Biddulph of the Royal Artillery (*Cleghorn Collection* Fig. 30), who had sketches published in contemporary British periodicals. One of Biddulph's originals survives at RBGE, as do some unused ones by Margaret Prinsep. The final point of interest of the conifer paper is ecological, a quotation of some work by Brandis on the differing lifespans of the needles of different species, which was found to be correlated with the conditions of light and shade in which the species grew (those of dense forests, spruce and fir, lasting eight to ten years; those of open slopes, *Pinus roxburghii* and *P. gerardiana*, being replaced every two or three).

Return to Calcutta

In early November, with the rest of the Indian Government, Cleghorn returned to Calcutta for a final winter. Brandis's return from Europe was repeatedly delayed, so Cleghorn continued to officiate as Inspector-General and some of his reports from this period have already been discussed. Brandis eventually returned to India in early April 1867, but his time away had been profitably spent, laying the foundations of a training system for young men for the Indian Forest Service.

9. FURLOUGH & LAST INDIAN DAYS, 1865-7

Brandis and the start of Continental forest-training

On his leave Brandis had studied forests in Scotland (doubtless at Cleghorn's suggestion, though why he also went to the forests of Corsica is altogether stranger). He also visited forest schools in France (Nancy) and Germany (Eisenach, Tharandt and Neustadt Eberswalde) and in Britain he had discussions on the subject of the training of forest officers not only with the India Office, but with individuals including Joseph Hooker, J.H. Balfour and Charles Lawson.[47] Everything that Brandis did was with Teutonic thoroughness and by means of voluminous, in-depth reporting: the question of education was no exception. Grandiose initial schemes for what might be termed an 'undergraduate' training in sciences, especially maths, in Britain were to be followed by specialist training in France or Germany.[48] However, as it was intended that the India Office would pay for this education (the probationers were paid an annual stipend of £120 and as the courses were of more than two years, by 1872 each student had cost the Government £450 by time he reached India), it was realised that training in languages and science could be made a prior requirement, examined by the Civil Service Commission. It proved possible to run what was effectively a trial scheme as two of his Burmese foresters, Captain William John Seaton and Lieutenant William Stenhouse (like Pearson, both recruited from the Madras Army; perhaps another indirect influence of Cleghorn?), happened to be on leave in Britain at the time and in October 1866 Brandis obtained permission to extend their leave for a period of training in Scotland and France. This had ended by July 1867, when the pair returned to Burma, and though it is not known where they went in Scotland (Scone is the most likely), in France they spent time in the forests of Villiers Cotterêts (in the Aisne), Haguenau (Bas-Rhin) and Remiremont (Vosges) and at the Nancy forest school.

Scotland had been the traditional training ground for gardeners for India and the Colonies, and was starting to do the same for foresters – one called Davidson had been sent to the Central Provinces on Balfour's recommendation in 1866, and W.L. Grahame arrived for the same department in early 1867. So Brandis's idea of sending students to Germany and Nancy incurred the wrath of part of the British horticultural/scientific community. In December 1866 Balfour read a paper to the Botanical Society of Edinburgh, later published in *The Farmer*, along the insular, if not xenophobic, lines that training in Scotland was good enough and attacking the idea of sending men to the Continent. More reasonably he argued that teaching probationers Hindustani would be of more use in India than French or German. In reply Brandis stated that Scottish training, while excellent of its kind, was not an appropriate model for what he had in mind for India. His line of argument, backed up by Cleghorn, was that British forests were too small-scale, that they were owned by private proprietors, and that the method was in nearly all cases that of clear-felling and replanting, which on grounds of scale would be impractical in India, where natural regeneration had to be relied upon – as it was in the State-owned European forests. Furthermore, Britain had no tradition of training in 'scientific forestry' (German 'working methods' with mathematically worked-out standing crops and yields), or of practical training schools such as had been developed in various German states since the 1770s, and at Nancy since 1824. In Britain there was effectively no such thing as a forestry profession, and certainly no forest literature worth speaking of (hence the need for language skills in French and German).

CHAPTER 9

The insular criticisms were answered with tact by Brandis, who convinced the Secretary of State (Viscount Cranbourne) and the India Office; he had, in any case, already chosen the first batch of students. The seven young men went to Europe in 1867, where they studied until 1869, reaching India the following year. The two for Germany (John Kipper Hume and Albert Edward Wild, both aged 20) were sent to Heinrich Christian Burckhardt, Civil Director of Forest Administration in Hanover, where they spent a year (in later years extended to 1½) of practical training, with excursions to areas such as the Hartz Mountains, followed by a year (later 1½) either at the Royal Saxon forest school at Tharandt near Dresden, or the Forest Academy of Sachsen-Weimar at Eisenach in Thuringia. The latter already had an international reputation, taking students from Scandinavia, Russia and Italy. The five men destined for France (Framjee Rustomjee Dasai, son of a Bombay merchant; Walter Henman; Alfred Pengelly; Edward Moir and Louis Gavin, aged between 17 and 23) spent two months (in later years extended to eight) doing practical work in the forests of Haguenau in Alsace under Clément de Grandprey and then two years, from November, at the École Nationale des Eaux et Forêts at Nancy under Henri Nanquette (Fig. 48). This pattern became established and by the year 1874, no fewer than 27 men had trained in France and 23 in Germany; however, from that year onwards all the training was done at Nancy, the last batch for Germany being that of 1873/4. Almost inevitably there were problems – lack of commitment on the part of the students, language difficulties, and, in some cases, lack of physical stamina. It was always realised that training in Britain would be more efficient (and cheaper). Nonetheless, Continental training continued until 1885, when, under the direction of Wilhelm Schlich on his retirement from India, a school of forestry was opened at the Indian Civil Engineering College at Cooper's Hill in Surrey (Fig. 49), until moving to the University of Oxford in 1906. It should be noted that at this point there was no question of training the upper ranks of the service in India, as none of the forests were yet exemplars of 'scientific' management, though in 1878 Brandis inaugurated a training school for junior (i.e., largely Indian) ranks at Dehra Dun, with Frederick Bailey as its first director.

Fig. 48. Pavilion Nanquette, École Nationale des Eaux et Forêts, Nancy. (Collection AgroParisTech – Nancy Centre).

9. FURLOUGH & LAST INDIAN DAYS, 1865–7

Fig. 49. Cooper's Hill Engineering College, where senior Indian foresters were trained from 1885 to 1906. (Author's photo, November 2014).

It wasn't only the first batch of students whom Brandis picked on his 1866 European visit; he also chose two highly qualified German foresters to leave for India immediately. Wilhelm Schlich and Berthold Ribbentrop were to be 'Special Assistant Conservators' for Burma and the Punjab respectively; they arrived at Bombay on 16 February 1867 from where they must have gone to Cleghorn in Calcutta. These two men were destined to have a major influence not only on the development of forestry in India, but on forest education in both India and in England, which would have enormous influence over the whole of the British Empire and America until well into the twentieth century. Cleghorn's own view on international co-operation in Indian forestry were balanced and in June 1866 he wrote: 'The Germans understand Forestry on a large scale better than we do – but an equal admixture of Scotch & Germans in the Dept. will be good'.[49]

Preparations for Leaving India

By March 1867 Cleghorn's thoughts must have been focused ever more closely on his return home. News came of progress on the rebuilding of Stravithie, which by now had a new roof and James McNab, Curator of RBGE, had been sent by Balfour to advise on redesigning and planting the garden.[50] The tradition of great international exhibitions was continuing in Europe and from 1 April to 31 October 1867, the second *Exposition Universelle* was held in the Champ de Mars in Paris, envisaged by Napoleon III as a showpiece of the Second Empire. Once again extensive collections were sent from the Indian Presidencies, but Cleghorn's contribution on this occasion was restricted to a single object, though one of spectacular proportions – a section of an enormous deodar trunk which he had had felled in Lady Canning's Grove near Simla – with a girth of 11 feet 11 inches at four feet from the ground and 210 annual growth rings.[51]

CHAPTER 9

A Government's thanks

On 22 April 1867, on his return to Madras, the Government of India published a Resolution entitled 'Services rendered to the State by Dr. Cleghorn'.[52] This did not go so far as the arguably exaggerated claims of the 1865 one, but summarised his work for the Supreme Government between 1861 and 1867, and ended: 'The Governor General in Council [still Sir John Lawrence, Fig. 50] desires to convey to Dr Cleghorn the thanks of the government of India for his long and successful labours in the cause of Forest conservancy in this country.'

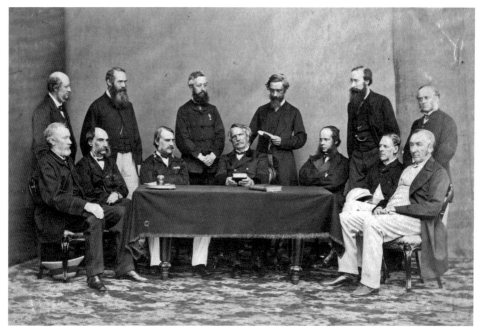

Fig. 50. The Viceroy, Sir John Lawrence, with his Council and the Government Secretaries, Simla, July 1869. Standing second from left is Colonel Craven Hildesly Dickens, Secretary of the Public Works Department (Cleghorn's boss); standing third from right (reading a paper) is his friend Sir William Muir. Albumen print by Samuel Bourne.

Farewell to mulligatawny

Cleghorn returned to Madras in mid-April 1867 to prepare for leaving India. It must have been a time of reflection after a richly fulfilling professional life of twenty-five years, which had covered the fields of medicine, horticulture, economic botany, botanical education and above all the setting up of forest conservancy both for Madras and (with Brandis) for the Government of India – added to which were his private efforts in the promotion of medical-missionary work. Personal touches emerge in the letters from this period such as his sending a box of 'equal proportion of Curry Paste, curry powder, Mulligatawny' for Mabel and for Mrs Balfour,[53] but there was still official business to be done. There must also have been farewells to his loyal Indian assistants and he asked Govindoo to make a few final drawings, on a large scale – of the sausage tree (now *Kigelia africana* – Cleghorn Collection no. 50) growing in the Agri-Horticultural Society garden, and of three other trees – teak, dillenia and baobab.

9. FURLOUGH & LAST INDIAN DAYS, 1865–7

In August the Madras Government sent him on one last field trip, to investigate the effects of forest clearance in Coorg that have already been discussed (p. 186). On 6 September Cleghorn submitted his final report to the Madras Government, which amounted to a decennial review of the Forest Department under himself and latterly Beddome.[54] After this (19 October), his last Government task was a written opinion on the question of whether or not the 1865 Forest Act should be introduced to Madras. The two documents and the response of the Government are best treated together as they provide insights into the beliefs of Cleghorn and the Madras Government, both of which were to have long-term consequences.

The report ended with some interesting items relating to Cleghorn's commissioning of botanical drawings – recommending that the large collection by the 'Draughtsman attached to the Forest Office' (i.e., Govindoo) be published as a supplement to Wight's *Icones*; and that an identification manual, a 'Flora Sylvatica' along the lines of Evelyn's *Sylva*, and more recent French (by Mathieu) and German (by Hartig) works, with descriptions, illustrations and notes on uses, be published. Both suggestions were approved by the Government and would be achieved by Beddome.[55] However, the bulk of the report related to the 'present and prospective operations of the [Forest] department.' Cleghorn was, unsurprisingly, proud of the profit that management of the 20 forest ranges (i.e., Government Forests) had brought – 1,858,038 Rs over ten years. His concern, however, was not only for short-term revenue, and he stressed the fact that some areas would not become profitable for many years, as restoration was required. He was committed to the benefits of plantations, not only for native species such as teak, toon, sandalwood and red sanders wood, but Australian exotics such as eucalyptus and acacia, and the need to increase the number of plantations for supplying railway fuel. He was also keen on legal control, and the need to strengthen the authority of forest officers by giving them the powers of 'subordinate magistrates'. A deep interest of Cleghorn's (going back to Great Exhibition days, and even before that) was in non-timber forest products such as dyes and medicinal plants, which had been depleted by the destruction of the forests. He did not, however, recommend their uncontrolled use by 'natives', rather, that their protection would result from levying a 'seigniorage' that would also yield revenue.

The Madras Government's response to the final report included a glowing tribute to Cleghorn's work, but also made significant and revealing comments on their own priorities – they looked on conservancy for:

> the development of the sources of supply, and the careful protection of those resources from waste, as the primary object to which the department should devote its attention, and that the securing a large revenue from the forests ... is a comparatively secondary consideration.[56]

Their response also stated that the 'demands, present and prospective, on the timber and fuel resources of the country are so enormous, that only by unremitting attention and care ... can ultimate disaster be obviated'. On plantations and reserves they observed that 'Too much importance cannot be attached to the restoration and extension of sources of supply'. But they differed slightly from Cleghorn on the matter of use by the 'natives' of non-timber products, which they viewed as 'a delicate one' that needed 'careful treatment ... the interest of the jungle tribes, who mainly collect this produce, shall be carefully protected ... The revenue to be derived from them is altogether a minor matter'.

CHAPTER 9

When asked for his opinion on applying the 1865 Act Cleghorn's conclusion was that its introduction 'generally speaking ... would have beneficial results'.[57] However, his views had not really changed since the empirical methods he had developed in Madras in the 1850s and he had not taken on board the criticisms of the new generation of foresters now emerging. In his lack of expressing a distinction between Reserved, Protected and Village forests, there is nothing to show that he understood the limitations of the Act as perceived by this new generation. By contrast, the Madras Government, under the Governorship of another liberal Scottish peer, Francis Napier, 10th Lord Napier of Merchiston and Ettrick (though under the influence of his Board of Revenue), had very strong views on such subjects, which have been hailed as being much more enlightened in terms of the respecting of 'native rights' by modern environmental historians[58] – as has just been shown over the question of non-timber forest products. On the question of the application of the 1865 Act, in their deliberations on Cleghorn's memo and those of the Board of Revenue, the administration considered that for Government Forests the Act was unnecessary as they were already protected by existing laws on trespass, mischief and theft.[59] The other forests were 'within village boundaries', to which villagers had immemorial rights that had been 'repeatedly recognised by Government, and ... scrupulously respected'; so even if the Government had wanted to alter their access this would be so repeatedly challenged on the grounds of existing rights as to be unenforceable. Furthermore it would lead to oppression: 'It will be in the power of the pettiest 'peon' in the [Forest] Department to harass the people at his own pleasure'. In 1869 Napier therefore decided not to introduce the Act. It should be added that Beddome, who had originally agreed with Cleghorn's views on the need for more extensive control, backtracked, which later drew down upon him the opprobrium of professional foresters,[60] though perhaps Beddome was always more interested in his own botanical and zoological research than in forest administration. It was only in 1882 that a change in attitude of the Madras Government came when they made their own Forest Act based on the 1878 national one that was by then in force.

In reply to his final report the Madras Government, when formally thanking Cleghorn for his immense work, this subtle difference in their attitudes is at least hinted at:

> Dr Cleghorn may well look with satisfaction on the results of his labors, and the Government desire to record here the high estimation in which they hold his services, and those of his able Deputy, Captain Beddome, and his other zealous subordinates, and their full appreciation of the extent to which the country is indebted to him and them for the advantages secured, both past and prospective. They have an additional satisfaction in the assurance they entertain that these results have been attained without any infringement of privileges which deserve to be respected, and that the real interests of individuals, no less than those of the community, have been promoted by the labors of the special department.[61]

On 11 October 1867 Cleghorn was granted sick leave for 20 months, so that although he left India shortly thereafter, his official Indian service would not end until November 1869 when he retired with the position of Deputy Inspector-General of Hospitals, the penultimate stage of the Medical Service hierarchy.[62]

10
Conjugal Reunion & a Grand Tour

For the fifth and last time Cleghorn returned to Europe from India, his third by the 'overland' route via the Red Sea and Egypt. In Egypt he crossed paths with William McIvor returning to India from leave bearing a consignment of fish for the Ooty Lake.[1] On the delightful island of Malta, after a separation of two years and two months, he met up with Mabel who had been 'in weak health in Edinburgh'. They planned to spend six months in Europe as Mabel, no doubt relieved to be missing an East-coast Scottish winter, required to travel in 'easy stages by land'. This was by way of a latter-day Grand Tour, on some of which they were following in the footsteps of his grandfather 80 years earlier. Like his grandfather Cleghorn also wrote a report on his observations in Malta, though not on its politics; Cleghorn was still on half-pay and while in Europe he made notes on natural resources and other aspects relevant to Indian forestry that he would in due course report to the Secretary of State for India.[2] It was not all play but there were certainly none of the boozy Masonic jollifications such as the one in which his grandfather had participated in Syracuse in 1790; and on Malta they were free to explore rather than being quarantined in the Lazaretto. On his own European tour Cleghorn assiduously visited universities, libraries, botanists and botanic gardens, but also missionary establishments and took the opportunity to attend significant meetings relating to Indian forest matters.

Malta

In Valletta the reunited couple stayed throughout December and January at Morells' Hotel, 150 Strada Forni, now Old Bakery Street, one of the three major streets of the fortified city, only two blocks away from three of the great buildings erected by the Knights of St John – the

Fig. 51. Temple in the garden of Villa Frere, La Pietà, near Valletta, visited by Cleghorn in 1867. (Author's photograph, February 2015).

CHAPTER 10

Grand Master's Palace, the Library and the Cathedral.³ Although this small Mediterranean island (only 17 by 9 miles) was a long way geographically from the Subcontinent, links with India were there to be found. They visited the terraced garden of Lady Hamilton Chichester, the woman who introduced lace-making to Malta from Genoa (Fig. 51). She lived in a house called the Villa Frere in the suburb of La Pietà previously owned by John Hookham Frere, her aunt's husband, who was also uncle both to Sir Bartle Frere and to Hatley Frere whom Cleghorn had known in Malabar (*Cleghorn Collection* Fig. 27). Cleghorn hoped to meet up with Lord Harris who was also travelling for the sake of his health, and the Governor and Commander-in-Chief of Malta for the last year had been Sir Patrick Grant, whom, when in a similar position at Madras, Cleghorn had accompanied for his forestry-exploration expedition to Burma of 1857. Grant was concerned to improve the agriculture and impoverished forestry of Malta and offered support for the report on the agriculture of Malta into which Cleghorn stuck his teeth. At this stage the possibility of a return to India was still not entirely ruled out, though his thoughts were turning more towards a botanical chair in one of the 'minor universities' (Oxford, which had just come up through the death of Charles Daubeny, he not unreasonably considered beyond his reach). As a staunch adherent to Reformation principles Cleghorn (and doubtless Mabel) took a lofty view of the colourful 'Romish festivals' they encountered on the island, as also of the very idea of:

> a large statue representing a man in the fire of purgatory and on the opposite side of the road ... a tablet indicating that if the traveller repeated so many Ave Marias and placed so many coins into a box under the tablet that he would receive <u>Indulgenzia [sic] plenaria</u>. We may bless God for our privileges in England & that we have been taught that salvation is the free gift of God, obtained without price!⁴

But their hotel was close to two potential sources of Protestant comfort: at the western end of the street was the Church of Scotland Kirk of St Andrew and if this offended their Free-Church principles the couple could have attended the much closer Anglican cathedral, opened in 1844 and paid for by the dowager Queen Adelaide. Cleghorn's report, which he read to the

Fig. 52. *Narcissus tazetta*, Island of Gozo. (Author's photograph, February 2015).

10. EUROPEAN GRAND TOUR, 1867–8

Fig. 53. 'Phoenician Ruins in Malta'.
The temple of Hagar Qim (3600–3200 BC), Malta, visited by the Cleghorns. Anonymous wood engraving, *Illustrated London News*, 21 November 1868.

Botanical Society soon after returning to Scotland on 11 March 1869,[5] is by a long way the most interesting paper he ever wrote, one that would have given pleasure to his father had he still been alive to read it. It shows Cleghorn at his best, doubtless inspired by his joy at being once again with Mabel, but still anxious to work. It is in the old Scottish tradition of statistical surveys, recording soil, climate, water supply, agriculture and botany, but revealing his own particular slant using a bibliographical approach, citing Classical authors including Ovid and Cicero up to contemporary botanical writers. It was too early in the year for the most colourful flowering season, but they saw almond blossom and native Mediterranean species such as the fragrant, gold-and-white *Narcissus tazetta* (Fig. 52) and the yellow celandine-like flowers of *Ranunculus bullatus*. As ever, Cleghorn had an eye for introductions, 'which obscure the original Flora Melitensis' and he noted two South African species – the then recently naturalised *Mesembryanthemum deltoides* and the colourful yellow *Oxalis cernua* (now *O. pes-caprae*, the Bermuda buttercup). The latter now almost completely carpets the fields and waste places of the island and had been introduced in 1811, a date Cleghorn learned from the *Flora Melitensis* of 1853, which records 716 species, the work of the only local botanist, Giovanni Carlo Grech Delicata, who combined the profession of 'avocat' [lawyer] with lecturing on Natural History to 10 or 12 students in the University. The Cleghorns visited the major horticultural sights of the island – the Mall and Argotti gardens in Floriana, a suburb of Valletta, and at Attard the St Anton gardens around the Governor's summer palace. The Mall was where the Knights of St John had in the mid-seventeenth century played an early form of croquet called 'palla a maglio', and had been turned into a garden by the island's first British Governor Sir Alexander Ball in 1805. In 1855 Stefano Zerafa, the professor of Natural History had turned the eighteenth-century

CHAPTER 10

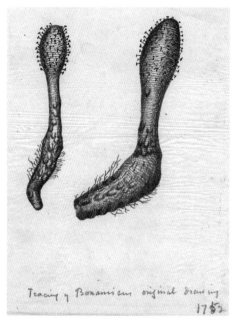

Fig. 54. 'Fungus melitensis' (*Cynomorium coccineum*).
Copy of Buonamici's manuscript made by Cleghorn in the Public Library, Valletta (above).
The Fungus Rock, once guarded by the Knights of St John off the west coast of Gozo. (Author's photograph, February 2015) (below).

Argotti Gardens into a botanical garden.[6] At St Anton Cleghorn noted Australian trees sent by Hooker from Kew and measured trees including stone pine, cypress, olive and Persian lilac that already had girths of six feet in only fifty years. There were expeditions to St Paul's Bay, where the Apostle landed after a shipwreck in AD 60, and Cleghorn recorded plants of the coastal zone including the medicinal squill *Urginea scilla*; here his interests broadened to include the zoological and he made a collection of shells.

198

Following the Grand Tour tradition, antiquities and architecture were not neglected and they visited the remarkable Megalithic temples (dating to around 3500 BC, but then believed to be Phoenician) of Hagar Qim on the south coast of the island near Qrendi (Fig. 53), and in Valletta the handsome Renaissance Palace of the Grand Master of the Order of the Knights of St John. Here, in the Sala del Piccolo Consiglio, Cleghorn particularly admired the set of 1697 Gobelin tapestries known as the 'Tenture des Indes'. Showing lush tropical scenes, he made the intriguing speculation that the plants depicted could only have been based on plates in the *Hortus Malabaricus*. A nice thought that Rheede's engravings might have been used as a visual source for the decorative arts, but unfortunately not the case. It was not then known, but these tapestries (of which at least eight sets were produced, over several decades) were based partly on paintings made in Brazil by Albert Eckhout, which accounts for the inclusion of a tapir and a llama; but as a zebra, a rhino, an elephant and some ostriches are also woven into the designs, it is evident that the French artists who prepared the cartoons added to Eckhout's drawings from other sources, with a regrettable disregard for biogeography. In Valletta Cleghorn also visited the University museum and in the Public Library took particular note of early botanical works and a fine shell collection.

The botanical curiosity most closely associated with Malta historically is the bizarre parasitic angiosperm *Cynomorium coccineum*, which, in fact, is widespread in the Mediterranean and occurs as far east as Afghanistan. The plant has no chlorophyll or leaves and resembles nothing so much as a reddish-brown turd thrusting vertically from the ground; it grows on saline soils and parasitises members of the families Cistaceae and Chenopodiaceae. Cleghorn studied descriptions of the plant in the Public Library – in a late seventeenth-century manuscript by Giovanni Francesco Buonamici (who called the plant *Fuco spicato coccineo Melitensi*) and in the published work of Paulo Boccone who named it the *Fungus melitensis*. Its *locus classicus* was a large rock off the west coast of the adjacent island of Gozo, once jealously guarded by the Knights of Malta for the plant's supposed medicinal properties, amongst which, based on the doctrine of signatures, was its use as a Medieval and Renaissance viagra, though this had ceased by Cleghorn's time (Fig. 54). Cleghorn appears to have visited Gozo himself and in 1882 he presented a specimen of the plant 'gathered in a cave on the seashore of the Island of Gozo' to the Botanical Society of Edinburgh; his transcript and tracing of the drawings of the Buonamici manuscript have survived in his illustrations collection, but not the specimen.

Sicily

Another old India hand was within easy reach of Malta: Henry Yule, who with his invalid wife Anna Maria, had been living in a house in the Giardino Inglese at Palermo for several years. On 11 February 1868 the Cleghorns therefore made the nine-hour voyage, on a postal steamer *Cariddi*, from Valletta northwards to Syracuse for a twelve-day visit to Sicily, a land he had known of from childhood.[7] He had probably heard tales from his grandfather, but he also knew it from schoolboy readings of Cicero, Thucydides and Strabo, and the botanical works of Dioscorides and Theophrastus, and as an adult from the travel literature of writers including Patrick Brydone.

The paper on the agriculture of Malta was therefore extended to include Sicily. The Cleghorns remained on *Cariddi* as it steamed its way to Palermo, with only short pauses at Syracuse and Catania, and an overnight stay at Messina. But there was enough time to pay the usual visits to cultural and educational establishments. In Syracuse he pondered on St Paul's visit, where, having survived shipwreck in the Mediterranean, the Apostle recuperated before continuing his journey as a prisoner to Rome to plead his case before Nero. The antiquarian Protestant in him regretted

CHAPTER 10

Fig. 55. 'Scilla & Charybdis', the Straits of Messina, through which the Cleghorns sailed on their way to Palermo. Steel engraving by A.H. Payne, 1840.

the adaptation of the ancient Greek temple of Minerva into a Catholic cathedral and though he visited the botanic garden there was no time to pay his respects to the papyrus at either of its well-known localities, the Fountains of Arethusa or the banks of the river Cyane. Sailing up the east coast there were views of the elegant volcanic cone of Etna, and at Catania they visited the Museo Biscari (filled with Classical antiquities), the library of a Benedictine convent, and Cleghorn was shown round the university's Orto Botanico by its director, Francesco Tornabene. The garden was little more than a decade old, but Cleghorn was impressed with the handsome, Neo-Classical, semi-circular classroom (designed by Mario Distefano, where 84 pupils were being instructed in agriculture and botany), attached to which was a library, herbarium and seed room. Back to the boat and on to Messina, which they reached late at night. The following morning they visited the Cathedral, the University and the Nunnery of San Gregorio, and then sailed through the two-mile-wide Straits of Messina with their infamous rocks and whirlpools of Scylla and Charybdis before reaching Palermo – appropriately, for what must have seemed like a second honeymoon, on St Valentine's Day (Fig. 55).

Cleghorn and Yule had been at school together at the High School in Edinburgh. Since then Yule had had a distinguished career in India, as a Bengal Engineer, best known for his work on irrigation with Richard Strachey, and for his part in Colonel Arthur Phayre's mission to Ava in Burma in 1855, on which he had published a book illustrated with his own drawings. Yule was a brilliant scholar and linguist and his reasons for living in Sicily were not only for the sake of his wife's health, but for access to manuscripts of early Italian travellers to Central Asia and the Orient. For several years he had been editing such works for the Hakluyt Society (*Mirabilia Descripta*, 1863 and *Cathay and the Way Thither*, 1866). At the time of the Cleghorns' visit Yule was working on his magnum opus, which appeared three years later in 1871 in two substantial volumes as *The Book of Ser Marco Polo* (for which Cleghorn provided a single footnote on the

10. EUROPEAN GRAND TOUR, 1867–8

Mysore name of *Gossypium religiosum*). While in Palermo the men must have discussed George Marsh's *Man and Nature*, as there are annotations in Cleghorn's copy of the book representing observations by Yule (though not in his hand). In Sicily the Yules received other distinguished visitors, not least among whome was John Ruskin, who took a particular interest in their spirited daughter Amy Frances, who herself became a distinguished linguist and bibliophile. The year after Ruskin's visit of 1874 Yule's wife died and he returned to England, and it was there that, with A.C. Burnell, he compiled the Anglo-Indian dictionary for which he is best remembered – *Hobson-Jobson* (1886). Another connection is that Yule was both painted and drawn by T. Blake Wirgman, which may have been one reason for Cleghorn using the same artist when he later came to have his own portrait drawn.

The timing of the visit coincided with a period of considerable political instability in Sicily – it had been conquered by Garibaldi only eight years earlier and became part of the Savoy Kingdom of Italy under Vittore Emmanuele II the following year, 1861. In 1866 there had been a rebellion, firmly crushed by the Italian army, and the island was still under martial law. To protect them from potential (though by this stage unlikely) guerrilla attacks on excursions from Palermo, Yule had arranged a military escort of Bersaglieri, whose hats were fetchingly decorated with long black plumes (traditionally taken from unfortunate capercailzies). In the environs of Palermo Cleghorn was shocked at the dilapidation of the palaces of the old aristocracy, though La Favorita, built 'under the precipices of Monte Pellegrino' by Ferdinand I (of the Two Sicilies) in 'grotesque Chinese style' in 1799, still had a fine garden. In it grew two of the plants in which the economic botanist took greatest interest – the sumac (*Rhus coriaria*) and the manna ash (*Fraxinus ornus*). Yule's linguistic skills were enrolled to translate a paper on the cultivation of the former, which was used for tanning (most famously for the leather of Cordoba and Morocco). Cleghorn made a detailed study of the cultivation of the ash – not the ground-hugging manna of the ancient Israelites, but a sugar-rich sap obtained by tapping the tree trunk like a maple. Cleghorn was always interested in educational establishments and was impressed by the achievements of the botanist, agronomist and mycologist Giuseppe Inzenga, who, since 1847, had been director of the Instituto Agrario at Castelnuovo. In a handsome Doric-colonnaded building, on a 30-acre site, 32 students (eight of whom were taught gratis) studied 'geometry, botany, physical science ... drawing, and ... practically in the fields the science of agriculture and the management of farm stock, as well as arboriculture in the woodlands'. Also in the environs of Palermo Yule pointed out the 'Papireto' where papyrus had once grown, though the swamp had been drained for many centuries. Cleghorn realised that in Sicily the giant sedge must have been an ancient introduction, being mentioned for the first time only in the tenth-century writings of Ibn Haukal, as recently revealed in a translation by Parlatore.

There was plenty to see in the city of Palermo itself. The tiny Orto Botanico (a mere 270 x 90 yards), founded in 1779, was under the direction of Agostino Todaro, though at this time the huge Moreton Bay fig (*Ficus macrophylla*) that now dominates the garden, with its huge buttresses and aerial roots, was only just getting established, having been planted in 1845. Here again Cleghorn was impressed by the class-room and, as he had at Catania, appreciated the busts of ancient botanists. In Palermo, as well as including the local Boccone, the presence of the Englishmen Ray and Dillenius came as a pleasant surprise. Also on display was a photograph of the contemporary Melbourne botanist Ferdinand von Müller – Todaro was keyed into the global plant-exchange network, and was receiving plants not only from Australia, but from New Zealand, the Cape of Good Hope and, while Cleghorn was there, a consignment arrived from Thomas Anderson in Calcutta. In Palermo Cleghorn loved the shops of the dealers in *erbi medicinali,* which chiefly sold members of the family Labiatae (*Origanum, Thymus, Nepeta*) and Malvaceae (*Althaea, Malva*) but also some obsolete relics including the *Fungo di Malta*. In the city visits were also made to the university and the public library. The Regio Studiorum Universita, 'a massy pile of a building', had a fine library and natural history collection in which the fossils 'from Sicilian caves' had been named by another Scottish EIC surgeon-naturalist, Hugh Falconer. The Librario del Commune held 75,000 volumes and Cleghorn compiled a bibliography of works relating to Sicilian botany and geology, but he was also pleased to see a copy of the *Hortus Malabaricus* and a run of Curtis's *Botanical Magazine*.

Italy

After ten days with the Yules it was time to start the journey north and on 24 February, on 'Florio's steamer' *Napoli*, Hugh and Mabel sailed from Palermo to Naples, which they reached in time for Mabel's thirtieth birthday on the 25th and were treated to a display of natural pyrotechnics, Vesuvius having been in a state of eruption since the previous month.[7] In Naples Cleghorn was impressed by the Orto Botanico, which had been founded under French rule in 1809, and was shown to him by Professor Vincenzo de Cesati with his assistants Giuseppe Pasquale (who the previous year had published a catalogue of the garden enumerating 6738 species) and Gaetano Licopoli. The garden received a generous annual grant of £400 from the Italian Government and the Neapolitan municipality. The climate was harsher than that of Sicily, and some of the Indian plants including *Caryota urens, Cerbera odollom* and patchouli that in Palermo could be grown outdoors had here to be grown under glass. As usual Cleghorn was interested in introduced plants and there were two areas for acclimatization – one for palms and bananas, the other for exotic conifers (North American – *Sequoia*; Japanese – *Cryptomeria*; and Australasian – *Agathis* and *Araucaria*).

Next stop on this Grand Tour was, inevitably, Rome, which was still (with French assistance) under the rule of the Pope, as part of the Papal States; not for another two years did it become part of the Kingdom of Italy, and the national capital not until 1871. Meanwhile, with the southern migration of the House of Savoy from Turin, it was Florence that was the Kingdom's temporary capital. The Cleghorns made the seven-hour train journey from Naples on 2 March and stayed a night in the Hôtel d'Amérique before transferring to the 'comfortable boarding house' of Madame Sopranzi at 25 via dei Cappuccini for three weeks. Hugh was able to catch up with his sister Isabella, who was by now more or less based in the Eternal City. Armed with Murray's *Handbook*, reminiscences from Classical authors from 'High School days', and Byron's *Childe Harold*, they visited the sights. The tower of the Capitol was out of bounds so that extensive views over the city were obtained from the summit of the Pincian Hill and the top of the Colosseum. They visited the recently excavated Palace of the Caesars, and in the Colosseum wallflowers and fennel were in flower, but the scraping of the ruins had already started and there can have been nothing like the 420 species within its six acres that Richard Deakin had recorded in 1855. Rome is renowned for its gardens and the couple visited those of the Vatican, and of the Palazzi Colonnae and Raspiglioni. The lawns of the Villa Borghese were attractively spangled with *Anemone coronaria* and there were 'merry fountains playing charmingly'; those of the Quirinal Palace, despite the wonder of an 'organ played by water', were deemed 'very stiff and formal'. In fact Cleghorn was generally underwhelmed and found the city 'deadening wearisome', being more impressed by the 'concourse of cardinals & Swiss guards in full dress' than by the Pope, a 'trembling old man' of 76. Pius IX remains to this day the longest-reigning pope, and it was in his pontificate that two of the Roman Catholic doctrines to which Cleghorn, as an evangelical Protestant, would have taken the greatest of exception were pronounced – the Immaculate Conception of the Blessed Virgin Mary and, at the First Vatican Council that began the year after Cleghorn's visit, that of Papal Infallibility.

Florence & G.P. Marsh

From Rome Cleghorn had written proposing a visit to George Perkins Marsh, the American plenipotentiary in Florence, introducing himself with the statement that he had carried *Man and Nature* (Marsh's great conservation polemic of 1864) 'along the slopes of the Northern Himalaya, and into Kashmir and Tibet'.[9] Marsh (1801–1882) was a polyglot scholar, a linguist and philologist, familiar with 20 languages; he was also a diplomat, first to the Ottoman Court, and since 1861, appointed by Abraham Lincoln, to Italy.[10] He was initially based in Turin, from where on sojourns to the Riviera and Piedmontese countryside he had composed *Man and Nature*, but in 1865 he had moved with his wife to the Villa Forini in Florence where he remained for the rest of his life, though he spent winters in Rome after the capital moved there in 1871. When they met the two men and their wives got on well and they would correspond from time to time over the following five years. In 1875, when Cleghorn returned to Italy to visit the forest school at Vallombrosa he would again see Marsh, who was in Rome at the time. Introductions to mutual contacts were made on this first visit. Caroline Marsh gave Mabel an introduction to the American gynaecologist Dr James Marion Sims, then based in Paris, and it was doubtless through her that five young American ladies were placed under Mabel's care; more significant, with his recent visit to Sicily in mind, Cleghorn gave Marsh an introduction to Henry Yule: 'I felt sure that you were kindred spirits & would value one another'.[11] The linguist Marsh shared many interests with Yule and this was the start of a more frequent and substantive correspondence than that with Cleghorn. Marsh provided Cleghorn with literature on Italian forestry who in return said that he intended to send copies of *Man and Nature* for the use of forest assistants in India. It was Yule, however, who later went as far as preparing an Indian edition of *Man and Nature*, published in Madras in 1882. It wasn't only the presence of Marsh that for Cleghorn made Florence a much more interesting place than Rome – they stayed in the Hotel de Milan,[12] and found the city a 'fine, growing, bustling town' in which he spent more time in the 'Halls of Science than the Galleries of Art'. Perhaps Mabel devoted more time to the latter, though Cleghorn was terrified for her health, which required 'extraordinary care ... & my great business is to prevent her from writing & standing or sitting in galleries – Sightseeing & health go together to a very limited extent'. The leading Italian botanist of the day was Filippo Parlatore, who in 1842 had been appointed by Leopold II, Grand Duke of Tuscany, as Professor of Botany at the renowned Museo di Storia Naturale, founded in 1775 and known as 'La Specola'. Here Parlatore established what amounted to a national herbarium and was at the time engaged in writing his *Flora Italiana*. Cleghorn attended some of the professor's lectures on geographical botany, those on the Swiss Alps calling to mind the Himalayas. He also attended a Conversazione of naturalists at the Museo, today best known for its remarkable collection of wax anatomical teaching models, which must also have been of interest to Cleghorn the medic. Another attraction of Florence was the cheapness of its antiquarian books.

CHAPTER 10

Switzerland & France

On 6 April, having taken a steamer from Livorno, the Cleghorns were in Marseille from where they travelled to Switzerland via Lyon. They visited the botanic gardens of Geneva (but failed to meet Alphonse de Candolle), Berne and Basel, but in the latter city, by now safely back in a Protestant country, they attended the Église Consistoriale Évangélique and Cleghorn's missionary interests resurfaced.[13] This was the headquarters of the Basel Evangelical Missionary Society (founded in 1815) whose Inspector, the Rev. Joseph Josenhaus, he visited. Here too were Indian links as the Society had, since 1834, run a mission station in Mangalore with which Cleghorn was familiar, and which had been the starting point for Sebastian Müller, his forest assistant in Canara. In Basel it was thus possible to reminisce with Dr Hermann Gundert, editor of *Missions Blatt*, over the days when they had 'cooked our rice together in Malabar'. In Switzerland Cleghorn also studied the communal forests around Berne and Geneva, and observed the spring activities where (as in the Himalayas) pine logs were brought down in torrents by the power of melting snow and where, as in the 'deep chasm valleys of the North-West Himalayas', it was necessary strictly to preserve the forests on the ridges 'as "boulevards" against avalanches and cataclysms'.[14]

Fig. 56. Gateway of the École Nationale des Eaux et Forêts, Nancy, through which Cleghorn must have passed in April 1868. (Collection AgroTechParis – Nancy Centre).

En route for Paris Cleghorn studied the forests between Basel and Nancy, stopping at the latter for six days on 19 April. Here he visited Henri Nanquette, director of the École Nationale des Eaux et Forêts, his assistant Auguste Mathieu, and the five British students chosen by Brandis who were already studying there (Fig. 56).[15] This appears to have been Cleghorn's only visit to an institution that would play a leading role in the training of probationers for the Indian Forest Service over the next 17 years. Cleghorn noted the strict discipline imposed from which the British students were exempted only from the wearing of uniforms (Fig. 58); Nanquette told him that the students were 'bien satisfait', that they were happy, and he alluded to 'the "bonté" shewn to them by the director and professors'. On the other hand Cleghorn considered that the stringency of the French rules on trespass and forest offences would not be applicable in India due to the 'mild provisions of the Indian Forest Act'. In Paris, where they stayed from 20 to 25 April, Cleghorn indulged in his usual combination of intellectual visits and retail therapy – he renewed the acquaintance of Decaisne at the Jardin des Plantes, visited a bookseller called Savi, and Vilmorin the seedsman. The eighteenth-century firm of Vilmorin-Andrieux specialised in agricultural crops and hybridization, so Cleghorn may already have been planning what he would do with the fields more than the gardens of Stravithie. At the l'Office d'Administration des Forêts he met Jules Clavé, author of the influential *Étude sur l'Économie forestière* (1862), which curiously was not in Cleghorn's well-stocked library, but was quoted by Marsh. Mabel, meanwhile, took up Caroline Marsh's introduction to Dr Sims, but he was unable to treat her unless she stayed for two or three months.[16]

11
Retirement in Britain

The Cleghorns returned to London at the beginning of May 1868. As usual Hugh reastablished himself with the botanical community, visiting Kew, and attending a Linnean Society soirée where he met the new doyens of Indian botany Joseph Hooker and Thomas Thomson. There were meetings at the India Office with Sir Stafford Northcote, Secretary of State for India under Disraeli's Conservative government,[1] and with Sir James Fergusson of Kilkerran.[2] Although at this point Fergusson was an Under-Secretary in the Home Office, he had until recently been in the India Office where he had been succeeded by Lord Clinton to whom Cleghorn submitted the report on his European tour.[3] These meetings almost certainly concerned the selection and training of probationers for the Indian Forest Department, for which examinations were to take place the following February (by which time there had been a general election, and the Duke of Argyll had replaced Northcote as Secretary for India in Gladstone's Liberal administration). Also while in London Cleghorn was, with Forbes Watson, a member of a committee for 'purchasing return presents for Indian Princes who [had] sent contributions to the Paris Exhibition', which met at the Crystal Palace.[4]

The Glasgow Chair Debacle

After a week in London Cleghorn returned north where the serious illness of George Walker-Arnott (impacted gallstones, severe jaundice, and a swelling in the throat), just as he was about to start his summer botany course in the University of Glasgow, offered a new opportunity.[5] Cleghorn stood in for Arnott, and between 25 May and the end of July delivered the course to more than 90 students, using Sir William Hooker's teaching diagrams, which Arnott had purchased for £200 on taking up the chair in 1845. When Arnott's illness proved terminal, Cleghorn applied to succeed him. Such appointments were messy affairs, involving patronage and politics as well as ability on the part of the candidate. Part of the application process involved soliciting testimonials, which the candidate had to get printed and circulated. The process was distasteful to Cleghorn but the list of references he collected was an extremely impressive one.[6] The 30 authors include virtually all the leading British botanists (Hooker, Babington, Bentham, Oliver *et al.*), senior academics (including Christison, Maclagan, Lyon Playfair and Balfour in Edinburgh, and the London zoologist Richard Owen), and an impressive tally drawn from the highest ranks of the Indian administration (Sir Bartle Frere, Sir Robert Montgomery, Sir Walter Elliot) including one who had acted as Viceroy (Sir William Denison). Others had Indian links notably Forbes Watson of the India Museum and Clements Markham. With several years of teaching experience in Madras, in-depth experience of economic botany, a creditable list of publications, and public testimonials from the Government of India, Cleghorn's application would appear to have been extremely strong. Glasgow at this time was one of Britain's greatest ports and industrial centres, and that several of the supporting statements stressed the value of his economic experience in a great 'commercial centre', suggests that Cleghorn himself suggested this as his trump-card when requesting the references.

Other strong candidates at least rumoured to be applying were the Rev. John Croumbie Brown (of whom more anon), Dr George William Davidson (lecturer in Botany at the Edinburgh School of Arts), Dr Robert Beveridge (pathologist at the Aberdeen Royal

Infirmary, formerly botany lecturer at Marischal College) and Dr William Ramsay McNab (son of James the RBGE Curator).[7] The post, however, went to the 32-year-old Alexander Dickson, who had held the Trinity, Dublin botanical chair for only two years. Dickson was a skilled morphologist, with post-graduate German experience, which suggests that Glasgow was looking for a research scientist and that by this time academics had more say in the appointment than patrons or outsiders.

British Association Meeting at Norwich

Between 19 and 26 August 1868 a large body of savants descended on the ancient cathedral city of Norwich in East Anglia – not only for the 38th annual meeting of the British Association, but also for an International Congress of Pre-historic Archaeology. The Presidents of the two gatherings, Joseph Hooker and Sir John Lubbock respectively, happened to be amongst Darwin's closest personal friends and supporters. With various Scottish colleagues including Walter Elliot, Cleghorn attended his fourth BA meeting. Although never a life member of the Association, nor entitled to a copy of the Annual Report, he was a 'Subscriber' for the rest of his life. As will be seen later, he was certainly a committee member of a BA-funded project on Scottish deforestation/rainfall in 1873, and in 1885/6 and 1887/8 was a member of the committee on W.C. McIntosh's work on food fish. Given his sociability it seems likely that he attended the annual meeting on many more occasions than these, especially when it was held within reach of Edinburgh or St Andrews, but for which no record exists in the Annual Reports because he did not speak.

Hooker's Presidential address in 1868 was a remarkable one,[8] which started with a nostalgic look back to his own first BA meeting of 1838 (which lay behind his great Antarctic adventure); it proceeded to a subject that had arisen at the previous year's meeting in Dundee – the recommendation of a large-scale survey of Indian anthropology, which had ended up being limited to an investigation of the tribes in the Khasia Hills who still constructed megaliths – dolmens and cromlechs – a subject on which Cleghorn's friend Henry Yule had published in 1844, and which Hooker had himself seen in 1850. Then followed a discussion of museums (local ones, the British Museum, and a plug for his own one of economic botany at Kew) and their role in biological education, after which came a review of the most important strides in botany over the previous decade, which he considered to have been in the fields of 'physiology' and palaeobotany. In the latter the revelation had been that some (extinct) fossils could be referred to extant plant families, though Hooker dubbed the field as 'the most unreliable of sciences … [in which] we do but grope about in the dark'.

The review of 'physiology' was an enthusiastic summary of Darwin's botanical work on cross-fertilization (in orchids and, by heterostyly, in various other families); his theory of pangenesis; his work on climbing plants and his latest work on domestication. Viewed *in toto* these were 'serried ranks of facts in support of his theories which … may well awe many a timid naturalist into swallowing more obnoxious doctrines than that of Natural Selection'. However, they had not only theoretical but also practical uses: 'What Faraday's discoveries are to telegraphy, Mr. Darwin's will assuredly prove to rural economy'. Hooker took to task a recent article in the popular periodical *The Athenaeum* which had claimed that – a decade on from the appearance of *Origin* – Darwin's theories 'were a thing of the past … rapidly declining in scientific favour'. According to Hooker 'so far from Natural Selection being a

11. RETIREMENT & THE INDIA OFFICE, 1868–88

thing of the past, it is an accepted doctrine with almost every philosophical naturalist'. It was possible to say this in the light of two recent and influential public acknowledgements – the acceptance of the theory by the famously cautious George Bentham in his Linnean Society presidential address of May 1868, and Lyell's re-writing of his *Principles of Geology* in its tenth edition (vol. 1 1867, vol. 2 1868). Hooker did, however, have to acknowledge a significant criticism of Natural Selection – its 'cleverest critique' – that had recently appeared in the *North British Review*, but was not yet known to be the work of Henry Fleeming Jenkin, Professor of Engineering at Edinburgh University. Somewhat cheaply Hooker accused the author of being hoist by the petard of his own self-admitted ignorance of biological facts, and took the author's most serious criticism to be that of the age of the earth. Being still too uncertain, Hooker would not allow this as valid (and Lord Kelvin's oldest estimate of 400 million years would, indeed, prove to have been wrong by a factor of ten), but he failed to pick up on Jenkin's most serious and astute criticism – that natural selection could not work under the operation of a blending inheritance such as pangenesis.

In many respects the most remarkable part of the address was the concluding section on the relation between science and religion. Hooker was notoriously cautious about expressing views on religion in public – and even his private statements are ambivalent. This speech seems to come the closest of any of his writings to revealing the nature of his belief. This is shown by his quoting, approvingly, a longish poem by his own cousin Francis Turner ('Golden Treasury') Palgrave, which appears to support a Cartesian dualism of mind and matter, both of which are under the control of an ultimately unknowable Creator. Before this poem, which forms a coda followed by no further discussion, Hooker had indicated that he considered science and religion to be separate realms, each with its own legitimate area of investigation that should be undertaken in a mutually respectful manner, not by turning the telescope the wrong way round, for each to diminish the role of the other by making it look small (a favourite tactic of the modern television atheist). Hooker also had something fascinating to say about the limitations of Natural Theology – that straw at which many have clutched in attempts to reconcile science and religion – which he denoted as 'that most dangerous of all 2-edged weapons ... to the scientific man a delusion, and to the religious man a snare, leading too often to disordered intellects and to atheism'. So perhaps Hooker, even although he certainly did disagree with some of the dogma of the Anglican church, was slightly less agnostic than authors such as Jim Endersby have suggested.[9] Cleghorn must certainly have pricked up his ears at this part of the address, though, unlike his friend J.H. Balfour, he himself never did play the Natural Theology card, content to use Biblical authority as the chief, if not the only, source of Divine revelation. Balfour's views on Darwin (which may or may not have been similar to Cleghorn's unexpressed ones) have been discussed on p. 120.

There were several Scottish botanical contributions to Section D. J.H. Balfour reported a new *Hieracium* (now known as *H. caespitosa* subsp. *colliniforme*) discovered between Selkirk and Philiphaugh; Thomas Archer, Director of the Edinburgh Museum of Science and Art (a keen economic botanist), described the germination of seeds of the crucifer *Erysimum* (now *Conringia*) *orientalis* in a hotbed, in soil taken from a garden in the Old Town of Edinburgh and Alexander Dickson spoke on 'some of the principal modifications of the receptacle, and their relation to the "insertion" of the leaf-organs of the flower'. The longest of these, however, was Cleghorn's own paper relating to his concerns of recent years, on the 'Distribution of the principal timber trees of India and the progress of forest conservancy', which summarised the work undertaken in the various regions of the Subcontinent in the wake of the 1865 Forest Act.[10]

It was not only a case of listening to papers and addresses, there was also a significant social side to such gatherings. The biological section went on an excursion to Kings Lynn, on which they saw one of the East Anglian fen rarities, the fern *Dryopteris cristata* (Fig. 57). After a luxurious luncheon in the Assembly Rooms Cleghorn, rather curiously for such an unmilitary figure, replied to a toast 'on behalf of the Army'. The party then went to the Guildhall to view a collection of antiquities and to visit some of the churches, after which the botanists went to the town museum before returning to Norwich by the 6.30 train.[11]

Fig. 57. Specimen of *Lastraea* (now *Dryopteris*) *cristata*, collected near Kings Lynn, from J.H. Balfour's herbarium (RBGE).

Work for the India Office

In London in the summer of 1868, before the start of his retirement and while he was still on sick-leave and half-pay, Cleghorn's experience was exploited by Sir Stafford Northcote, Secretary of State for India, in helping to select young men for training in France and Germany prior to their departure as junior officers in the Indian forest service. Cleghorn would perform the same task each spring for the next twenty years. Although mostly undertaken in 'retirement', this can be seen as one of Cleghorn's major contributions to Indian forestry. Cleghorn's combination of personal experience of what was required of individuals proposing to live in tough environments, and of the methods developed by Brandis, gave him a unique insight that allowed him to choose the very best candidates.

The system of Continental training for senior forest officers for India had been devised by Brandis on his European leave of 1866 and the first batch of students were by now half-way through their training in Germany and at Nancy (Fig. 58).[12] Details of Cleghorn's involvement in the selection process are known both for his early days,[13] and his final year in the role two decades later. In 1868 he selected eight men to start training in 1869, four in France and four in Germany, and in 1869 seven for France and five for Germany. The studies of both these batches would suffer an unexpected disruption with the outbreak of

11. RETIREMENT & THE INDIA OFFICE, 1868–88

the Franco-Prussian War in July 1870, which meant that the French students had to return to Britain. After completing their Continental studies, six of Brandis's class of 1867–9 returned to Britain for two months of practical training in Scottish forests – perhaps as a sop to critics like Balfour. Making use of his Scottish contacts Cleghorn arranged placements for these men at either of two of the great Scottish forestry estates – under William McCorquodale, forester to the Earl of Mansfield at Scone Palace in Perthshire, or with Grant Thomson, forester to the Earl of Seafield at Castle Grant in Strathspey. Cleghorn had probably first visited these estates with Brandis in July 1865 and he certainly visited Strathspey the following July (1869) with J.L. Stewart.

In the case of the Nancy refugees further intervention was called for and Cleghorn clearly felt a sense of responsibility for these young men at the start of their intrepid careers. He had already been asked to lecture on botany at St Andrews University in the winter term of 1870,[14] and suggested that the seven students he had selected that year (who were still doing practical work at Haguenau), should come to St Andrews to attend these and other science lectures, then be sent to McCorqudale or Thomson the following spring. Three of the senior year (including James Sykes Gamble) he offered to put up at Stravithie, where they could attend the lectures but also help practically on his own estate, and make use of his amply stocked library and herbarium.[15] This scheme was eagerly accepted by the Duke of Argyll (by now Secretary of State for India), and Cleghorn was paid a hefty £500 for his services; in December he duly delivered his 20 lectures on forestry for the refugees and other interested students. The ten displaced students returned to their studies after peace was re-established in the spring of 1871, though after this only one further batch was sent to Germany after

Fig. 58. French forestry methods in action. ?Lithograph by Théophile Schuler from the cover of Lucien Boppe's *Exposé des Faits Généraux Relatifs a la Production Forestière sous le Climat de la France* (Nancy, 1882-3) (RBGE: Major Frederick Bailey's copy).

which training was entirely at Nancy. In 1871 there were 67 applicants, 24 of whom got as far as taking the Civil Service exams. Of these only 11 passed in French, from whom Cleghorn chose three for France, one of whom was an Anglo-Indian called Fernandez who had previously studied at Deccan College, Poona.

James Sykes Gamble (1847–1925)

Gamble (Fig. 59), who had been selected for Indian service by Cleghorn, seems particularly to have benefitted from his hospitality in the winter of 1870/1 and might be counted as his most important protégé.[16] Over a long career Gamble was to make a major contribution not only to practical forestry in India, but also to the literature of forestry and to the taxonomic botany of the Subcontinent. Having been a fine mathematician at Oxford he only just failed the Indian Civil Service exam, but if the turn to forestry seems surprising it should be remembered that in German scientific forestry, mathematics was, in Brandis's words, essential for 'an officer who is expected to spend his life surveying land, measuring timber, and the like'. There might also have been an element of family tradition, as Gamble's Irish father Harpur had strong zoological interests and was (like Cleghorn, though 17 years earlier) a product of the Edinburgh medical school. After serving as a forest conservator (especially in Bengal – including Sikkim/Darjeeling – and from 1882 to 1890 in the Northern Circle of Madras), Gamble directed the Forest School at Dehra Dun 1890–9, and contributed substantially to botanical/forestry literature, both as an editor of the *Indian Forester* (1878–82, 1891–9) and as author of the standard *Manual of Indian Timbers*. This work, densely filled with information, was first published in 1881 and sprang from an interest of Gamble's kindled by the timber collection at Nancy, but, more immediately, by his work in selecting timbers to be sent to the 1878 Paris Exposition Universelle.[17] Among all his other activities Gamble found time to amass 'probably the largest private collection of Indian plants ever owned in India' – 50,000 sheets (now at Kew),[18] which he put to productive use in retirement in writing his authoritative *Flora of Madras*.[19] His other major contribution to systematics was his 1896 monograph of the taxonomically challenging bamboos of British India. The beautiful set of specimens of Indian forest trees that he made in Madras for the 1884 International Forestry Exhibition form another memorial to Gamble, linked as they are with Cleghorn's specimens in the RBGE herbarium (Fig. 60).

Fig. 59. J.S. Gamble. Photographer unknown, from Hill, 1926 (RBGE).

Fig. 60. Specimen of sal (*Shorea robusta*) collected by J.S. Gamble for the 1884 Edinburgh Forestry Exhibition – the generic name is in Gamble's hand and the printed cuttings from his *Manual of Indian Timbers*. After the close of the Exhibition the specimens were given to Cleghorn. (RBGE).

CHAPTER 11

Annual visits to London

Cleghorn's work for the India Office gave him an annual excursion to London, which he used to maintain links with metropolitan botanists and institutions, and was doubtless a welcome break from provincial life and lairdly/uxorial duties. In November/December 1869, as on later occasions, he stayed at the United Hotel in Charles II Street – a pleasant walk through St James's Park to the India Office in Whitehall.[20] On this occasion he took the opportunity to undertake a rare and final foray into botanical taxonomy – investigating two species of caper that he had known in South India and considered to have been confused – *Capparis divaricata* and *C. heyneana*. On 2 December, following a research visit to Kew, he read a paper on the subject to the Linnean Society, though this was not published in the Society's *Transactions*, surviving only in manuscript.[21]

The timing of Cleghorn's visits to the Metropolis, linked with the Civil Service examination system, seems to have changed over the years: in 1871 he was in London in March, and in 1888 in June. The report resulting from the latter, his last visit, is revealing,[22] not only in showing the considerable rigour of the selection process, but also for a charming vignette that says much about Cleghorn's kind nature, not to mention a picturesque scene on Hampstead Heath: two bewhiskered men, a gym instructor, and 32 boys, two of whom clearly verged on the effete – as his obituarist Andrew Taylor put it, Cleghorn 'loved his young men'.[23] The serious purpose of relating details of the selection process, however, is that it demonstrates the immense efforts made to select foresters of the highest possible quality for service in India. Advertisements were put in the press, after which Cleghorn personally interviewed each of the applicants (35 in 1888). The men next had to undergo a medical examination and a 22-mile walk (introduced a few years earlier at the suggestion of Cleghorn and G.F. Pearson).[24] Those who survived these ordeals (in this year 32) then had to take written exams, which took place at Cannon Row, Westminster over two days in the following subjects:

Group I: Handwriting, Orthography, English Composition,

Group II: Arithmetic, Geometry, Algebra & Logarithms, Plain Trigonometry and Logarithms, Mensuration

Group III: Mechanics, Physics (omitting Electricity & Magnetism), Botany, Mineralogy & Geology

Group IV: Chemistry

Group V: Geometrical Drawing, Freehand Drawing

Group VI: French: Oral, and Translation; German: Oral, Translation, Handwriting.

In 1888 Cleghorn reviewed the exam results and recommended twelve men (four Scots, one Irish, and eight English) for Continental training, on each of whom he had to write a report. Whatever the post-colonial questioning of the aims and methods of the Imperial Forest Service (let alone unknowable questions of 'motivation'), it should never be forgotten that Cleghorn and his colleagues were passionately committed to supplying the very best in the way of staff. A pastoral side to this process is to be seen in Cleghorn's account of the walking test, for which Wilhelm Schlich came over from Cooper's Hill to witness:

11. RETIREMENT & THE INDIA OFFICE, 1868–88

The walking excursion took place on Saturday, 16th, 32 candidates were present. The party met at the "Castle Hotel", Hampstead, and, after breakfast, started under the direction of Mr Claude Scott, of the Gymnastic Club, St John's Wood. The day was favourable, and the candidates, with two exceptions, completed the whole distance (22 miles) in a satisfactory manner. Of the two exceptions, one was much fatigued and went home in a cab, the other ... accomplished the first half of the distance well, but when returning and about a mile short of Hampstead thoughtlessly got up on a cart, and was thus disqualified. He mentioned the matter to me at once and expressed regret, whereupon I asked Mr Scott two days after to take him again over the same ground; this he kindly did, and was satisfied with his power of physical endurance.[25]

Although Cleghorn's help in selecting forest officers was his most significant work with the India Office, he also made an important contribution in pushing for financial support for Indian floristic publications with the upper echelons of the administration. He played a major role in promoting the completion and publication of J.L. Stewart's work on a tree flora of NW India,[26] and in 1871 actively canvassed for the financial support of Joseph Hooker's work on what became the *Flora of British India* with Mountstuart Elphinstone Grant-Duff, to the extent of passing on a private letter of Hooker's for Grant-Duff to show to the Duke of Argyll.[27]

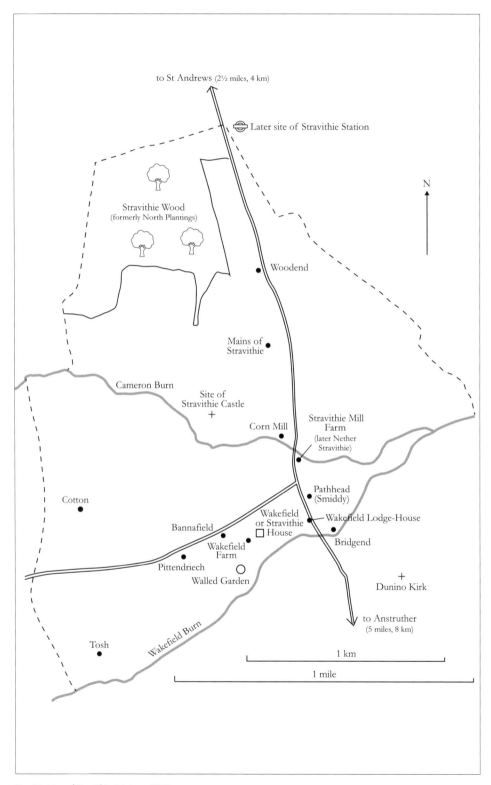

Fig. 61. Map of Stravithie Estate, c. 1860.

12
Laird of Stravithie

Cleghorn's retirement was spent largely in Scotland. However, a major limitation for a biographer lies in the dearth of documentation for this period – almost no letters in his hand and no diaries, though there are extensive records relating to the running of the estate in the Cleghorn Papers that would repay closer study.[1] Fragments of his life have to be pieced together largely from published sources that are not only extremely gappy, but almost entirely devoid of the personal and anecdotal. Only a single leisure activity is known of – golf. On 7 October 1868 he was elected a member of the famous Royal and Ancient Golf Club in St Andrews.[2] This body had developed from the Company of St Andrews Golfers, which his grandfather had captained in 1802/3, and in 1854 had merged with the Union Club founded by Major Playfair in 1835. Like most St Andrews schoolboys the young Hugh is likely to have learned to golf in youth and in old age may have taken at least an occasional turn around the links. However, he won no medals and served on none of the Club's committees, though he did use its Union Club House from which to write the occasional letter. Considering Mabel's health it seems surprising that (so far as is known) the couple didn't spend more time abroad, especially in the winters – but this was the season for Hugh's attendance at the meetings of the various societies in Edinburgh that became one of his major activities, as will be described in the next chapter. Edinburgh continued to be a major focus, not only for these activities, but because of his Cowan in-laws and friends such as J.H. Balfour; in later life he was a member of the Caledonian United Services Club (at 14 Queen Street).

Only three foreign visits made by Cleghorn in the last 27 years of his life are known though there may have been others that have left no archival trace. He is known to have visited the Tharandt Forest School near Dresden,[3] and as no Indian forest students went to Germany after 1874, the visit must have been before that date. In May 1875 Cleghorn visited Italy: he started in Rome, where George Marsh gave him some books, but the primary reason for the visit is likely to have concerned his sister Isabella, perhaps in one of her periodic personal or financial scrapes. It must have been Marsh who suggested that Cleghorn visit the forest school of Vallombrosa housed in a beautiful Benedictine monastery in the Appenines,[4] where it had been established under Adolofo di Bérenger (and where, curiously, Marsh would die on a similar visit seven years later). The last recorded European visit that Cleghorn made was to Germany, referred to incidentally as recent and on 'urgent business' in his August 1886 Arboricultural Society Presidential address,[5] which might also have concerned Isabella who certainly in her earlier years had spent time at German spas. If this trip was to see her, it may have been the last time they met and the siblings do not appear to have been close. One of the most moving folders in the Cleghorn Papers is the one relating to the affairs of the elder of his two sisters.[6] She was the classic Victorian invalid spinster whose final years were spent between Rome and Naples, eking out a £200 annuity from the Stravithie estate left in her father's will. She may have been feckless, or simply trusted the wrong people and been taken advantage of – in the letters is talk of being swindled out of money and a lawsuit (of which her younger sister, sitting pretty in Mayfair, disapproved). A possibility that does cross the mind for the distance between brother and sister, though for which there is no evidence, would be if Isabella's fondness for Italy had led her to convert to Catholicism. Whether or not this is the case, she certainly cadged money from the English chaplain to Rome, Canon Henry Wasse, who at the time was forking out money to get G.E. Street's Anglican church of All Saints built, and she also borrowed from a Mackenzie cousin. Her end came on 7 November 1888 with none of her family in attendance; the Canon telegraphed her brother a notably heartless message to the effect that unless he replied immediately, with an offer to pay, she would be given a pauper's burial. This, of course, did not happen: she may even have got a simple marble cross that was offered for between £10 and £20 and Hugh sent his faithful accountant John Dovey to sort out her embarrassed financial affairs and put flowers on her grave. Whereas Cleghorn and his younger sister Rachel seem to have been models of Victorian prudence and rectitude, the middle pair of siblings, Allan and Isabella, present a more bohemian counterpoint – though whether this was genetic or a result of the disturbances of their early childhood, who can say?

Financial Affairs

Cleghorn worried a great deal about money. The topic frequently crops up in his letters to Balfour – the concern being to have an adequate income to run, and if possible improve, Stravithie on succeeding his father as laird, a sense of responsibility that must have been instilled into him in childhood by his grandfather. The spectacular success with which he achieved this is apparent from his will and inventory,[7] which concerns only his moveable assets (and not the landed property): that is, his shares, and the incomes from his Indian pension and his tenant farmers. Despite the pain it must have caused him to stay on for so long in India (leaving Mabel an invalid in Scotland and the running of the estate to a factor), the result was a substantial annual pension of £676 from the Indian Government. By the time of his death the annual income from his tenant farmers amounted to £882. But what comes as the greatest surprise is the enormous value of his shareholdings and the income these yielded, showing how much he must have scrimped and saved from his salary throughout his career and invested this, and the money inherited from his father,[8] with considerable skill. At the time of his death the moveable part of Cleghorn's estate was valued at £38,190. For comparison, of the estates left by his friends,[9] colleagues and Indian contemporaries, only a single one left more – this was Sir William Muir, who in 1905 left £44,390, but in India Muir had held a much more senior post as a civil servant, followed by a well-paid job after his return as Principal of Edinburgh University. Those who left less included civil servants of similar or even higher rank to Muir including Lord Harris (under £35,000), Sir Walter Elliot (£36,855), Henry Yule (£29, 365), Sir Robert Montgomery (£24,431) and Sir Donald McLeod (under £10,000). Fellow medical-service colleagues left very significantly less than Cleghorn: Alexander Hunter £12,132 and Hugh Falconer £16,000; poor old Thomas Farquhar died £467 in debt.

The names of the business advisors who made this accumulation of wealth possible are also to be found in the will. His legal requirements were simple: there was no need to waste money on expensive Edinburgh lawyers, so he stuck with his grandfather's solicitors, the firm of Stuart & Charles Stuart Grace in St Andrews. Financial advice, however, was a different matter for which he did go to Edinburgh. His accountant was John Edward Dovey, whose recommendations were clearly shrewd, and who seems to have been regarded as a friend. Dovey shared evangelical and bibliographical interests, and it was he who was sent to Rome to sort out Isabella Cleghorn's affairs after her death and who in the same year was among the audience at the launch of the Forest Library. By far the largest proportion of Cleghorn's money was in debentures – investments for a specified term that yielded a fixed rate of interest (mostly four or five per cent). Most of the capital was invested in businesses operating in the Colonies, including insurance and loan companies, banks, governments and civic authorities, in Canada, New Zealand, India and Australia, and also some in the United States. A £7000 investment in the paper-making firm of Alexander Cowan & Sons doubtless represents Mabel's dowry. The railway investments are discussed elsewhere, but on the question of 'ethical investment' it is of interest to note that Cleghorn did not object to having shares (if only a small number) in the British Public House Company, and he can therefore not have been nearly so concerned with temperance as many of his evangelically-minded contemporaries. Cleghorn banked with the Royal Bank of Scotland with which he had accounts in branches in St Andrews and Edinburgh.

Stravithie Estate

Cleghorn had inherited Stravithie, the 1058-acre estate, originally purchased and developed by his grandfather, on the death of his father Peter in 1863 (Fig. 61). At the time he was heavily committed in the Punjab and it had taken him two years even to make a reconnaissance visit to ascertain the state of its finances. What he discovered necessitated a return to India for the sake of a higher pension, even though this had meant leaving Mabel behind. During this six-month visit he had made arrangements for essential rebuilding works on the mansion-house that were undertaken in his absence (in 1867) under the supervision of his factor Charles Stuart Grace, and his father-in-law. After retirement the estate inevitably became one of Cleghorn's major preoccupations – it was not only the source of a substantial income, but he had responsibility for a population amounting to more than a hundred souls – tenant farmers, their families and servants, agricultural labourers, tradesmen and artisans. The Laird also had financial responsibilities – in days before social security and the benefits system, support for the destitute came from the parish-based system of 'Poor Relief', paid by a rate calculated on the rental value of the property – so, for example in 1866 Cleghorn paid £7/2/2 (on a rental value of £1357/13/11), but the following year this went up to £11/9/11.[10] As a Heritor of the parish church of Dunino he had to contribute to the salary of the school-master,[11] to the stipend of the minister,[12] and to make periodic contributions for general parish purposes.[13]

The rebuilding of the house by John Chesser has already been discussed, but what of the garden? In the time of Cleghorn's grandfather and father the policies and woods seem to have been of the simplest sort – there was the large Stravithie Wood (139 acres) at the northern end of the estate and a few shelter belts; close to the house the most attractive feature had been provided by nature – the Wakefield Burn flowing through a heavily wooded den. There may have been a few flower beds around the house, but the only substantial horticultural feature was a traditional Scottish kitchen garden beside the burn a hundred yards upstream from the house. Even this was of a vestigial nature – a single wall to shelter the ground from the east winds rather than the square enclosure usual for a Scottish estate. Planning for his return Cleghorn clearly had an intention of growing at least some specimen trees, but, as with architecture, any landscaping or garden-making was undertaken with an eye to economy and in later years he never employed more than a single gardener. In March 1867, as the house was being completed, James McNab, Curator of RBGE, made a two-day visit to Stravithie,[14] to plan some terraces on the south-facing slope between the house and the burn, and also some other 'embellishments' though these must have been minor (Fig. 46). The main supervision of the grounds on a daily basis during the rebuilding was in the hands of George Sinclair who appears to have been recruited from RBGE,[15] with a budget 'limited to £250 ... for working operations',[16] who seems to have stayed on for a number of years as a general factotum. Bills

CHAPTER 12

from the period 1865–7 from the Kirkcaldy nursery firm of Edward Sang & Sons amount to a total of £61/17/4 for young trees of 2 or 3 feet,[17] which must have been planted in the den, in hedgerows (more than 2000 thorns) and plantations, which in 1892 were recorded as occupying 70 acres. The main species were beech (more than 6000), with several hundred each of elm, spruce, silver fir, larch, sycamore ('plane') and five types of poplar (black Italian, Ontario, woolly, balsam, and hoary). What are likely to have been specimen trees included 25 each of purple beech, lime and Spanish chestnut, and 50 ash. For under-planting there were the usual gloomy Victorian evergreens – Portugal and bay laurels, and hollies both green and variegated – doubtless as cover for game. In 1868 some more expensive shrubs were purchased from the Edinburgh firm of Dickson – more hollies and laurels, English yew, and a dozen *Aucuba japonica*.[18] Sang & Sons also provided 500 yards of box hedging, probably for the borders of the kitchen garden.

The only contemporary account of the grounds is to be found in the report of a visit made by the Arboricultural Society to Stravithie in 1892,[19] when the house was covered in roses, and the kitchen garden was producing the flowers and seasonal fruits that were on offer in a marquee erected on the lawn for the occasion. The trees planted since his return from India were listed, and Cleghorn provided measurements for those of which he was proudest, but it was pointed out that the climate and soil were far from ideal for tree growth, beech performing the best of the (semi-)native species. Of particular interest are those Cleghorn brought back as seed from the Himalayas: deodar, *Abies pindrow*, *A. webbiana*, *Picea smithiana*, *Pinus wallichiana* and, more surprisingly, *Cupressus torulosa*. This was the heyday of the introduction of exotic conifers, especially from North America, and these were represented (using modern nomenclature) by *Sequoiadendron giganteum*, *Thuja plicata*, *Cupressus nootkatensis*, *Chamaecyparis lawsoniana*, *Pseudostsuga menziesii*, *Calocedrus decurrens* and *Abies procera*, and, from South America, that Victorian favourite the monkey puzzle. There were also the European conifers *Pinus cembra*, *Abies alba* and *A. pinsapo*, and from Asia *Ginkgo biloba* and the specimen that was planted to commemorate the visit, a golden form of the Japanese cypress: *Chamaecyparis pisifera* cultivar 'Aurea' (Fig. 62).

Fig. 62. Plaque commemorating the tree planting at Stravithie by the Royal Scottish Arboricultural Society on their visit in 1892. (Author's photograph, 2015, with permission of David Chalmers).

As a forester, Cleghorn's most surprising action on inheriting the estate appears to have been the felling of Stravithie Wood to turn into farmland; but on the highest part of the estate, at an altitude of 300 feet above the cold North Sea, its timber must have appeared feeble compared with the forest splendours of the Himalayas or the Western Ghats. Cleghorn would probably be disappointed were he to see his policies today – the Den is occupied by a scrub of self-sown beech and the odd Scots pine, and of his specimen trees only a magnificent Douglas fir, an Atlas cedar and an undistinguished Wellingtonia certainly date back to his time, though making allowance for poor soil, so perhaps might a western red-cedar and a deodar.

12. LAIRD OF STRAVITHIE

The farms

A map of the Stravithie estate shows a rectangle with sides sloping northwards from a southern boundary formed by the Wakefield Burn (known upstream as the Kinaldy Burn; downstream it becomes the Kenly Water).[20] Another burn, the Cameron, which powered the estate's corn mill, crosses the estate from the north-west, to join the Wakefield in the south-eastern corner. At an altitude of 300 feet the northern boundary forms the highest point along which runs a disused railway; the eastern and western boundaries are the arbitrary divisions of a surveyor. The land is largely covered by glacial drift and now entirely arable, though looking much more prairie-like than it did in the nineteenth century: the forty or so fields of Cleghorn's day are now reduced to about 25 and with fewer hedgerows. Apart from a volcanic intrusion in Dunino Den the underlying geology is of Carboniferous sandstone, the only exploitable physical resource, which was briefly quarried at the northern edge of the estate, and which for centuries has provided the building material of the houses and cottages. Golden when first quarried, it weathers to ash grey; the cottages pantiled, the farms and mansion slated.

Much can be gleaned about the estate from the records of the national censuses made every ten years – three in his own and two from his father's periods as laird (the first being in 1851).[21] These provide rich detail about land-holdings and tenancies, buildings, and the occupations of their inhabitants. Originally there were two major farms on the estate, Stravithie Mains in the north and Wakefield in the south. The latter name is derived from a common Scottish place name 'Waukfield' – a place for the retting of flax (though no longer cultivated by the later nineteenth century, out-competed by Bengal jute imported into the nearby port and city of Dundee), but the name gradually fell out of use, doubtless to prevent confusion with the Yorkshire city, though even in the 1880s it was still sometimes used by 'the County' as a territorial designation for 'Dr Cleghorn of Wakefield'. The etymology of Stravithie is disputed, but although the 'strath' is more likely to be a corruption of 'rath' (a garden or enclosure) than its usual meaning of a valley, romantic Victorians saw in it 'Strath Beithean', a valley of birches.[22] A further two substantial farms evolved during the course of the three Cleghorn generations – one attached to the mill, Stravithie Mill Farm, later known as Nether Stravithie; the other called Woodend, created by Cleghorn between 1861 and 1871 by the felling of Stravithie Wood.

The lands of Wakefield farmed personally by Peter Cleghorn included a home farm (with charming, single-storey Gothick buildings, probably built by his father), and the smaller farms of Bannafield, Pittendreich and Tosh, but in Peter's and Cleghorn's era the buildings of all of these farms (17 in total by 1881) were used for housing agricultural workers and tenants of various sorts, until Tosh was made a farm in its own right, and (by 1881) the fields of Bannafield were added to Stravithie Mill Farm. From the occupations named, considerable changes in farming practice are also revealed in the censuses: Cleghorn seems to have replaced beef and dairy cattle with sheep, while increasing the amount of arable land, not least, as already noted, by decreasing the area of unprofitable forestry (no forester or wood-cutter was employed after 1861).

The 'big house' itself (with a split personality in the censuses, as Stravithie or Wakefield) was relatively modest in terms of both accommodation and indoor staff, having 16 rooms with windows both in Peter's time and after the 1866/7 rebuilding, though another five rooms had been added by Hugh by the time of the 1891 census. Family needs were attended to by a housekeeper, a cook and, variously, two or three maids. The housekeeper for most of Cleghorn's

CHAPTER 12

time was Anna Maria Galloway, six years his senior, who had retired by the time of his death – she came of respectable, middle-class stock in Edinburgh where her father was an accountant and member of the Highland Society, and in gratitude for her services she was left an annuity of £30. There are no known photographs of the interior of Stravithie in Cleghorn's time, though the valuation of its contents at £561/17/3 at his death in 1895 suggests that it cannot have been extravagantly furnished, an inference that can be drawn from a consideration of the library. Although books are notorious for becoming worth far less than their purchase price, and their value either not recognised by a local valuer or deliberately given a low probate value, even at a low valuation Cleghorn's library must have represented a substantial part of this total (for example in the three years between 1852 and 1855 he had spent £300 on books). Beyond the books there must have been the usual souvenirs of the Victorian Raj (leopard-skin rugs, blackwood furniture and the like). Still in the Stravithie dining room, purchased back from the Sprot family by the present owner, are two handsome pieces of mahogany furniture, both dating from Peter's time, an architectural sideboard and a dining table at which it is easy to imagine a late-Victorian spread hosted by Hugh and Mabel for a mixture of neighbouring gentry and friends from their Punjab days such as Edward and Elizabeth Lake who stayed at Stravithie in 1871. In the drawing room are two handsome mirrors that may once have reflected the figures of Hugh and Mabel – a tall, pillared pier glass and, over the fireplace, an octagonal one with heavy ormolu mouldings. Silver is mentioned in the inventory and of pictures some still survive in the family.

For these a visit to Wiltshire was necessary to view the three framed prints/drawings of the wreck of the *Sutlej* commemorating the young surgeon's escape from peril on the sea (Fig. 11). The Cleghorn portraits have, appropriately, been kept on ancestral lands at Stravithie by Geoffrey Cleghorn Sprot, Cleghorn's great great great nephew. Finely framed, with gilt identification plaques, these allow a glimpse of the interior decoration of Stravithie in Cleghorn's era, and one can only speculate on what he would have thought of their removal to more manageable properties better suited to modern living – first to his Mill Farm and more recently to his Mains Farm. The finest of these, appropriately, is an exquisitely finished and coloured pastel of Hugh Cleghorn senior in his full vigour by Archibald Skirving (Fig. 1), and of equal interest if lesser quality (its artist disowned it) is a tiny oil portrait of Peter by his friend David Wilkie (Fig. 1). Of the next generation, in masculine rosewood frames, is a pair of fine chalk portraits of Cleghorn and his unfortunate brother Allan, both aged about 18 by the Glasgow-born Alexander Blaikley then at the start of his career (Fig. 1). The sisters Rachel and Isabella, also drawn in chalk but gilt-framed, are shown slightly older, in their early twenties, dressed in fine gowns: handsome girls leading one to wonder why Isabella never married. The drawing of Rachel, with her husband Alexander Sprot and their first child, has already been discussed, as has her oil portrait in mature widowhood. The last of the portraits is the presentation charcoal of the Grand Old Forester in bearded old age by Wirgman (Fig. 66).

At the entrance to the drive, beside the Anstruther to St Andrews road, is a picturesque, two-roomed lodge, with diamond-leaded windows and a central chimney stack. This was used to house a married couple, of which the husband at different times worked either as the gardener or as the laird's coachman. Cleghorn must have been particularly fond of the coachmen as two successive ones, both kept on in tied cottages after retirement, were the only individuals other than Miss Galloway and his nephew to be named in his will: William Morris (who had been Peter Cleghorn's valet) and William Duff.

12. LAIRD OF STRAVITHIE

Stravithie Mains was one of the two large farms of the estate, originally the home farm for the long demolished Stravithie Castle that had once stood on the north bank of the Cameron Burn to the west of the Mill. The farmhouse eventually had nine rooms, and by 1881 the area farmed was 365 acres, run by six men and two boys or women. From 1851 to 1891 the tenant was William Hood, but by 1895 he had been replaced by William Waddell who paid an annual rent of £340. The Mains is today the home of the Sprot family, who still farm the estate, 210 years after it was purchased by the first Hugh Cleghorn. The second large farm was the eight-roomed Stravithie Mill Farm, now called Nether Stravithie, where a monkey puzzle forms a conspicuous landmark beside the St Andrews to Anstruther road; it became home to Hugh and Elizabeth Sprot after they sold Stravithie House in 1979. The acreage of this farm changed substantially over Cleghorn's period, from 78 acres in 1861 to 356 in 1891 – initially by taking over a smallholding farmed from Pathhead, then with the more substantial addition of the Bannafield/Wakefield lands, and by purchase of land from his neighbour Walter Irvine in 1869. This farm stayed in a single family throughout Cleghorn's period, Colin Berwick having succeeded his father William by the time of the 1861 census, and by the 1890s, when it yielded an annual rent of £375 it was farmed by nine men and eight boys or women. Even more dramatic than the increase in acreage of Stravithie Mill Farm was that of Woodend. In 1861 there were two cottages on the site, but ten years later one of these had been enlarged into a three-roomed farmhouse with 175 acres, though some of the land was probably only grazing, as the tenants in 1895 (Alexander Scott and Thomas Kay) paid only £90 in rent.

Most of the houses on the estate, apart from the farms and the smiddy, were tied properties for which the occupants paid no rent, but at the time of Cleghorn's death there were five let cottages (two at Bridgend, beside the Wakefield Burn, two on the farm of Bannafield, and one at Stravithie) whose occupants paid rent of either £2 or £3.

The tenantry

The decennial censuses represent a goldmine for the social historian, giving, as they do, the trades, professions and occupations of the people enumerated. For Stravithie between 1851 and 1891, out of a population that varied between 92 and 131, the largest number (between 28 and 39) were employed on the land as labourers (both men and women), ploughmen, farm servants, cattlemen or shepherds – attached to the farms, and living in bothies or groups of cottages known picturesquely in Fife as 'Cottons' (i.e., Cottar's touns). The next largest group comprised domestic servants (11 or 12 throughout the period) who worked in the larger farms and at Stravithie House. In what was largely a self-sufficient rural economy the miller, with an assistant or apprentice, ground the barley at the mill driven by the waters of the Cameron Burn; and at a time when horses were used for haulage, ploughing and drawing the laird's brougham, the blacksmith who shod the beasts was also a major figure. The smiddy was at Pathhead, almost opposite the Lodge, and for the whole of Cleghorn's period was occupied by the Donaldson family (who in 1895 paid £7 4 shillings in rent). For his smiddy work Mr Donaldson had two assistants in 1851, but none by 1891 by which time public transport had improved and some of the ploughing was probably being done by traction engine. In addition to these artisans, the remainder of the Stravithie population comprised a fluctuating assortment including a cartwright, up to two grocers, a woollen draper, a tailor, two shoemakers, a joiner, a road surfaceman, a letter carrier, a brewer's clerk, a sheep dealer and a seaman.

CHAPTER 12

Stavithie lies in the parish of Dunino and it is therefore necessary to consider what lies to the south of the estate, beyond the Wakefield Burn: the small and scattered group of buildings that form the hamlet of Dunino, around the kirk, the manse and the school. The primary education of Cleghorn and his siblings had been provided at home, though the parish school catered for the offspring of his tenants and those of neighbouring estates. When Cleghorn returned from India, as a committed member of the Free Church, his church-going was probably in St Andrews. Nonetheless, as the Laird of Stravithie was one of the parish's heritors, *noblesse oblige* surely dictated that the family must, at least on occasion, have attended the kirk. In bad weather they would have gone by pony chaise but on a fine spring or summer morning they must sometimes have walked from the house down the Wakefield Burn and up Dunino Den, filled with snowdrops followed by a carpet of bluebells, to the pretty little Gillespie Graham church with its crocketed bell-cote (Fig. 71). The kirkyard, with its distant glimpse of the sea beyond Boarhills is where, in due course, the family bones were all laid to rest. The site is an ancient one, and to the west of the church, neatly cut into the flat top of a sandstone cliff is a remarkable basin, possibly a font; in the graveyard stands the lower part of the shaft of a Pictish cross. On an unseasonably wet day in July 1883 Cleghorn and his nephew Alexander Sprot showed these antiquities to a group of Edinburgh worthies consisting of members of the Society of Antiquaries of Scotland and the Royal Physical Society.[23] Cleghorn, with his horror of 'primitive' religions and idolatry, would be appalled at the sight that would greet him were he to return today – the natural beauties of the den, over which the cliff with its basin towers, are not enough for the New-Age hippies who have colonised the site. Against the trunks of the trees are placed bunches of repulsive artificial flowers; tied to their branches are squalid ribbons and rags, toy frogs and plastic dragonflies; one of the hazel bushes has been covered with a web of coarse red wool, as though the product of a genetically modified spider. Even the stump of the Pictish cross in the graveyard has been defaced, its top used as a receptacle for rusting coins.

Cleghorn & Railways

Railways were one of the major causes of deforestation in India and it is therefore somewhat ironic to discover that Cleghorn, the father of Forest Conservancy in India, not only chopped down his own grandfather's wood, but invested serious amounts of money in Indian railways. It is, of course, inappropriate to apply modern concepts of 'ethical investment' retrospectively and Cleghorn undoubtedly saw the development of mechanised communications as part of his belief in 'improvement'; this was the age of railway mania, in which speculation in the new companies carried with it at least the possibility of handsome financial returns. In Madras, as early as 1859, he had purchased 50 shares in the Great Southern Railway, an investment that he must have built on, as at the time of his death he had £3014 worth of stock in the Company: a reasonable investment as he had paid £2740 for it.

What is more surprising to discover is that late in life Cleghorn himself became a director of a railway company, providing one of the more fascinating episodes in his sparsely documented retirement in Fife.[24] Since his undergraduate days the Edinburgh & Northern Railway Company (later the North British) had been pressing northwards, reaching the county town of Cupar in 1847 with extensions to Leuchars in 1848, and to Tayport, the southern end of the first Firth of Tay crossing, in 1850. In 1852 the St Andrews Railway Company had opened a branch line from Leuchars to the 'city of golf', with a terminus on the West Links.

12. LAIRD OF STRAVITHIE

Fig. 63. Cleghorn's railway station at Stravithie, looking east. Photographer unknown, ?1930s.

The Leven & East Fife Railway Company had later launched a line from Thornton Junction that had reached the fishing port of Anstruther in 1863, leaving a gap of only 13½ miles between Anstruther and St Andrews to complete the coastal loop. A preliminary meeting in 1865 estimated the cost of this line at £65,625, but nothing further happened for more than a decade. In 1879 a Provisional Committee was established, of which Cleghorn was a member, but to set up a new railway company an Act of Parliament was required. This was duly enacted on 26 August 1880, authorising the raising of share capital to the tune of £57,000 in £10 shares – the rest of the capital was to be borrowed – Cleghorn himself purchasing 277 shares. On 30 September 1880, with the abolition of the Provisional Committee, Cleghorn found himself elected a Director of the Anstruther & St Andrews Railway Company, along with the wonderfully named Colonel Moneypenny of Pitmilly as a fellow Director, his neighbour John Purvis of Kinaldy as Chairman, and Philip Oliphant, an Anstruther solicitor who had been the scheme's most active promoter, as Company Secretary.

Despite the fact that the line has been described as 'possibly the most fascinating and least visited part of the whole [British railway] system', the business was 'one of the least prosperous and most ill-fated of the small companies',[25] which must have caused Cleghorn and his fellow directors many a sleepless night. Difficulties concerned not only the raising of the required capital, but the bankruptcy of the original contractor, a faulty locomotive and the human tragedies of three unfortunate navvies who died, in separate incidents, as a result of intoxication. The trouble was that, beautiful though the scenery might be, the 'business case' for the line was tenuous in the extreme – the population it served was tiny, consisting of the village of Crail, the hamlets of Kingsbarns, Boarhills, and a scattering of estates and farms. The inhabitants were already well served by road transport, though the carriage of agricultural produce was a slightly more credible justification. Other than getting tourists to Crail it is hard not to see the whole enterprise as an act of enormous vanity on the part of the landowners whose estates it linked: Erskine of Cambo, Moneypenny of Pitmilly, Purvis of Kinaldy, Cleghorn of Stravithie and

CHAPTER 12

Whyte-Melville of Mount Melville. Such cynicism is substantiated by a change in the original route of the line to include stations at Stravithie and Mount Meville, which not only added to the length of the line by taking it further inland, but involved the addition of two substantial physical obstacles – a steep incline between St Andrews and Mount Melville, and a three-hundred-foot summit at Stravithie, both of which required the provision of shunting sidings. This all resulted in rising costs, which reached an astronomical £125,000, for which another share issue of £35,000 had to be launched, loans borrowed, and deals done with the North British company. The only resources exploited by the line other than agricultural produce lay in the form of a stone quarry at Stravithie, which helped to justify a private siding and a station for Cleghorn's estate (Fig. 63) – but, ever the canny businessman, he obtained this facility on the cheap – he had to pay £187/0/9 for the siding, but the Company paid him £180 for the value of the stone extracted to build it. Such deals must have been given help when John Dovey, Cleghorn's own accountant in Edinburgh, had replaced Oliphant as Company Secretary in November 1883.

The line was built in two stages, the first sod being cut at Anstruther on 13 June 1881, the line reaching Boarhills on 1 September 1883. Finally, on 30 June 1887, Cleghorn received a telegram from the Board of Trade giving permission for the opening of the final section, which took place the following day with a ceremony at a new station in St Andrews (the original terminus of the line from Leuchars having been moved in order to join with the new line). Even when complete the line was not for those in a hurry: to get to Edinburgh it was still quicker to go to the main line at Leuchars. When the Arboricultural Society visited Stravithie in 1892,[26] it took them 3½ hours by train using the coastal route from Edinburgh. The Company was never profitable, though in 1893 it was able to issue a 2% dividend. At the time of Cleghorn's death in 1895 the shares for which he had paid £2770 were worth a mere £346 5 shillings and only two years later the Company was absorbed into the North British Company. The stations of Kingsbarns, Boarhills, Stravithie and Mount Melville were closed in 1930, and the whole line in 1964. The station building at Stravithie still survives and a railway coach parked in its garden operates as an upmarket B & B for latter-day railway enthusiasts.

13
Scottish Societies

For intellectual stimulus in his later Scottish life Cleghorn relied on participation in the activities of various societies. Locally there was the St Andrews Literary and Philosophical Society but the major ones were based in Edinburgh. Of these he was a fellow of the Society of Antiquaries of Scotland, and as already noted he hosted a visit of its members at Stravithie in 1883. As a fellow of the prestigious Royal Society of Edinburgh he must surely have attended its meetings from time to time, but of this there is no record as he took no part in its administration and read no papers. He was also a member, and served on the council, of the Scottish Geographical Society. The bodies in which he was most active and to which he contributed most significantly were more practically orientated, what might perhaps be termed 'semi-learned societies': the Highland & Agricultural Society of Scotland, the Botanical Society of Edinburgh and the Scottish Arboricultural Society.

The Highland & Agricultural Society of Scotland

The differences of opinion already described between certain Scottish interests and the India Office over the question of training of Indian foresters, first raised by Balfour in 1866, had rumbled on and led to a development in which, rather curiously, Cleghorn became involved. The Highland Society of Edinburgh was founded in 1784 by an influential body of Scottish gentry with the aim of improving the Highlands, not only in terms of agriculture and fisheries, but also (slightly surprisingly, a mere four decades post-Culloden) in preserving its Gaelic culture.[1] The cultural role faded rather rapidly: agricultural improvement became its major aim and in 1834 it was renamed the 'Highland and Agricultural Society of Scotland'. The Society, which had offices and a museum in the premises of the Lawson seed company on George IV Bridge (now part of the City Library), became extremely active nationally, by means of meetings, publications, the organisation of shows and the awarding of large numbers of prizes ('premiums') relating to a wide variety of topics – practical (including livestock, implements and forestry), and the writing of essays and reports. From 1856 the Society had also been active in education, granting diplomas in agriculture, and by 1870, the year in which Cleghorn was admitted to membership (at the meeting on 19 January),[2] it had the enormous number of 4033 members under the presidency of George Hay, 8th Marquess of Tweeddale, a former Governor of Madras. Members were based throughout the country (and beyond), showing to what degree Scotland was still a largely agricultural nation, with two levels of annual subscription – gentry paying £1 3s 6d, tenant farmers and those of similar rank ten shillings.

In June 1870, at Balfour's instigation, the Society had petitioned the India Office along the lines of his earlier arguments on Indian forestry training – what was the need to send boys to the Continent, and at a stage 'most inconvenient to parents, removing ... [the young men] entirely from their parental control at a critical age'?[3] This letter was sent to the Duke of Argyll as Secretary of State for India, followed up in July with the visit of a delegation under the Duke of Buccleuch (this was high-powered stuff, the group included, among others, an earl, two baronets and five MPs), which was joined by like-minded representatives from the Horticultural Society and the Institution of Surveyors. Argyll was not to be brow-beaten, but although the Society got nowhere in this matter it was the spur to the formation of its own Forestry Department under Balfour (departments devoted to Chemistry and Veterinary Medicine already existed). There also arose the idea of awarding First and Second Class Certificates in Forestry by examination, the first of which was to be held in November 1870. Despite Cleghorn's active support for Continental training, he agreed to be a member of the new department along with John Wilson, Edinburgh Professor of Agriculture, four landowners, and Charles Lawson (head of the seed merchant's firm, effectively the Society's landlord). Cleghorn clearly saw no conflict of interest and doubtless considered that the status given by these certificates, at a more elementary level than what was offered on the Continent, might be useful in raising standards in Scottish estate forestry.

The exam covered five subject areas, given here with the examiners for the early years:
1. 'Science of Forestry and Practical Management of Woods' – examined by Cleghorn (in later years sometimes taking on others, including McCorquodale and Thomson) until 1891 (when Frederick Bailey and William Somerville, the Edinburgh University forestry lecturers took over).
2. Elements of Botany – Balfour (until 1883, when replaced by Cleghorn who continued until 1894, after 1889 jointly with Isaac Bayley Balfour).
3. Relating to soils, drainage and climate – John Wilson, Professor of Agriculture.
4. The measuring and surveying both of land and timber – A.W. Belfrage.
5. Accounting and book-keeping – Kenneth Mackenzie.[4]

In this era of the auto-didact, candidates were responsible for preparing themselves for the exams, which consisted of a three-hour written paper, followed by an oral the following day, which started with 'practical forestry', and proceeded no further if this was unsuccessful. The Society, generously, made no charge for the exam or certificate: the candidate simply had to get himself to Edinburgh. The difficulties, however, were immense and one of the examiners, John McGregor (forester to the Duke of Atholl), in his evidence to the 1887 Select Committee, drew attention to the lack of a decent textbook from which to learn. In his own 1885 evidence Cleghorn had stated that as preparation for the certificates 'we find that two seasons are generally sufficient' and referred rather grandly to 'what books they have been able to get in their cottages'.[5] Such difficulties, together with the fact that proprietors seem not to have been impressed by the qualification – doubtless in case successful candidates expected a wage-rise – explains the low take-up rate, which meant that Cleghorn's role as examiner was hardly onerous. In the whole of his period as examiner (1870–94) only seventeen First Class certificates and ten Second Class were awarded. It may originally have been hoped by Balfour that some of the certificands would have gone to India or the Colonies, but this seems not to have been the case. Two did, however, go on to have distinguished academic careers in Britain – John Hardie Wilson (1884) and William Somerville (1886), whose stories come later.

The Society had also played a major role in the development of the teaching of rural sciences in Edinburgh University – in its earliest days (1790) it had backed the chair of Agriculture, and in 1840 a chair of Veterinary Studies, and at this point it was actively lobbying and fund-raising for a chair in Forestry. Cleghorn was also involved in these activities.

13. SCOTTISH SOCIETIES

Botanical Society of Edinburgh (Part 4)

Cleghorn had been a keen member of the Botanical Society since 1838 – participating in its activities during furloughs and by correspondence when abroad. This connection was in part due to his friendship with J.H. Balfour, who at meetings would read letters, reports and papers that Cleghorn sent back from India. But in retirement Cleghorn took a more active role and during the period 1868–93 there was only a single year (1870/1) when he did not act as an office-bearer.[6] The first year of his retirement (1868/9) he was elected President; for ten years he acted as Vice-President; for 12 as a Councillor; and from 1870 he was on the publications committee that oversaw the editing of the *Transactions*. In this period the summer monthly meetings (April to July), consisting of a committee meeting followed by a public one, took place, as they had from the society's 1836 foundation, at the Botanic Garden. By this time winter meetings (November to March) were held at 5 St Andrew Square, the headquarters of the National Bible Society of Scotland where other societies, including the Scottish Arboricultural and the Edinburgh Photographic, also met. Cleghorn attended whenever possible, sometimes standing in for the President as chairman. Through its meetings and publications the society was very active at this period, its dominant personalities being Balfour and his teaching assistant John Sadler, the horticulturist Isaac Anderson Henry, and two men from Cleghorn's past – Sir Walter Elliot and Sir Robert Christison. By this time two of the Society's aims from the days of Cleghorn's earliest membership had, however, disappeared – the herbarium had been combined with that of the University in 1838 (and moved to RBGE at least by 1862), and in 1872 their valuable library had been handed over to the Government (the Board of Works) where it joined the herbarium at RBGE for use by academics, students and, to some degree, the public.

Cleghorn's election as the Society's President took place on 10 December 1868, succeeding Charles Jenner, owner of a well-known Princes Street department store, but also a keen horticulturist and algologist. On 11 March 1869 the new President read a fascinating paper on his recent observations on the botany and agriculture of Malta and Sicily.[7] At the end of his year in office, on 11 November 1869, on being succeeded by his old friend Walter Elliot, Cleghorn delivered an address on the current state of botany in Britain.[8] The talk started in nostalgic vein, with reminiscences of the time he had joined the Society, which brings to mind Joseph Hooker's similar speech that Cleghorn had heard at the British Association in Norwich the previous year. It was not, however, Hooker's talk that Cleghorn cited as a precedent but, rather, a thirty-year old one by Greville. Beyond the theme of reminiscence, another more particular, but unacknowledged, similarity to Hooker's address is Cleghorn's recommendation of the use of local museums in natural-history education. But altogether more remarkable is the deafening omission of one of Hooker's chief topics.

CHAPTER 13

The nostalgia allowed Cleghorn to acknowledge his early mentors including Sir William Hooker and Sir David Brewster, and more recent botanists; he made an uncalled for, and therefore all the more generous, tribute to Alexander Dickson who had just beaten him to the Glasgow chair. He then went on to talk about some of the great changes in botany over the previous 30 years, including the decline in the number of private herbaria, and the great increase in publicly funded ones in Britain (Edinburgh and Kew), on the Continent, and (since 1861) in Calcutta.[9] He also noted the declining contribution to botanical study from medics and what would today be called 'amateurs' (specifically Indian civil servants), due to the increasing intensity of medical training, a narrowing of focus as reflected in a new breed of specialist journals, and what would now be termed the professionalization of science – with significant contributions now coming from foresters, and employees of the topographical and geological surveys. The great series of Colonial Floras being produced from Kew was noted with approval, as was the Duke of Argyll's recent go-ahead for J.L. Stewart's 'Flora Sylvatica' of NW India, though Cleghorn regretted that there had been no further progress on a comprehensive Flora for the Subcontinent. The bibliographer also revealed itself in his anticipation of the Royal Society's encyclopaedic *Catalogue of Scientific Papers* (1800–1863).[10] The subject of Indian forestry was discussed, in particular a justification for sending young Indian foresters to France and Germany for training; there was, however, nothing about the matter of forestry *per se*, as usual preferring to talk about applied botany (in particular cinchona and ipecacuanha). Cleghorn discussed additions to the British flora, including wool aliens at Galashiels, and recent local Floras, and then summarised recent work and literature on cryptogams, fossil botany and physiology. As always he had prepared his work conscientiously, as can be seen from surviving letters soliciting information for the palaeobotanical part of the talk from William Carruthers, an evangelically minded pupil of Balfour, who since 1859 had been Assistant Keeper of Botany at the British Museum.[11]

Before ending with an extensive bibliography of botanical works published during the previous year, and a list of recently departed botanists, there was a section on 'Physiology', which is significant in terms of one subject that is treated, and one that is not. Recent work on the morphology of conifer leaves by Meehan, Dickson and Carrière was discussed, which Cleghorn showed to have practical relevance. In the Himalayas he and Brandis had discovered that the needles of different conifer species differed widely in lifespan and that this was correlated with their ecology. The deafening silence was over the subject of evolution: the only research of Darwin cited was his work on the fertilization of orchids, with not a single word about transmutation or natural selection. Cleghorn could, perhaps, have been said to have covered himself at the start of the address, pleading his recent Indian sojourn and that 'I may omit to allude to some points which appear more prominent to you, and which may have altogether escaped my notice'. But the great question of the 1860s simply cannot have escaped his notice and he clearly felt reticent about discussing it in public.

By now Cleghorn had exhausted his stock of original Indian material and undertook no further botanical research so his contributions to the Society resulted in no publications other than obituaries. He did, however, occasionally exhibit botanical curiosities at the monthly meetings. Two of these were interesting examples of long-distance dispersal through the agency of man – an eight-foot sapling of the New Zealand tree *Aristotelia racemosa*, whose seed had hitched a lift on the stem of a tree-fern translocated to his father-in-law Charles Cowan's greenhouse at Wester Lea, Murrayfield;[12] and a plant of the cereal *Sorghum vulgare* grown from a seed that the Marquess of Lothian (clearly a hands-on farmer) had found in the fleece of a sheep.[13] One exhibit was home-grown at Stravithie – a stem of the broom (*Cytisus scoparius*) only eleven years old but 12 feet long and eight inches in diameter at one foot from the base.[14] In 1890 he showed the original specimens from which William Coldstream had illustrated his *Grasses of the Southern Punjab* (1889), which were presented to the Society and added to the RBGE herbarium.[15] But Cleghorn's most significant contribution was as an editor and obituarist, which gave scope for his long-standing interest in bibliography.

The *Transactions of the Botanical Society* [*of Edinburgh*] was a learned scientific journal devoted to botany (with some horticulture). It had an extensive distribution both nationally and internationally and covered not only local subjects such as accounts of field excursions, but – given its close links with the Botanic Garden – interesting plants that had flowered there. There was also generous coverage of topics such as plant anatomy, physiology, phenology and geographical and cryptogamic botany. On two occasions in Cleghorn's period a major monograph occupied a whole part – Richard Spruce's account of the liverworts of the Amazon and Andes, and a work on the economic botany of W Afghanistan and NE Persia by J.E.T. Aitchison (at £140 the most expensive part they had published up to that time). The Society had employed an artist from its foundation – initially James McNab and in the 1880s Dr David Christison (son of Sir Robert) – so the *Transactions* were well illustrated. In 1870 Cleghorn, with Balfour and Elliot, was appointed to form a publications committee. The following year Elliot wrote to Cleghorn that they should 'exercise a much more rigid censorship than we did last year' and that it should not be made 'a vehicle for petty discussions on … trivial matters' which should be placed 'in the pages of "Nature" or other ephemeral papers' – which shows the lability, over time, of what are now known as 'impact factors'.[16] Cleghorn remained on the committee, which always consisted of senior academic members of the Society, for the rest of his active life, but his editorial hand is invisible.

What is visible are the obituaries, of which Cleghorn wrote no fewer than 14 between 1868 and 1889. By far the most substantial, forming standard biographical sources for their respective subjects, are those of George Walker-Arnott (read 9 July 1868)[17] and of 'dear old' Robert Wight (read 11 July 1872).[18] Cleghorn took these works seriously and, as for the shorter ones, compiled bibliographies of the deceased's publications, but in these two cases used sources available to him as a personal friend – including letters, diaries and reports. Other Indian botanists for whom it fell to him to perform this task were two colleagues from Punjab days, John Lindsay Stewart (1875)[19] and William Jameson (1882),[20] and, from Madras days, Sir Walter Elliot (1887).[21] There were also write-ups for a less well-known Indian medical contemporary, William Traill, (1886)[22] and for George Thwaites (1882)[23] with whom he had corresponded in Ceylon but never met. For distinguished foreign botanists such as Carl von Martius and Antonio Bertoloni (both in 1869)[24] Cleghorn had to rely on published sources.[25] British botany was not neglected and there were tributes to George Dickie and Richard Parnell (both in 1882, the latter a founder member of the Society and graminologist),[26] and the last of these contributions (1889)[27] was for J.T.I. Boswell (*né* Syme). Boswell Syme was son of the artist Patrick Syme (artist to the Caledonian Horticultural Society and William McIvor's drawing master at Dollar Academy), who, before moving to London, had acted as the Society's Curator. Latterly Boswell had inherited his mother's family estate of Balmuto (near Kirkcaldy) and so became a fellow Fife laird, though he is best remembered for his substantial third edition of 'Sowerby's *English Botany*'.

CHAPTER 13

Like other academic societies (such as the Linnean and the Royal), the Botanical Society had a dining club which, on 25 July 1890, was invited by Cleghorn to visit him in St Andrews and Stravithie.[28] In the ancient city they visited the College Museum (still in Upper College Hall at St Salvators), the ruins of the Cathedral and the Castle, and under the guidance of his friend Professor McIntosh, the six-year-old Marine Laboratory on the East Links. John Hardie Wilson showed them round the even newer Botanic Garden in the grounds of St Mary's College (Fig. 69) and the nearby University Library before the party boarded a train to Stravithie, where a walnut tree was planted at the south-west corner of the house to commemorate the visit. There is still a walnut tree on the grassy slope to the south-west of the house, but if it is the same one, the soil must be extremely devoid of nutrients. Also present were Alexander Christison, Robert Lindsay (curator of RBGE), the anatomist John Bell Pettigrew and William Somerville the Edinburgh Forestry lecturer.

Cleghorn's last official involvement with the Botanical Society, by which time at the age of 73 his energies must have been flagging, was as Vice President for the year 1892/3, having on 11 December 1890 been elected as one of its only six Honorary British Fellows.

The (Royal) Scottish Arboricultural Society

The third Scottish body in which Cleghorn played a major role during his retirement was the Scottish Arboricultural Society, of which he served as President for two spells – the first for the two years 1872/3 and 1873/4, the second a decade later, for three successive years from 1884 to 1887.[29] The Society had been founded in 1854 for:

> the promotion of the science of Arboriculture in all its branches, by periodical meetings of the Members for the reading of Papers; by offering Prizes for Essays, and Report on the Practical Operations of Forestry, and publication of the same; and by such other means as may be found advisable.[30]

Its role was therefore somewhat similar to that of the Highland Society, though it never had anything like such a large membership, which rose from 600 in Cleghorn's first period of presidency to 738 by the end of his second. As with the older society, it was strongly hierarchical, with substantial input from the aristocracy and gentry, though by means of four levels of subscription, and doubtless with a view to raising standards, it was also open to those of less privileged status. Annual subscriptions were half a guinea for Proprietors (5 guineas for life); five shillings for Factors and Nurserymen (3 guineas for life); 3 shillings for Head-Foresters and 2 shillings for Assistant-Foresters (2 guineas for life). In fact according to John McGregor's evidence to the 1887 Select Committee, the proprietors didn't attend meetings, and it was the practical foresters who were the active participants. Despite this, and due to its

13. SCOTTISH SOCIETIES

Fig. 64. The Dunkeld larches, visited on the Scottish Arboricultural Society's excursion to Perthshire in autumn 1884. Photograph by George Washington Wilson, reproduced as a Woodburytype in C.Y. Michie's *The Larch* (Edinburgh, 1882).

still deeply feudal nature, Scottish forestry was then exclusively the province of landowners rather than of the State; it was largely plantation based and (with the exception of Scots pine in Speyside and Deeside) already dependant largely on exotic species, notably the European larch (Fig. 64) and the North American Douglas fir.

The Society, like the Botanical Society, had offices at 5 St Andrew's Square, but in its early days its annual meetings had been held at the Freemason's Hall in George Street, where Cleghorn gave his first Presidential address in 1872. The following year and thereafter, doubtless under the influence of Cleghorn, the annual meeting was held at RBGE. Here Balfour and his assistant John Sadler were key players with Cleghorn in the Society and this year, with the appointment of a Museum Committee, the Society started to accumulate collections (including cones, photographs and wood samples). Forestry being something of a Scottish speciality, and through the involvement of widely travelled men like Cleghorn, the Society had both a national and, increasingly, an international role (as may be seen from the papers in the *Transactions*, which, from the late 1870s, included subjects from as far afield as India, South Africa, Australia and the USA). In addition to meetings, from 1866, there was an Annual Dinner and, from 1878 onwards, excursions, including an annual one in August the first of which, significantly, was to Scone, one of the centres of modern Scottish forestry. For the years 1881–3 there was an additional autumn excursion in October.

The excursions were considerable social events, not least for Cleghorn who attended nearly every one between 1881 and 1888, in some of which forces were joined with the corresponding English body. Up to 40 or 50 members, who styled themselves the 'Jolly Foresters', took part, and on at least one occasion they bedecked themselves with sprigs of Scots pine. The national tree had been

CHAPTER 13

taken as the Society's logo, along with a couthie motto devised by Sadler: 'Be aye stickin' in a tree, it will be growin' while ye're sleeping'. The railway network enabled a wide geographical range to the excursions, which often extended to two or three days – all the great Scottish policies and estates were visited and among those Cleghorn is known to have attended were Dalkeith Palace and Newbattle Abbey (Midlothian, 1879), Morayshire and East Lothian (both in 1881), Argyll and Fife (both in 1882), Strathearn (1883), Riccarton and Perthshire (1884 – Fig. 64), Mountstuart and Inverary Castle (Argyll and Bute, 1886), and Ferniehurst, Montaviot and Minto (Roxburghshire, 1888). In Galloway in 1891 Cleghorn was already staying with the Maxwell family at Munches and accompanied the party on to Castle Kennedy where he must have been particularly interested in the Sikkim rhododendrons, which, forty years on from their first introduction by Joseph Hooker, had already attained huge stature. Cleghorn also took part in at least one of the Society's forays south of the border when in 1885, after visiting Dumfries-shire (Langholm, Canonbie, Netherby) they went to Carlisle (the 30-acre Knowefield Nurseries) and Westmorland (Lowther Castle). Hearty breakfasts, luncheons and dinners were enjoyed, including the ritual drinking of toasts, provided by hosts such as the Duke of Buccleuch at Drumlanrig, or in local inns such as the Spread Eagle at Jedburgh or the Weem Inn near Aberfeldy. The 'big house' and its gardens were usually inspected, guided by the proprietor, his factor or wood manager, as well as any notable local antiquities such as castles or abbeys, though the main focus was, inevitably, on trees. Notable specimens were measured with the aid of 'Kay's dendrometer';[31] these were usually planted specimens, including recently introduced North American conifers, though in some estates there were also ancient pines, oaks and Spanish chestnuts and near Jedburgh, on the Ferniehurst estate, were two ancient oaks, survivors of the primeval Jed Forest – the 'Capon Tree' (which still stands) and the 'King of the Wood'. Accounts of the excursions were published nationally in the *Journal of Forestry* and the *Gardeners' Chronicle* and by the Society in pamphlet form;[32] these contain much of interest – descriptions of forests and trees, the afflictions of foresters including devastating gales, and pests and diseases. With the onset of extensive introduction came the beginnings of outbreaks of *Phytophthora* in larch, which in the case of Spanish chestnut had started as far back as the late eighteenth century. There were also reports of new technologies and at Dalkeith the party saw a new and improved way of making charcoal in a retort partly fuelled by the gases released. Altogether more unexpected was the third Marquess of Bute's early attempt at naturalising colonies of beavers and wallabies at Mount Stuart.

Two of the Annual Excursions are of particular note – a two-day visit to Balmoral in the last of Cleghorn's presidential years, by invitation of Queen Victoria (the Society's Patron since 1869) in the year of her golden jubilee (Fig. 65).[33] 1887 was also the year in which the Society was given a royal charter and became the *Royal* Scottish Arboricultural Society. The party was shown round the royal estates of Deeside by her factor ('Commissioner') Dr Alexander Profeit and here, unusually, it was native forests that provided the greatest interest. The party of over 40 visited Birkhall, the royal shooting lodge of Alt-na-Giuthasach at the eastern end of Loch Muick, then travelled through Glen Gelder to Balmoral. The following day, 29 July, they visited the forest of Ballochbuie on the south side of the Dee to the west of the castle, which had been purchased nine years earlier by Queen Victoria to preserve its ancient Scots pine from logging and which remains to this day as a substantial remnant of the ancient Caledonian Forest. In addition to the magnificent trees Cleghorn the botanist must also have examined the ground flora, noted for rarities such as *Linnaea borealis* and *Pyrola secunda*; they must also have heard the sharp calls of Scottish crossbills as they tweezered seeds out of pine cones, and the occasional crashing through the trees of Britain's largest game bird, the capercailzie. The Jolly Foresters examined the stump of the remarkable 270-year-

13. SCOTTISH SOCIETIES

old tree with a girth of 25½ feet, of which the Queen had exhibited a cross-section at the 1884 Forestry Exhibition (Fig. 65), which they considered to be the result of the fusion of several trunks. Also noted was the fact that no regeneration of trees was occurring except within fenced areas near the Castle, due to intensive grazing by deer. In the private chapel at Balmoral they admired panelling made from finely-figured Ballochbuie pine and to commemorate the visit a specimen of the Arolla pine (*Pinus cembra*) was planted in the grounds.

The second excursion that calls for notice took place on 10 August 1892,[34] when Cleghorn invited the Society to Stravithie, though he did not accompany the party onwards to Balbirnie, Falkland, Scone, Murthly and other Perthshire estates – at 72 his energies were clearly on the wane. The party left Waverley Station, Edinburgh, at 6.35 a.m. and by means of the 2½-year old Forth Bridge, which spanned the estuary 'with cyclopean strides', and the new (if slow) coastal line, they reached the railway station at Stravithie at 10 o'clock. The Jolly Foresters affectionately dubbed Cleghorn the 'Grand Old Forester' and presented him with an 'oak walking-stick, which combined with it an ingenious steel band-saw arrangement'. After an elegant *déjeuner* in a marquee on the lawn the party inspected the property and made the valuable record of Cleghorn's garden and estate that has already been described.

Fig. 65. Balmoral forest, as sent by Queen Victoria to the International Forestry Exhibition, Edinburgh, 1884. Wood engraving by John Swain, after drawing by Horace Morehen based on sketches by 'W.D.' from *Illustrated Sporting and Dramatic News*, 12 July 1884.

CHAPTER 13

Presidential addresses

Cleghorn performed various roles in the Arboricultural Society – he was Convener of the committee responsible for its *Transactions* from at least 1878 up to his death (and, as with the Botanical Society, contributed money to its illustrations fund), and from at least 1878 to 1885 was Convener of the group that judged the annual competitive essays. The Presidency also carried with it the duty of an annual address. If, at times, somewhat rambling in structure, the frequent references to India can be taken to be not so much as a case of nostalgia on Cleghorn's part, as an indication of the great interest of the Subcontinent to the people of Scotland (a large number of whom had relations working there), and of the international reach of the Society. The six talks contain much interesting material regarding Cleghorn's mature views on arboriculture and education (like the Highland Society, the Arboricultural was a major force in lobbying for an Edinburgh Forest School), and, as with the similar Botanical Society address, reveal his interests in bibliography (including current literature) and the paying of tribute to departed colleagues.

The first address of 1872 is the most substantial, and represents a tactful and understated contribution to the argument of why the Continent rather than Scotland provided the most suitable training for Indian forest officers.[35] It also contained an important statement about the reasons for the beginning of forest conservancy in India – that 'the introduction of railways, and the rapidly increased demand for timber for railways and fuel, at length forced the attention of the Government to the vital question of forest management'. At this stage he was still being realistic. Seven years later, as discussed below, this had changed.

Cleghorn started with the question of the importance of forests for the environment, the differences between British and Continental forests, the history of forest exploitation and the fact that Britain was the exception to the rule: he believed that there was no need for State control because of its coal resources and ability to import timber. In this lies an irony as it meant that forestry could be left to private landowners who had the luxury of being able to think of aesthetic matters, even although:

> few private individuals can afford to take that higher view of forest conservancy which wishes to make provision for generations yet unborn, and fewer still, perhaps, realise the extent of their ... national responsibilities to maintain the forests which they possess.[36]

There were, however, a few points of similarity across the continents – for example the alpine forests of Italy, France and Germany provided important lessons for the Himalayas, and there was even a tenuous link between the unique system of floating logs down the River Spey and the use of much more powerful (and un-dammable) rivers for similar purposes in India. But there were vast differences, not only in such matters as elephants, but the much larger scale of operations in India, where forests could not be enclosed, where foresters had to know much more botany due to the large number of species involved (and hence the need for Floras and Manuals), and the use of forests for a wide range of products other than timber. These matters were reverted to in several of the later addresses, and in the third, in 1874,[37] he quoted Brandis's statement on the importance of forest conservancy for public benefit and future need, which was to be achieved by a broad and 'comprehensive view of the whole forest vegetation, and not only what might *initially* [emphasis added] appear to be important [i.e., the timber]' – a philosophy that came to be known as the 'household of nature' view. Cleghorn,

13. SCOTTISH SOCIETIES

as had Brandis, stressed the value of the extensive French and German forest literature as another strength of the Continental system, and frequently referred to recent publications, either relating to Europe or India (on which he kept himself up to date, for example Baden-Powell's and Gamble's Report on the 1874 Allahabad Conference).

The most interesting topic in the 1873 address was Cleghorn's statement on progress with a project that went back to earlier interests of his own and had a link with the British Association – though relating to Scotland rather than the Tropics.[38] At the 1868 annual meeting of the Arboricultural Society, in the year of his return and perhaps at least partly at his suggestion, a deputation had been named 'to wait on the British Association in testing the influence of Forest on climate'.[39] The convener of the group was J.H. Balfour and the other members appear to have been Cleghorn, Robert Hutchison of Carlowrie (an active member and several-times President of the Society), John Sadler and Alexander Buchan (Secretary of the Scottish Meteorological Society). This resulted in the BAAS awarding Balfour grants of £20 in each of the years 1870, 1871 and 1872 towards investigations into the effect of trees (or their lack) on rainfall in North Britain.[40] In his address Cleghorn paid credit to Marsh's historical review of the topic, but noted that there was still a great need for quantitative data. At the Bradford meeting in September 1873 Balfour's group was able to read a progress report,[41] as on 11 July of that year Cleghorn and Buchan had chosen three stations at Carnwath in Lanarkshire, one inside, and two outside, a forest, at which accurate recordings started to be made two months later.

Cleghorn frequently urged junior members of the Society to self-improvement through observation; for example of insect pests, or of the facilities offered by RBGE – its museum and the Pinetum that in 1873 had recently been laid out by James McNab. Scope for observations of exotic trees was greatly increased after 1877, when the policies of Inverleith House, adjacent to RBGE, were purchased by the City for £20,000 as an arboretum. Related to this was growing pressure for a School of Forestry, a major subject at the 1885 Select Committee to be discussed later. Cleghorn's 1884 address was mainly devoted to a review of the spectacular Forestry Exhibition,[42] one of the most impressive sections of which, somewhat surprisingly, had been sent by Japan, a country where the importance of forestry to general welfare had been realised long before it had in Britain, and where forestry formed 'an important feature of national education'. The last address in 1886 was the shortest,[43] in which Cleghorn talked about his recent visit to the Indian and Colonial Exhibition in South Kensington, where, as a reminder of his interest in botanical illustration, he had been especially impressed with photographs and drawings of Australian eucalypts and Canadian conifers. He also spoke about the growing collections of the Society – it had been decided to give the 'Specimens illustrative of forestry' to the Edinburgh Museum of Science and Art (but some specimens were kept that, with the library, were donated to RBGE in 1895).

Cleghorn's devotion to and efforts on the Society's behalf were recognised with his election as an Honorary Member in 1885.

Fig. 66. Charcoal portrait drawing of Cleghorn by Theodore Blake Wirgman, dated 6 July 1888, presented to Cleghorn by the Royal Scottish Arboricultural Society at a dinner in the Waterloo Hotel Edinburgh on 7 August 1888. (Sprot family collection).

14
Grand Old Forester

The sobriquet bestowed on him during the Jolly Foresters' visit to Stravithie in 1892 aptly describes Cleghorn's status in Scottish and Indian forestry circles during his later years (Fig. 66). It is the more surprising, therefore, that the India Office never saw fit to bestow on him one of the orders that were now routinely awarded to men of distinguished Indian public service. In the last two decades of his life he continued to publish and to contribute to institutions that kept him at least to some degree in the public eye, though the only public accolade that he ever received was an honorary doctorate of laws (LLD) from St Andrews University in 1891.[1] These 'late works' form the subject of this penultimate chapter – his contributions to the scholarly ninth edition of the *Encyclopaedia Britannica*, his involvement with the ground-breaking Edinburgh International Forestry Exhibition of 1884, the evidence he gave to Sir John Lubbock's Select Committee on forestry the following year, and his involvement with the two universities he had attended as an undergraduate nearly half a century earlier.

The Ninth Edition

Other than his Arboricultural Society annual addresses, Cleghorn's last significant forestry publications were a pair of articles published in the ninth edition of the *Encyclopaedia Britannica*. Since its first edition of 1768–71 (written largely by William Smellie, a botanical pupil of John Hope), this work had been something of a Scottish national institution. Two of Cleghorn's friends had written the article on Botany for the previous two editions – Walker-Arnott for the seventh (1831) and Balfour for the eighth (1854). The ninth edition, the third to be published by Adam & Charles Black, was to be on an altogether more ambitious scale. It was to be edited by Thomas Spencer Baynes who since 1864 had been Professor of Logic, Metaphysics and English Literature at St Andrews – and hence, probably, Cleghorn's invitation to write the articles on Arboriculture (vol. 2, 1875)[2] and Forests (vol. 9, 1879)[3]. Known as the 'Scholar's Edition', Baynes commissioned articles from more than a thousand authors, including well-known figures from Robert Louis Stevenson and Patrick Geddes to the Indian philologist Max Müller. Several St Andrews academics known to Cleghorn also wrote articles (in the second volume these included W.C. McIntosh on Annelida and, in the ninth, John Bell Pettigrew on Flight), as did individuals from his Indian past. Henry Yule wrote numerous geographical articles and Richard Strachey wrote the one on Asia for the second volume; Balfour once again authored Botany (vol. 7, 1878). Baynes wanted the work to be both 'an instrument as well as a register of scientific progress', which he achieved by taking advice from James Clerk Maxwell and Thomas Henry Huxley.[4]

Cleghorn's two articles are substantial works (each running to about ten double-columned pages), essentially reviews, in which he dipped widely, if not particularly deeply, into his extensive library, supplemented by his own rich experience. The first is on Arboriculture – the culture of trees, which he considered the most recent branch of agriculture, necessary due to historical deforestation, achieved by means of plantations. Although Cleghorn stated that the best trees for this purpose were those native to the region (in Britain: oak, ash, elm, Scots pine, etc., or those from similar latitudes, especially larch and silver fir), a large part of the article was, in fact, devoted to one of his greatest interests – exotics, of which he noted the importance of the role of botanic gardens in their distribution from the time of John Evelyn in the seventeenth century onwards. Cleghorn treated, very briefly, various general topics such as planting for avenues and hedges, raising trees in nurseries, propagation from seed, the transplanting of trees (which he favoured over direct seed sowing in creating plantations), pruning, and the use of nurse-trees to encourage growth. The longest section, however, consisted of individual species accounts – timber trees (conifers, followed by broad-leaved – both 'hardwoods' such as oak, and 'softwoods' such as poplar), followed by ornamental trees, in which non-natives again figured largely. The article ended with an apology for its brevity, and, as usual for Cleghorn, a bibliography of works where further information could be found (including literature not only on North America – the richest source of introductions – and Europe, but, inevitably given the author's experience, India).

An intriguing speculation is what Cleghorn might have thought about one of the other articles encased in the same handsome binding of Volume 2 – the one on 'Angel' by William Robertson Smith that, with successors from the same pen, became *causes célèbres*. Robertson Smith was Professor of Hebrew at the Free Church College in Aberdeen, but having studied in Germany was an exponent of the 'higher criticism' – the application of modern techniques of literary analysis to the books of the Bible. While not questioning their unique value, Robertson Smith accepted (without apology or gloss) that this showed that many were neither the unimpeded, nor indeed unimpeachable, Word of God, but the product of widely varied authorship, date and authenticity. This was a step too far for the Free Church authorities, who tried Robertson Smith for heresy over a period of several years, sacking him in 1881. After this he became joint (and, later the sole) editor of the ninth edition of the *Encyclopaedia Britannica*, then an Arabic and Semitic scholar at Cambridge (where he held a fellowship at Christ's, Darwin's college).[5] Unless he had mellowed substantially in late middleage, Cleghorn is likely to have viewed such developments in biblical scholarship with, at the very least, disquiet (though Henry Yule was a supporter of Robertson Smith as was Cleghorn's father-in-law Charles Cowan). It did not, however, stop Cleghorn from publishing in a later volume of the *Encyclopaedia* – Smith's most notorious article – on the Bible – appeared in the third, but Cleghorn's next article, on Forests, was in the ninth, in 1879.

This second article concerned the 'management of forests or sylviculture'. In his evidence to the Select Committee given six years later, Cleghorn would make an interesting distinction between sylviculture and arboriculture. Although the question was put to him in an overtly leading manner, his reply was that 'No country in the world has such fine specimens of trees in point of arboriculture, but as regards sylviculture we are deficient; the one is what is called *jardinage*, and the other is professional forestry'.[6] The article can therefore be seen, along with his active promotion of the establishment of a Forest School, as a step towards raising standards in such matters. But it is also much more interesting than the article on Arboriculture in another respect, as it reveals a development in Cleghorn's ideas on the value and importance of forestry – away from the utilitarian stance of his professional Indian period, towards a greater emphasis on the 'secondary benefits' of forests. It starts with an explicit statement of their importance: 'in the general economy of the globe, influencing humidity of the air and the soil, mitigating the extremes of heat and cold, affording shelter to man and beast, and enriching the soil on which they grow'. A humanitarian concern is also present when he firmly attributes the 'causes of the terrible famines in India and China' to the 'denudation of mountain slopes'. The main part of the article is arranged geographically on a continent-by-continent basis. For the major forested countries from Europe to America, Asia, Australia and (briefly) Africa, Cleghorn described the forest types, their acreage and species composition, 'the systems of conservancy adopted or the preliminary measures adopted for the better management of *state* forests [emphasis added]' and, for European countries, the systems in place for the education of foresters.

That the section on India is especially interesting is hardly surprising, but what does come as a shock is Cleghorn's claim that 'attention was first directed to conservation in India by the appointment of a committee by the British Association in Edinburgh in 1850'. The general reader was hardly to know of Cleghorn's dominant role in this committee and its report, but it is not so much the immodesty, as the questionable accuracy of the claim, that calls for notice. As noted earlier there is little evidence to show that the report had a direct influence on the Indian government, whereas there is strong evidence coming from pressure from engineers and users of timber. After his global survey Cleghorn ended with something that is hard not to read as a blatant, if uncharacteristic, act of vanity – an almost verbatim quotation of the by now antiquated conclusions of the 1851 BAAS report. Of the seven items only one superfluous conclusion was removed, substituted by one that forms some sort of a balance to this 'new Cleghorn' – an assertion of his old belief in the utility principle: 'that it is a duty to prevent the excessive waste of wood, the timbers useful for building and manufactures being reserved and husbanded'. Also mentioned is the concept of 'reservation' that did not exist in 1851.

The reason for this change in emphasis is unknown,[7] though several possibilities come to mind. It could be a veteran's view, long after the predominant financial needs of his employer the EIC had faded into the background. Another influence must have been the recent renewed interest of the British Association in the effect of 'denudation' on rainfall, and the committee on which Cleghorn had worked with Balfour in the early 1870s. But it could also be the result of his readings of, and friendship, in the post-Indian years with G.P. Marsh (whom he had visited for a second time in 1875). Marsh's work, however, is quoted only very briefly and yet, duly summarised and given more prominence, would have been a far more powerful argument for forest conservancy than the limp repetition of the conclusions of a 28-year old report. Whatever the reasons for Cleghorn's change in emphasis, it is worth quoting the peroration at the end of the article, which can be taken as his final, and strongest, statement on the 'climate issue', while remembering that his priorities had not always been thus:

> It is the climate and physical importance of a due proportion of wooded land, independent of the utility of forest products ... that has at length awakened most of the civilised Governments to the necessity of protecting forests from ruthless spoliation .... Necessity has caused the adoption of these principles [those of the 1851 report] in many lands .... and forestry will henceforth be studied as a science as well as practised as an art. It is manifestly of the greatest importance that in the progressive development of great countries, just and enlightened principles should influence the views and actions of those who are charged with the duty of advising Governments in regard to the material resources comitted to their care for behoof of present and future generations.

The 'Ninth Edition' was extremely widely read (not only in the original, but also in various pirated American versions). Cleghorn's mature opinions were therefore widely circulated, and at least one case has come to light in which these had a direct influence in contentious questions of conservation. In a parliamentary debate in New South Wales in 1881, when the question of killing trees by ring-barking (for conversion of forest into grazing lands) was controversially being discussed, Cleghorn's article was cited as an authority.[8] And it is likely that this is not the only such case.

CHAPTER 14

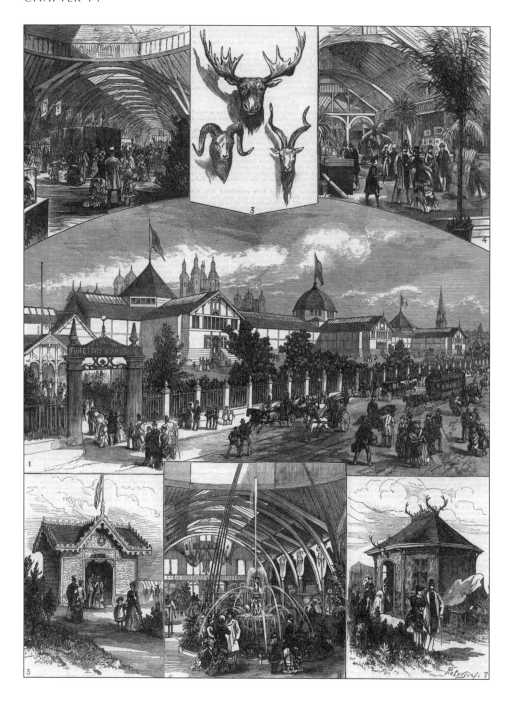

Fig. 67. The International Forestry Exhibition, Donaldson's Hospital, 1884.

Exterior view (above, middle), Indian Collection with Central Fountain and palms provided by RBGE (lower centre) and Queen Victoria's Balmoral Châlet (lower right). Anonymous wood engraving, *Illustrated London News*, 12 July 1884 (above).

Interior, with Cleghorn (standing, second from left) manning the Scottish Arboricultural Society stand. Lithograph (based on a photograph) by McFarlane & Erskine. *Transactions of the Scottish Arboricultural Society* vol. 11, part 1, 1885. (RBGE) (right).

14. GRAND OLD FORESTER

International Forestry Exhibition 1884

When standing in West Coates, the main thoroughfare leading westwards out of central Edinburgh, facing Donaldson's Hospital it is impossible to imagine the scene had one stood on the same spot on 1 July 1884. All but a fringe of ogival pepper-pot pinnacles, in copper and stone with gilded-pennant weather-vanes, of William Henry Playfair's version of an Elizabethan prodigy house (built as a school for deaf children), would have been obscured by a long, low, wooden building, twice the length of the Hospital, crowned by three domes and with three projecting transepts (Fig. 67).[9] On that day, to the martial strains of the band of the Royal Scots Greys, the Marquess of Lothian, Schomberg Henry Kerr, declared the International Forestry Exhibition open. Although forestry had been represented in earlier international exhibitions, this was the first one ever to be devoted to the subject exclusively. The brain-child of the Scottish Arboricultural Society it had been two years in the planning,[10] during which Cleghorn had played a major role as the Society's representative on the Exhibition's Executive Committee (styled an 'Honorary Secretary', under the chair of Lord Lothian). His obituarist Andrew Taylor considered that this event was probably 'the greatest break' in the 'systematic routine' of Cleghorn's retirement.[11] Appropriately, this triumph was realised in one of Cleghorn's presidential years and the only kill-joy note, in a typical example of East-Coast-West-Coast rivalry, came from the *Glasgow Herald*.[12] Rather than waiting for the event itself, the edition of the opening day gloatingly predicted that there was 'little probability of the majority of the sections being anything else than a mass of interesting confusion'. This was because not all of the exhibits had arrived in time – notably the large Japanese contribution and, on a much smaller scale, J.S. Gamble's one from Madras. Two of the more crowd-pleasing features – electric lighting and an electric train – were also in a state of hiatus due to technical problems. All of these,

however, proved but minor and temporary glitches and between 1 July and 11 October more than half a million people visited the exhibition, which was open from ten in the morning to ten at night for the price of a one-shilling ticket.

The scale of the enterprise was enormous – the site extended to 15 acres, as to the west of the exhibition building was a field in which outdoor exhibits and machinery were displayed. The building itself was designed by the City Architect, Robert Morham, who already had experience in large display-buildings – including both a Greenmarket and a Wholesale Meat Market for the city, and a Winter Garden in West Princes Street Gardens. In addition to private loans, official ones came from all over the world, with national exhibits not only from British colonies, but countries including Venezuela, Denmark, Johore and Japan. The Arboricultural Society had its own substantial display area close to the main entrance, but one of the most spectacular features, at the heart of the building, was the Indian Court arranged with military precision by Colonel James Michael and decorated with palms and tropical plants lent by the RBGE.

The range of objects shown was breathtaking, and although largely didactic and serious in purpose – including large quantities of wood samples, machinery representing the latest technology, official publications and maps[13]– in retrospect it is hard not to be drawn to the more esoteric and picturesque items among the 309 individual collections (some very large) shown inside the wooden marquee. From a ladies 'Muff made from Thistle Down' lent by Agnes Cowieson and 'Wood from the staircase of the house where Adam Smith wrote the "Wealth of Nations"', to a series of skins of woodpeckers – friends of the forester for their entomophagous lifestyle. These birds were not the only taxidermy – by means of loans from the Queen, the Prince of Wales and others, displays of big game trophies gave the interior the appearance of a grand Highland shooting lodge. From Colonel Michael himself came a gaur and some elephant tusks, mementoes of his days in the Anamalais that Cleghorn must have viewed with nostalgia. The Colonel also lent two rather more curious artefacts – the 'Head of a Man-eating Tiger, mounted as a letter-box' and an 'Elephant's Foot – Footstool'.

The arts – both pure and applied – were heavily represented. Among the latter were carved wooden items of all possible sizes, shapes and functions – from walking sticks to substantial furniture, including a 'Kaffir Pillow' (headrest) lent by Cleghorn, doubtless a souvenir from his enforced sojourn at the Cape of Good Hope in 1848. At various points in the building were hung paintings by artists both professional (including Waller Hugh Paton) and amateur, among which numbered a 'Series of [watercolour] Forest Studies from Nature in the Himalayas, California, Ceylon, Fiji, New Zealand, Australia, Japan' by Constance Frederica Gordon-Cumming (Fig. 68). 'Eka' was an intrepid traveller, writer and artist, a Scottish answer to Marianne North, though less exclusively botanical. She came of an interesting family, her mother Eliza Gordon Cumming (who may or may not have borne a child of Shelley's) was a distinguished collector of Old Red Sandstone fossil fish from the area around the family seat of Altyre, Morayshire, in which she had been encouraged by Hugh Miller, but also by John Grant Malcolmson, a Madras surgeon who in India had greatly helped Robert Wight's career.

Cleghorn rented a house in the neighbourhood for the duration of the exhibition so that he could be on hand, and according to Taylor 'the spare lithe form of the doctor fleeting about its halls, was well-known to the general visitor' (Fig. 67). Much of this time must have been spent in the 'Indian court', but in the present context only a taste of the exhibits lent by Cleghorn

14. GRAND OLD FORESTER

Fig. 68. Simla Bazar, showing Christ Church, c. 1868. Watercolour by Constance Frederica Gordon-Cumming.

and his friends can be given. Cleghorn contributed eleven objects, the first one listed being a memento of his Himalayan travels: a wooden Tibetan printing block prepared at Keylong by the Moravian missionary the Rev. H.A. Jäshcke. There were also three exotic fruits – a baobab from Madras, the gigantic pod of the climbing legume *Entada* from Malabar, and 'vegetable ivory', the fruit of the South American palm *Phytelephas*. Unfortunately anonymous were four watercolours, one showing the effects of the 1864 cyclone on the Calcutta Botanic Garden, and three of 'Aborigines of the Nilgiris' (presumably Badagas and Todas). Most of Cleghorn's loans were more strictly forestry-related, showing the effects of insects or plant parasites on wood: a mango branch encircled by the mistletoe *Loranthus longiflorus* from Calicut; two items from Rangoon damaged by the wood-boring, marine mollusc *Teredo navalis*;[14] from Madras came a railway sleeper damaged by the carpenter bee *Xylocopa* and a stick of a coffee bush from Wynad damaged by the Coffee Borer. At Stravithie Cleghorn evidently had not only a library and herbarium, but an interesting collection of artefacts and naturalia.

Of items lent by friends or acquaintances, the Marquess of Tweeddale (whom as Lord William Hay Cleghorn had known in Simla) lent a model of an Indian temple in carved shola pith.[15] Sir Walter Elliot sent a carved portfolio stand made in Bombay from blackwood (*Dalbergia latifolia*), Henry Yule some leaves of the raphia palm from Zanzibar, and William Coldstream paper-making material of *Desmodium tiliaefolium*, familiar to Cleghorn from the Himalayas. Douglas Hamilton, now a Lieutenant General, who had been with Cleghorn in the Anamalais whose sketches had been used to illustrate *Forests & Gardens*, exhibited some drawings of 'Forest Scenery' and 'Sketches of Sporting Incidents'. More intriguing, however, was a set of 'Coloured Drawings of Indian Trees' lent by Sir George Birdwood – no details are provided, but given his Bombay connections (and the fact that the next item in the Catalogue, 'Botanical specimens', were from the Poona Botanical Gardens) might these, just possibly, have been Alexander Gibson's Dapuri Drawings? Alexander Hunter lent 17 smallish wooden objects, the most interesting of which were items that had been made in his own School of Art – including a carved satinwood picture frame with a 'Design from *Momordica charantia*' (i.e., the twining cucurbit, bitter gourd), which shows that at least some of the printed abstractions made by his students from nature were carried out in three dimensions (see *Cleghorn Collection* Fig. 41).

Typical of the much larger numbers of exhibits relating to more serious forest matters was Gamble's, which eventually arrived from the Northern Circle of the Madras Presidency.[16] In addition to a detailed account of conservancy in the area (which was making a small annual profit of 124,355 Rs, largely from the sale of timber (sal and teak), 'fancy timber' (red sanders and sandalwood) and railway fuel), Gamble sent a set of beautifully mounted herbarium specimens, leaves only, of 136 species of forest tree. Cleghorn must have claimed these after the Exhibition and they are now in the RBGE herbarium, mounted with the appropriate clippings from the pages of Gamble's *Manual of Indian Timbers* (Fig. 60).

In contemporary accounts of the exhibition great attention was paid to the contributions from the Royal Family. Queen Victoria had early on agreed to be Patron and her own most spectacular loan was located in the open-air section. This was arranged around a band-stand and refreshment room, the 'Machinery in Motion' being placed out of harm's way at the northern extremity of the site, and included trade displays by leading manufacturers. There were also more picturesque items including a Swiss Chalet, a model Manitoba Farm, and the Queen's octagonal chalet made entirely of Scots pine from the Deeside forests, its roof sprouting antlers and thatched with pine-bark, beside which was a cross-section of the trunk of a veteran pine (Fig. 65). Altogether more modern was the Electric Railway, which, after initial teething difficulties, opened on 17 July.[17] Designed by Henry Binko, an Austrian, the train plied 700 yards around three sides of the exhibition building, on which visitors could ride for a fee of thruppence. This pioneering piece of technology, powered by a fixed dynamo, had been demonstrated for the first time at Crystal Palace only two years earlier. At the end of the exhibition Binko used the same dynamo to connect to the horse-drawn tramway on the road between the Exhibition and Haymarket, proving for the first time in Britain that an electric tramway was a possible form of public transport, far cheaper and with far less fuss than the cumbersome system adopted in Edinburgh 130 years later. The train was miniature in scale, and in contemporary photographs the passengers dwarf the three carriages and the locomotive 'Ohm'. The exhibition was visited not only by the masses, but by royalty and the Prime Minister (himself an exhibitor, who, notable as an arboricide, had lent a 'Presentation Silver Axe and Splinters of Ash'). For the visit of the Prince and Princess of Wales on 22 August, Binko had a local carriage-builder make an elegant royal coach (the 'Alexandra'). This coach was thereafter put into general use, and was used by Gladstone when he visited a week later, accompanied by the Earl and Countess of Rosebery. On this occasion Mrs Gladstone and her daughter were presented with alarming-sounding 'electric bouquets', sparkling, but hopefully not sparking, with incandescent lights.

The exhibition also formed part of the Arboricultural Society's strategy to establish a School of Forestry in Edinburgh, and the small surplus made (£291 14 shillings) was earmarked for this purpose. Another legacy was the in the form of official publications – a general catalogue, one for the Indian collection, and a Calcutta-published bibliography of the publications and maps that had been sent from India. The *Official Catalogue* can be browsed with immense enjoyment for the ingenuity and sheer exuberance of the mixture of spectacle, education and eccentricity, not only of the exhibits themselves, but the interspersed advertisements, including one for 'Oldridge's Balm of Columbia' – 'the best and only certain remedy ever discovered for preserving, strengthening, beautifying or restoring the Hair, Whiskers, or Moustaches, and preserving them from turning grey'. The spectacle should not, however, distract from the didactic aims of the exhibition, which Lord Lothian (a Conservative politician and later Secretary for Scotland, but also a keen arboriculturist on his estates at Newbattle, Ferniehurst and Monteviot) had expressed in his opening address – the aim was "to show what Nature was, and to try to teach those who visited it how every tree might be fostered and preserved for the benefit of man in the future".[18]

Colonel Michael, curator of the Indian exhibit, asked Sir George Birdwood to write a preface to the catalogue devoted to this substantial section of the exhibition. In this Birdwood gave an account of the current state of the Indian forest service, which by now administered 46,000 square miles of Reserved Forest and was yielding a net annual revenue of £300,000. More remarkable is his account of the history of the development of the service, opening his preface with the extraordinary statement: 'It may be said that Forest Conservancy in India originated in Edinburgh'.[19] The service had by now reached a stage of maturity when histories of origin were being forged: and as is not unusual in such cases, these were not undisputed, and Birdwood's version is certainly open to question.

The reason for asking Birdwood for his contribution must have been his extensive experience with Indian exhibitions, of which it is worth saying more before returning to the 'origin myth'. Birdwood's background had many parallels with Cleghorn's own,[20] though born 17 years later, in the Bombay Presidency and of military parentage, he too had studied medicine in Edinburgh (with botany under Balfour) graduating MD in 1854. Birdwood returned to Bombay and held a similar position to Cleghorn, as Professor of Anatomy and Physiology, and later of Botany and Materia Medica, at the Grant Medical College. As curator of the Government Museum he had a major role in the establishment in 1861 of the botanic garden in the grounds of the exquisite Victoria & Albert Museum, which was based on the collections of the Agri-Horticultural Society of Western India.[21] After returning from India in 1868 Birdwood held various posts in the India Office, including charge of the India Museum and the curation of exhibits for several international exhibitions, being knighted in 1881. Birdwood was a great promoter of Indian handicrafts, while at the same time notorious for his denial of the existence of an Indian fine-art tradition. In 1910 he would claim that 'a boiled suet pudding would serve equally as well [as a sculpted Buddha] as a symbol of passionless purity and serenity of soul'.

While not quite so controversial, Birdwood's version of the origin of the Indian forest service in the 1884 catalogue calls for comment. In his account Birdwood distinguished 'practical' from 'scientific' forestry – but although he did mention the efforts of 'Daddy Gibson' in Bombay, he gave the palm to Colonel Michael as 'the pioneer of practical forestry' – relating to his efforts in the Anamalai teak forests, following pressure on the Madras Government from Frederick Cotton. But Birdwood also claimed that the 1851 BAAS report had 'attracted the attention of' the Court of Directors', which is one of the most explicit statements of any such influence of Cleghorn's report, but made 34 years after the event and as he was an Edinburgh undergraduate at the time it is hard to know how much to believe it.[22] The British Association meetings, including Cleghorn's report and its conclusions, were reported at the time, and in some detail, in the popular press (e.g., in *The Athenaeum*), and the full report in the Association's annual report in 1852, and two years later in Calcutta in the *Journal of the Agricultural and Horticultural Society of India*, but no *contemporary* evidence has yet come to light that the hard-nosed EIC Directors were influenced by any of this. They are surely as likely to have shared the anti-scientific view expressed by Charles Dickens in *Bentleys' Miscellany* from 1837 to 1839, in which he satirised the BAAS as the 'Mudfog Association for the Advancement of Everything'. Birdwood fell short of claiming that the report was the direct inspiration for forest conservancy in India, but allowed that conclusion to be drawn by saying that 'within a few years afterward regular Forest Conservancy Establishments were sanctioned for the Madras Presidency and British Burmah'. This he took as the start of 'scientific' forestry in India and hailed Cleghorn as 'the father of scientific forestry' – thus putting limits on Sir John Lawrence's claim of 1865 (see p. 171). Lawrence's claim was unfair to Alexander Gibson, but as Cleghorn was most certainly not a 'scientific forester', Birdwood's was certainly less than fair to Brandis.

In 1901 Birdwood would repeat his claim for the influence of the 1851 report, which he believed had drawn attention:

> to the need of extending forest conservancy in India, and to the fact that where supervision had been wisely exercised improvement had at once taken place; and undoubtedly this also contributed to the resolution arrived at by the Court of Directors on the basis of General Michael's operations to make the conservancy of the forests henceforth an object of primary solicitude in India.[23]

Lubbock's 1885 Select Committee on Forestry

The Exhibition was widely noticed and undoubtedly contributed to raising awareness of forestry issues in Britain in the 1880s, and a widespread concern arising in part from the pitifully low forest-cover in Britain compared with its Continental neighbours (1½ million acres in England, 750,000 in Scotland). There were also worries about the low standards of management, which, with the exception of the three Crown forests in England (Dean, New, Windsor) were entirely in the hands of private proprietors and therefore subject to vagaries of personal taste and market forces. Moreover, the men employed as foresters, though in many cases skilled on a practical level (especially in Scotland), had no scientific training. Such training would allow the making of better decisions in matters such as matching species to growing conditions (to avoid mistakes that would not become apparent for decades) and in observing matters that were starting to become problematic such as insect pests and what are now known to be fungal diseases (the larch disease being much discussed at the time).

To address such questions Sir John Lubbock, on 15 May 1885, tabled a question in the House of Commons that led to the appointment of a Select Committee to consider the question of 'whether, by the establishment of a Forest School, or otherwise, our Woodlands could be rendered more remunerative'. There were two other reasons for the timing of this Select Committee. The first was the visit in 1881 of a high-powered group from the Nancy École (the three professors Lucien Boppe, Eugène Reuss and Eugène Bartet, with some of their students) who had written a report that included remarks critical of British forestry,[24] and concluded with the recommendations that a 'National Forest School' should be founded in Britain, and that chairs of sylviculture should be instituted at Edinburgh and at the Indian Engineering College at Cooper's Hill. Cleghorn had accompanied this French party on the Scottish leg of its tour, which had been paid for by the India Office and arranged by Colonel George Falconer Pearson. Pearson was another retired Indian forest conservator (like Cleghorn also originally of the Madras Army), who had supervised the British students in Nancy 1873–84, and who also gave lengthy evidence to the Select Committee (in 1885 and again in 1886). The second related reason was that in September of this same year, 1885, the French recommendation had been carried out by bringing Wilhelm Schlich from India to start the training of Indian foresters at Cooper's Hill (Fig. 49) so that students formerly sent to France and Germany could be trained at home. In other words, the question of training foresters for India and the Colonies was in the air in 1885, for which reason Indian experience, including Cleghorn's, was extensively consulted in the witnesses examined by the Select Committee, which could be expected to have a bearing on training for local needs in Britain.

Due to parliamentary interruptions the Committee would meet over three years, involving ten days of evidence from 30 individuals. The discussions ranged widely and raised many issues that are still of interest (including competing land-use between sport and forestry, pathology and entomology, and the first stirrings of the question of State involvement in British forestry – for example, government loans to proprietors to increase the forest cover, especially by exploiting large areas of 'waste' in Ireland, Scotland and northern England).

It is not possible to go deeply into these discussions here. They were published at the time in three 'Blue Books',[25] but, given the great interest taken in Scotland, especially in Edinburgh, where discussions on these matters had already been initiated, the three reports were summarised in the *Transactions of the Scottish Arboricultural Society*.[26] Hardly surprisingly no clear conclusions emerged, but there was strong overall support for the better training of foresters, though opinions varied widely as to the best way to achieve this. For example, whether or not it was necessary to have a dedicated large forest as a training ground, and whether by a school in each of the three countries, by an extension of the remit of Cooper's Hill, or by using existing establishments such as the Agricultural Colleges of Cirencester and Downton or Edinburgh University. The final conclusion, in 1887, was to recommend the setting up of a Forest Board under a Government-appointed director 'to organise Forest Schools, or, at any rate, a course of instruction in forestry' and to draw up a syllabus and examine on the subject. This body would be made up of representatives of interested parties including the Scottish Highland and Arboricultural Societies, the director of Kew, the Agricultural Colleges, MPs, peers and landowners. But in the meantime, as will be seen, Edinburgh (with Cleghorn's help) had gone ahead on its own.

Evidence was given by Cleghorn on Friday 24 July 1885, having been summoned at a day's notice (for which he was paid £8 3 shillings in expenses, of which £5 was the trainfare from St Andrews). During this and the previous session, on 21 July, four of the six witnesses had Indian connections (three practical foresters and one India Office official, the others being the Assistant-Director of Kew, William Thistleton-Dyer, and a representative of the Surveyors' Institution). On Cleghorn's day, in addition to Sir John Lubbock in the Chair, ten other MPs were present of whom two, somewhat surprisingly, were medically qualified and it was they who asked Cleghorn by far the majority of the questions. Dr Robert Lyons, Liberal MP for Dublin, had a major interest in the reforestation of Ireland. Dr Robert Farquharson, Liberal MP for West Aberdeenshire, had a similar background to Cleghorn's own: Laird of Finzean with an Edinburgh MD. Cleghorn was strongly in favour of the establishment of a school of forestry, or preferably one each in England, Scotland and Ireland, over and above what was on offer at Cooper's Hill:

> I think immense benefit would result, both economically and otherwise, from a more systematic management of our woods, and the skilled training of wood managers and subordinates ... Without a Forest School I do not see how we are to make any further progress.

CHAPTER 14

This is hardly surprising – he had been called to London both for his role as president of the Scottish Arboricultural Society and for his lengthy Indian experience (in particular for his annual selecting of Indian students, as part of which he had personally visited the Forest Schools at Nancy, Vallombrosa and Tharandt). Furthermore, he had been heavily involved in the 1884 Edinburgh Forestry Exhibition, and was an examiner for the Highland & Agricultural Society the only body in Britain then awarding any sort of professional qualification to foresters. Cleghorn clearly favoured Edinburgh as the location for a pioneering British Forest School, as it already possessed numerous advantages – the newly established Arboretum (publicly funded and attached to RBGE); the recent experience of the Exhibition whose small profits had been used to start a fund for a forestry chair at Edinburgh University (for which £10,000 would be needed); museum collections and a library (those of the Arboricultural Society and RBGE, and books from the 1884 Exhibition deposited in the Museum of Science and Art); fine woodlands within easy reach; and a group of young men anxious for improvement but currently held back for lack of training courses, and limited to self-improvement from books. In the interview Cleghorn was led on to support the idea of extensive new tree planting, to make Britain more self-sufficient in timber when foreign supplies were either too expensive (India), or themselves declining through over-exploitation (Canada). The single note of (unintentional) humour in Cleghorn's grilling was the question:

> "Do you think those woods would have been better planted by a highly trained man skilled in inorganic chemistry?"

The sort of question that today would be asked by a bean-counting, non-academic – what relevance has academic study for a practical subject? But it came not from a businessman but from Dr Farquharson. Cleghorn, not to be browbeaten by such cynicism, held his ground and in reply to a further, less aggressive, question from Farquharson, "If they had had a little more theoretical knowledge, they could have turned their practical sagacity to better account?", he replied "I think so, for many of them do wish they had had that". In other words it was the foresters themselves who were asking for education, not only the 'powers that be'.

Two other witnesses are of particular interest, for different reasons. On the first day, evidence had been given by Colonel James Michael, who had been involved in the earliest of the Madras efforts towards forest conservancy in the Anamalai Hills in 1848 and, as seen above, had more recently been closely involved with Cleghorn over the 1884 Exhibition. Surprisingly, if revealingly, he took this as an opportunity to give his own opinion on the question of the origin of forest conservancy in Madras, which he firmly attributed to his one-time hunting companion, the engineer Frederick Cotton:

> Riding across the southern forests of the Madras Presidency ... he [Cotton] was struck by the bad order in which they were kept, being destroyed as they were by Government leasing them out to contractors instead of keeping them in their own hands. It struck him that if the Government would put a stop to that system, this terrible waste would be stopped, and he urged the matter upon the Madras Government.

That is, at least eight years before Cleghorn's appointment as Conservator, nothing to do with climatic concerns, and *predating* the 1851 British Association report. So, rather than Birdwood's, or the view of recent authors such as Grove, Colonel Michael stressed the inspiration as being purely practical and financial concerns.

Another of those who gave evidence, the Rev. Dr John Croumbie Brown, brought with him a rather different perspective. As has been shown by Richard Grove, Brown was a major figure in the linkage between forest conservation and the 'climate question'. A prolific author on forest matters, he had already made a detailed study of Continental forest training on which he had published a substantial pamphlet with the self-explanatory title *The Schools of Forestry in Europe. A Plea for the Creation of a School of Forestry in Connection with the Arboretum at Edinburgh* (1877) – a topic on which he had addressed the Arboricultural Society in November 1877. Brown's was an unusual academic background in theology and botany, combined with practical experience as Colonial Botanist at the Cape of Good Hope. Curiously when that post was abolished in 1866, he had applied to Cleghorn for a possible posting in India, which Cleghorn forwarded with a strong recommendation both to Travancore and to Brandis in Simla, though with reservations over Brown's age,[27] but this is the only proven contact between the two men who apparently had so much in common. Brown's evidence (in 1886 and 1887) to the Select Committee did not enter into questions either of climate or India but is of interest for showing his independence of mind. One of the major obstacles that had always been cited as a reason why no Forest School had yet been, nor easily could be, set up in Britain was a lack of a substantial forest managed on scientific principles as a training ground. Unlike virtually all the witnesses with practical experience Brown did not see this as a problem, as he thought that experience could be gained on short visits by students to Continental forests. He favoured Edinburgh as a site where, given all the resources already mentioned, 'we can at comparatively small expense establish a School of Forestry equal to the most celebrated schools of the Continent of Europe'. Brown (like several other witnesses) was concerned about accessibility of courses to poorly paid aspiring or practising foresters (an anti-elitism reveals itself – especially with regard to the grand, quasi-military, Nancy establishment), which led him to cite as a model, for its 'perfect freedom and liberal course of study', the little-known Spanish forest school at the Escorial. Brown favoured a Government-funded school under the Department of Education (at this time in the process of evolving from the Science and Art Department of the Committee of Council on Education), but, failing that, teaching through short courses (including evening classes) either at Edinburgh University or the Watt Institution.

Involvement with Scottish Universities

During his retirement Cleghorn was actively involved with his two *almae matres*, being a member of the General Councils of both the Universities of St Andrews and Edinburgh. The General Councils are part of the administrative set-up of Scottish universities, which had been established to allow graduates a role in university affairs following the Scottish Universities Act of 1858. They were sizeable bodies: in 1864/5, the first list on which Cleghorn's name appears as a member,[28] Edinburgh's had 2263 members, which by 1869/70 had risen to 3652, and 6376 in 1891.[29] The General Council was made up of members of academic staff and – on payment of a small registration fee – alumni. Edinburgh's met twice a year, in April and October, to discuss 'questions affecting the well-being and prosperity of the University', with powers to elect the Chancellor (who chaired the Council, an appointment for life), to elect Assessors to the University Court (the university's governing body, the Assessors being independent lay members who acted like trustees), and in the year of Cleghorn's return, as a result of recent electoral reform, to elect a Member of Parliament jointly to represent the Universities of St Andrews and Edinburgh. The first of these MPs, elected in November 1868, was Lyon Playfair, well known to Cleghorn from their early St Andrews days and their work for the

CHAPTER 14

Great Exhibition. Another benefit of membership of the Council, doubtless appreciated by Cleghorn, was that (for a fee of five guineas for life) it conferred borrowing rights from the University Library.

Cleghorn's main contribution to his universities concerned the establishment of lectureships in forestry and botany – but at Edinburgh he performed another role in the medical faculty. Given the large number of medical students, additional examiners were required, with up to twelve being annually appointed by the Court, who were each paid £50 a year from a parliamentary grant. In the University Calendars Cleghorn is recorded as having been appointed as one of the additional Medical Examiners in the years 1875 and 1880. As W.C. McIntosh stated that Cleghorn held the post for ten years,[30] these appointments must each have been for a period of five years, and almost certainly related only to botany. McIntosh (who was also appointed an examiner in 1875) noted that nobody had 'more assiduously or more conscientiously performed his duties' than Cleghorn.

The Edinburgh Forestry Lectureship

Edinburgh was the first British university to offer a course of lectures in forestry, having had a chair in agriculture since 1790. The lectureship arose from a growing awareness of the importance of forestry in Britain, but it had also been stimulated by developments in India, in which Cleghorn had been closely involved. There was a need to improve standards by professional training, and to avoid the expense (and wounding of national pride) of having to send students to the Continent. Pressure to involve Edinburgh University in forest education came largely from the Highland Society (which, as seen earlier, already offered its own certificate in the subject), and from the Scottish Arboricultural Society, in conjunction with the voices of influential individuals among whom were J.H. Balfour, J.C. Brown and Cleghorn. The acquisition of the policies of Inverleith House as an arboretum adjacent to RBGE in 1876 added weight to the arguments,[31] but it was not until July 1889 that William Somerville was appointed as the first lecturer. In 1886, examined among others by Cleghorn, Somerville (1860–1932) had obtained a first-class certificate in forestry from the Highland Society, followed by a BSc in Agriculture from Edinburgh two years later and, in 1889, a PhD in economics from Munich (where his special subjects were forestry and economic science).[32] Somerville was a high-flyer and held the lecturer's post for only two sessions (1889/90 and 1890/1) before resigning in April 1891 on his appointment to the Chair of Agriculture at Armstrong College, Newcastle (from where he went on to hold the agricultural chairs successively at Cambridge then Oxford). The course was of 100 lectures, given in Old College, but as there were only five students in the first year and two in the second, their fees of three guineas each per annum did not amount to much of a contribution towards the lecturer's salary. In these early days the Treasury, through the Board of Agriculture, provided a grant of £100 towards the salary, but the Highland and the Arboricultural Societies, while trying to persuade the Treasury to raise the salary to a more appropriate level of £300, also proved generous with additional grants. At this point forestry was still not a degree subject in its own right and the examinations (which were ranked by Class) must have counted towards the BSc in Agriculture – ordinary degrees in Forestry had to wait until 1911/12, and honours not until 1924.

It was during Somerville's tenure, on 31 January 1891, that Cleghorn offered the University Court, anonymously, a donation of £1000 'for behoof of [a] proposed endowment of Education in Forestry ... by the institution of a separate Chair or Lectureship'.[33] The Court, however, was cautious and seems not to have taken up the offer at this point: 'resolved not to come under any obligation as to the time at which they might consider that the funds available were sufficient for endowing either a Lectureship or a Professorship'. Meanwhile the Highland and the Arboricultural Societies were actively fund-raising for the post: in January 1893 they already had a fund of £2500, which they were trying to raise to £5000, which, if achieved, would be matched by the Treasury for the £10,000 required to establish a chair. This did not happen, but in the academic year 1893/4, for example, the Highland Society gave the University £631, and the Arboricultural Society £433, for the lectureship.[34]

In October 1892, following Somerville's departure for Newcastle, Colonel Frederick Bailey was appointed to the post of Lecturer in Forestry. Bailey (1840–1912), who had much in common with Cleghorn, would hold the post until 1910/1. He had started his career in the Royal (Bengal) Engineers, joining the Indian forest service in 1871, three years after Cleghorn had left the Subcontinent. His wife was Florence Agnes Marshman, daughter of John Clark Marshman, brother of Rachel who had been Brandis's first wife, so there were also strong evangelical/missionary links. Bailey was also interested in forestry education having, from 1879 to 1884, been the first director of the Forest School at Dehra Dun, following which, for three years, he had replaced Colonel Pearson as supervisor of the British students in Nancy. In 1887 Bailey returned for a final spell in India, initially as Conservator of Forests for the Punjab, and then for two periods as Inspector-General. Florence Bailey's salon in Drummond Place was well known in Edinburgh for its display of Tibetan artefacts collected by their son, the Great-Game player, Frederick Marshman Bailey, known to botanists and gardeners for his discovery of the blue poppy *Meconopsis baileyi*. The Edinburgh forestry course was greatly developed by Bailey, with additional specialist lecturers giving courses in botany, zoology and engineering, but it was his successor, Edward Percy Stebbing (appointed 1910, best remembered for his indigestible history of Indian forestry), who made the major strides, with a new building for Agriculture and Forestry in George Square in 1911, the awarding of forestry degrees, and the eventual establishment of a Chair in 1919 of which he was the first holder.

Before this ultimate realisation of his ambitions for the department Cleghorn, in May 1895 had died, and it was not until the year 1895/6 that his name appeared as a benefactor in the University Calendar. While it is possible that Cleghorn may have given the money immediately before his death, it seems more likely that, fearful of losing it, the University authorities belatedly claimed the £1000 promised four years earlier from Alexander Sprot, his nephew and executor. This was not, however, Cleghorn's only benefaction to the University of Edinburgh, and following his death, it received from Sprot a sizeable gift of books from the Stravithie library, to be discussed in the next chapter.

University of St Andrews

On Cleghorn's return from India the University of St Andrews was in a moribund state – in 1876 it had a mere 130 students. Nevertheless he tried to become involved and in 1869 was proposed (by A.K. Lindesay of Balmungo) as Assessor to the Court. As so often in university politics, shenanigans were clearly at work, as, despite the support of the locals, Cleghorn was beaten to the post by a Dr William Richardson of London.[35] This may have put him off as he did not become a member of the General Council of St Andrews until 1880/1 (at least 16 years after he had signed up for Edinburgh's). But things were about to get better – a period of growth for the older university, at least partly led by competition

CHAPTER 14

with a new institution on the other side of the Tay. In Dundee a University College was founded in 1881, which quickly attracted stars of the quality of Patrick Geddes and D'Arcy Thomson.[36] Following the Scottish Universities Act of 1889 it effectively became a college of St Andrews, with representation on the older establishment's Senate. These were stormy times among competitive academics, but the fights were not all between the older and newer establishments – there was an old-guard at St Andrews who rued the decline of theology and the Classics and wanted nothing to do with the growth of science. In 1882 William Carmichael McIntosh was appointed to the Chair of Civil and Natural History, and in the last two decades of the century McIntosh, with support from the holder of the Chandos Chair of Medicine, the anatomist James Bell Pettigrew, and the Marquess of Bute (from 1892 to 1898 the University's Rector – he of the Mount Stuart beavers and wallabies), managed to build up science and pre-clinical medicine at St Andrews to a previously unimaginable degree – including the establishment of a botanic garden and what, by a whisker, was Britain's first marine laboratory.[37] Cleghorn was able to witness and contribute to the early stages of these developments, both financially, and eventually, from 1890 until shortly before his death, as one of the Assessors on the University Court.

McIntosh was a key figure both in the development of science at St Andrews, and Cleghorn's role in it. He had been born in St Andrews in 1838, where his father was a prosperous builder and town councillor (he built the Town Hall, to which Peter Cleghorn had subscribed in 1856). McIntosh studied medicine at Edinburgh, then for 22 years devoted his life to psychiatric medicine in Perthshire (first in Perth at the Murray Royal, then at a new asylum at Murthly, which lay adjacent to a famous forestry estate), while applying his super-human energies to an active research programme largely in the field of marine zoology (especially annelid worms). After disappointment in applications for four Scottish chairs, his outstanding and well-known talents were eventually rewarded with the St Andrews one. The teaching of natural history at St Andrews had a rather chequered history – the Rev. John Gibson Macvicar, had been appointed as lecturer in the subject in 1825 and taught botany, but only for two seasons. In 1850 William Macdonald was appointed to the Chair of Civil History but in 1862 his personal interests in natural history resulted in the chair being renamed with the unlikely combination of Civil and Natural History. Macdonald's, and his successor Alleyne Nicholson's, interests, however, were in geology and zoology and he taught no botany. In the end, but not until 1897, Natural History triumphed when it was Civil History (the chair that Cleghorn's grandfather had held for so long) that got dropped from the title. Cleghorn must have been close to McIntosh who wrote by far the best obituary of his friend,[38] which includes much otherwise unknown anecdotal detail. Among this is the fact that it was this family association that had initially given Cleghorn the idea of donating some money to the chair; but McIntosh persuaded him that botanical teaching was a needier cause.

Under the new regime lectures in botany began in 1888 when for two sessions they were given by McIntosh's talented young assistant John Hardie Wilson, paid for from university funds.[39] Wilson (1858–1920) had been born in St Andrews to a family of nurserymen and had gone on to train both at Edinburgh and at St Andrews, graduating from the latter with a First Class BSc in 1887 (and DSc in 1889). In Edinburgh, while working at RBGE including a period in charge of the Rock Garden, Wilson attended university lectures; Cleghorn may well have known Wilson since childhood and encouraged him to present himself to the Highland Society, from which he received a First Class Certificate in Forestry in 1884. As McIntosh's prize-winning

14. GRAND OLD FORESTER

student and assistant Wilson was as at this stage as much interested in zoology as in botany, though it was the latter that he was asked to teach. At this point Cleghorn offered student prizes in the subject, and the run of Stravithie to young botanical explorers; he also donated 'a series of large structural drawings of plants for lectures, 12 volumes on botany, and a very valuable herbarium in a cabinet'. It is unfortunately no longer possible to identify these items, though the teaching diagrams could well have been the ones sent by Balfour for use in the Madras Medical College. The herbarium would later have been incorporated into the general St Andrews collection, which eventually ended up at RBGE, and one possibility is that it could have represented Robert Maclagan's Himalayan specimens.

One of Wilson's most visible and lasting achievements was the setting up of a small botanic garden (Fig. 69). St Andrews has the town plan characteristic of the older Scottish burghs and still apparent today – that of the 'guttit haddie' (*Anglice*: eviscerated haddock), with narrow wynds spreading at right-angles from a spinal high street (in the case of St Andrews there were three, diverging from a head represented by the Scottish national cathedral), the gardens forming 'long rigs' extending from the backs of the houses, between the wynds. The site for the garden was the bottom quarter-acre of the garden of St Mary's College, next to Westburn Lane,[40] where Wilson laid out a garden containing 828 species, 'its boundary wall, containing carved fragments from the old Cathedral, was hirsute with wall ferns'.[41] Cleghorn donated some plants and attended the opening on 28 June 1889. In the planting of the garden Wilson was helped by another young St Andrews botanist/horticulturist, Thomas Berwick (1860–

Fig. 69. St Andrews Botanic Garden. Photographer unknown, 1889. (Courtesy of the University of St Andrews Library, Ms-38491-1.1).

1957), who must also have been well known to Cleghorn. Berwick had trained in the Wilsons' Greenside Nursery, and at RBGE where he attended botanical classes; in 1885 he obtained a Highland Society Second Class Forestry Certificate, having the previous year won a gold medal for an essay on 'The formation and management of forest tree nurseries' in connection with the Forestry Exhibition.

Wilson, however, was a man on the move. In October 1890 he left St Andrews for a brief spell as curator of the RBGE herbarium and library, followed by a period at the Yorkshire College in Leeds. In 1900 he returned to St Andrews, to teach agriculture, where his best-known work was on potato breeding. It was at this point, shortly after Wilson's departure, that in March 1891 Cleghorn made his donation to allow the continued teaching of botany or towards the future establishment of a chair.[42] The meeting of the University Court on 6 April 1891 was notable not only for the announcement of Cleghorn's gift, but for the attendance of one past and one future Viceroy of India – the meeting was chaired by the Marquess of Dufferin & Ava, but also present was the (9th) Earl of Elgin, son of the Viceroy whom Cleghorn had attended in his final days in Dharamsala, and whose Viceregal footsteps the son would follow in three years' time. Principal Donaldson stated that £1000 had been donated by 'a gentleman who takes a deep interest in this University': as Cleghorn himself was also sitting as a member of the Court, some knowing looks may have been exchanged, or perhaps, in those more discreet days, the secret was known only to the donor and the Principal. The cash was deposited with the Royal Bank of Scotland and later put into railway stock, in which the University was investing heavily at this time – oddly, if appropriately, the company they chose was the Scinde, Punjab and Delhi Railway.[43] In 1892 this yielded a dividend of £26 13 3d, which was duly donated to the new lecturer.[44] Robert Alexander Robertson (1873–1935) had been appointed to succeed Wilson, having in 1890 graduated BSc at Edinburgh. Generous though Cleghorn's gift was (the same amount that he gave to Edinburgh for forestry) it was nothing like enough to endow a chair (which even then took £10,000); neither was the income from it enough to pay the lecturer a living wage, which had to be topped up with a combination of student fees and £150 from the university. The gift was doubtless intended as a spur to the university, which eventually responded in 1903 when a fund of the requisite £10K was put together as a permanent endowment for the botany lectureship, consisting of the railway stock purchased with Cleghorn's donation topped up with £6000 from the Carnegie Trust, and £3000 from a mortgage on the Earl of Dalhousie's Brechin estates. The post would not, however, be turned into a botanical chair until 1929.

A marine byway

In the list of Cleghorn's publications is to be found a curious 'late work', also attributable to his friendship with McIntosh – the subject concerns fish. The 1885 meeting of the British Association was held in Aberdeen: McIntosh was president of Section D and Cleghorn must surely have attended. The St Andrews Marine Laboratory had been established the previous year in a converted wooden fever hospital on the East Bents, funded partly by the Fisheries Board of Scotland. This was an exciting time for marine biology, but it was also a political hot potato. Soon after arriving in St Andrews McIntosh had been approached by T.H. Huxley to take part in a Trawling Commission, chaired very actively by Lord Dalhousie.[45] At this point there was tension between local fishermen (line-fishers and small coastal vessels) and large steam trawlers, but very little scientific evidence on crucial questions such as the life-

14. GRAND OLD FORESTER

Fig. 70. Shanny (*Blennius* [now *Lipophrys*] *pholis*) emerging from a tuft of *Halidrys siliquosa*. Chromolithograph by G.H. Ford (printed by Mintern Bros.), based on drawing by Roberta McIntosh from her brother W.C. McIntosh's *The Marine Invertebrates and Fishes of St Andrews*, 1875.

cycles of fish, on which to base informed policy. The fisher-folk of Fife believed that the trawlers were damaging stocks and destroying fish eggs, but McIntosh and others found, on the contrary, that the eggs of most food-fish were pelagic, floating on the surface of the water, and therefore probably unscathed by trawling. McIntosh concluded that there was no need to control the activities of the trawlers, which led to his being burnt in effigy outside his own front door in genteel Abbotsford Crescent. In fact the Fisheries Board ignored his advice and did close off areas of the Scottish coast to trawling – the start of ongoing disputes on such matters both nationally and internationally.

The Fisheries Board failed to provide anything like enough funding for the staffing and running of the marine lab and in 1896, the year in which it was rebuilt as the Gatty Marine Laboratory, withdrew its support entirely. In the meanwhile McIntosh had applied to the BAAS for supplementary funding for his work: £75 was awarded in 1885 to a committee under McIntosh, comprising five other members.[46] Of these three were holders of Scottish medical chairs (from Glasgow: John Cleland, Anatomy and John McKendrick, Physiology; from Aberdeen: William Stirling, Physiology), James Cossar Ewart was professor of Natural History at Edinburgh,[47] but curiously it also included two botanists – Frederick Orpen Bower of Glasgow and Dr Hugh Cleghorn. Doubtless their role was nominal, as McIntosh and the distinguished visiting scientists he was soon able to attract, were more than capable of doing their own thing. A second committee was appointed at the 1887 BAAS Manchester meeting, when McIntosh was given another £50.[48] Cleghorn was again on this committee, but Bower had been dropped and the following added: the zoologists E. Ray Lankester and G.J. Allman; the physiologist John Burdon-Sanderson (Waynflete

CHAPTER 14

Professor at Oxford); and Ramsay H. Traquair. Traquair was the Keeper of Natural History at the Edinburgh Museum of Science and Art, an expert in fish anatomy especially of the fossilised variety, and married to the artist Phoebe Traquair. He and Cleghorn were friends and in 1871, after fossil fish had been discovered in a paraffin-shale quarry at Pitcorthie close to Stravithie, Cleghorn had invited Traquair to visit and inspect them (General and Mrs Lake were also staying at the time); the accompanying plant fossils were given by Cleghorn to William Carruthers at the British Museum.[49]

The two short reports made by McIntosh's fishery committee count among Cleghorn's last publications.[50] They can still be read with interest – not only for the palpable sense of excitement at fundamental discoveries in matters that were still, literally, mysteries of the deep – the basic biology (from their floating ova through embryo and larval stages) of organisms that were of crucial importance to the feeding of nations, but also for the wonderful names of the fish. Greater and lesser weever, shanny (Fig. 70), ballan-wrasse, shagreen-ray, piked dog-fish and porbeagle-shark were among those studied in this productive year. During its course J.H. Wilson also investigated the biology of the common mussel, but the greatest excitement was the catching, dissection, and skeleton-preparation, of a nine-foot-long tunny fished from the mouth of the Forth.

Philanthropic & County Business

When presenting Cleghorn with his portrait Sir William Muir noted that:

> our esteemed friend is known for his interest in the welfare of all around him, and warm sympathy with every philanthropic movement having for its object the good of the people. He weeps with those who weep, and rejoices with those who rejoice.[51]

And in the obituary, written by the friend of his old age, Professor McIntosh similarly stressed how seriously Cleghorn, as a landed proprietor, took his 'philanthropic and county business'.[52] Of such 'county' activities, one was his appointment in 1870 to the Board of Lunacy for Fife and Kinross, under the chairmanship of David Gillespie of Mountquhanie.[53] This board advised on the running of the county lunatic asylum at Stratheden near Cupar, which had been opened in July 1866 under the enlightened medical superintendence of Dr John Batty Tuke. Three reports for this period show that the patients were well fed and humanely treated, the abler ones being given useful tasks including shoe-making, joinery, sewing and laundry, and the running of a farm and garden.[54] Another late-Victorian public-health concern was sanitation, particularly in cities. In 1881 Sanitary Protection Associations had been founded in both London and Edinburgh, the latter at the instigation of Henry Fleeming Jenkin, Darwinian

14. GRAND OLD FORESTER

critic and Professor of Engineering. These were membership organisations formed by the middle classes (a joining fee of two guineas, one annually thereafter), which complemented statutory and municipal work, providing advice to members on the sanitation of their own property, and those of dependents or of the poor in whom they took a charitable interest. The Edinburgh association had offices in South Charlotte Street, and Cleghorn was a member of what was an extremely distinguished council under the chair of his old friend Sir Douglas Maclagan, by now President of the Royal College of Physicians, and Professor of Medical Jurisprudence and Public Health – two other Edinburgh professors were councillors – Alexander Crum Brown (Chemistry) and Thomas Fraser (Materia Medica) as was Sir David Brewster's son-in-law Charles Brewster Macpherson.[55]

As a result of pressure from reformers earlier in the nineteenth century, steps had been taken to improve conditions in prisons and Prison Visiting Committees had been set up with responsibility for inspecting the county gaols. Cleghorn was on the Fife committee, responsible for the gaol in Cupar. As a caring laird with an interest in the local community he was particularly concerned that small-time criminals should not be 'sent to herd with the criminal population of Edinburgh, Dundee, or Perth [where] there would be less hope of their reformation; and the effect, on first offenders especially, would be very bad'.[56] The anonymous author of his *Scotsman* obituary stated that Cleghorn was 'especially interested in efforts to give discharged prisoners a start in life again',[57] which suggests that he was almost certainly active in the Discharged Prisoners' Aid Society in Cupar, one of six such bodies in Scotland. Another social concern of nineteenth-century Scotland was thrift – that the deserving poor should have inducements and mechanisms to enable the saving of their hard-earned cash. The Rev. Henry Duncan, minister of Ruthwell in Dumfries-shire, was a pioneer in the Savings Bank movement with connections to Cleghorn both direct and indirect. Close to Thomas Chalmers, Duncan had received a DD from St Andrews in 1823 and twenty years later became one of the founders of the Free Church. His 'enlightened' interests included research on the magnificent eighth-century Ruthwell Cross, and he was the first to describe (from Lochmaben) a fossilised dinosaur track. The direct connection is that Duncan's second wife was Mary Lundie (*née* Grey), mother-in-law to the Rev. Robert Henry Lundie, the husband of Mabel's sister Elizabeth who had officiated at Hugh and Mabel Cleghorn's wedding. It is therefore hardly surprising that Cleghorn took an interest in the local manifestation of the movement and at the AGM of the St Andrews Savings Bank following his death, tribute was paid to the interest he had taken in it by endeavouring 'to impress on those amongst whom he lived the benefit accruing from the Savings Bank'.[58]

Cleghorn's interest in tertiary education has already been discussed, but he also took an active interest in the younger generation and in 1884 he was nominated by his neighbour Sir Robert Anstruther of Balcaskie to the Board of Governors of the Waid Trust Education Scheme. In 1804 Lieutenant Andrew Waid had left money to build an establishment for orphans of poor mariners and fishermen, which belatedly, 80 years later, was used for rather broader purposes: to establish a free secondary school in Anstruther, which survives as the Waid Academy.[59] At the opposite end of the social spectrum Cleghorn's will reveals that he had five shares in the 'St Andrews School for Girls Company Ltd'. This female establishment had been founded in 1877, and occupied, among other buildings, the house once rented by his grandfather, by which name the school is known today – St Leonards. In addition to formal education for the young, the nineteenth century also saw the development of adult education by means of the public lecture, sometimes illustrated, sometimes not. Notable performer-educators included John Ruskin and Thomas Henry Huxley and, in a Scottish context, Cleghorn's friend John Hutton Balfour. In his retirement Cleghorn, with his wealth of experience and vast collection of illustrations, also contributed to the genre, for the benefit of his local communities. For example, in the Town Hall at Anstruther on 3 March 1873 he lectured on 'Plant Life' and:

> illustrated his interesting subject with many beautiful diagrams, and the lecture was evidently the production of one who has lovingly and successfully studied the marvellous arcana of Nature; but being delivered in so large a hall was rather indistinctly heard by the audience, which ... in numbers ... was very far from equal to the eminent merits of the occasion.[60]

The following year, at the Duncan Institute in the county town of Cupar, his subject was 'On Forest Life in India',[61] and there were doubtless other similar performances.

Justice of the Peace

Cleghorn's involvement with education, public health and similar activities can be seen as expressions of his charitable concerns, but it should not be forgotten that the maintenance of law and order – the transgressions largely of the 'undeserving' poor – was also a major concern of the landed class to which he belonged. Property owners without legal training played a significant role in this and among his 'county business' Cleghorn's position as a Justice of the Peace (JP) was perhaps the most significant. He sat as a magistrate on cases heard in the local court in St Andrews Town Hall, which dealt with matters such as awarding licences to deal in game or sell alcohol, and also minor criminal cases, including assaults, prostitution and vagrancy that could lead to the imposition of fines. Though no records survive for this local court one case heard by Cleghorn and four fellow magistrates in 1872 is known for the local interest it generated and the resulting press coverage.[62] Alexander Duncan, a baker of Castle Street, St Andrews, had been caught in possession of an illicit still and other equipment including a 'worm tub' and a 'mash tub', for which he was liable to a fine to the tune of the enormous sum of £2200 that would have bankrupted a much more prosperous man. There were extenuating circumstances as Duncan, without the knowledge of his wife, had been reduced to undertaking such activities only through desperation, having had to give up farming when his cattle contracted bovine pleuro-pneumonia and were compulsorily slaughtered.[63] The justices showed leniency and took the legal option of reducing the fine to a quarter, though this still amounted to a hefty £550. As his neighbours had doubtless been only too happy to take advantage of the cheap liquor (and the pleasure of allowing them to stick their thumbs up to the excise men) it is to be hoped that they rallied round and helped Duncan to pay the fine.

What do survive are the minutes of the statutory Quarter Sessions of the JP court held in Cupar every three months in March, May, August and October/November.[64] Cleghorn was conscientious in his attendance at these, missing only a single year between 1872 until illness caught up with him in 1894, nearly always attending the May session and usually the autumn one, but often also one or two of the others. By this date this court had virtually no real power and it is impossible not to get the impression that these meetings of the great and good of Fife (some were chaired by the Earl of Elgin) were largely an excuse for a social get-together: one can easily imagine the convivial dinners that followed the adjournment of the court. For such trivial powers the attendance was remarkable – in almost half the sittings during Cleghorn's period more than 20 JPs were present and on one occasion 51. At each May meeting the Chief Constable was rubber-stamped as the JP's 'Fiscal', a report was read from the committee on Weights and Measures (which dealt with the apparatus in the 'Steelyards' at toll stations on the roads), and, up to 1877, reports were read of those unfortunates (usually miners and labourers) who had been prosecuted in the local courts under the Day Trespass Act, in pursuit of game onto private property, though the JP court had no power of prosecution itself. In fact the only matter over which it had any power at all was the question of the licensing of alcohol. In 1876 a new Act came in and a County Licensing Committee was established to set rules and supervise the local courts – Cleghorn was a member of this committee from 1876 until 1891. But heard at the Quarter Sessions JP court were appeals from those who had been refused licences by the local courts, and this took up by far the largest part of their time – some of the refusals for licences ('Certificates') were confirmed; others dismissed. During this twenty-year period, only a single topic arose that is of any general interest, in showing an early awareness of environmental pollution. In 1878 the Fife JPs had received protests from members of the public about the nuisance caused by smoky traction engines on public roads. The Justices undertook a survey, which revealed that of the 29 machines in Fife (14 in the agriculturally advanced St Andrews district) 12 were 'constructed partly to consume their own smoke' and 17 had no such control mechanism. All that the JPs were able to do, however, was to issue a letter recommending to owners the desirability of using the smoke-consuming variety of the species.

Of Cleghorn's politics nothing is known until his later days, when evidence of his allegiance to a new political party emerges from reports of meetings in local newspapers. Not least from his marriage to a member of the Cowan family, leading supporters of William Ewart Gladstone, Cleghorn's original outlook is likely to have been staunchly Whig. However, loyalties changed with Gladstone's conversion to the principle of Home Rule for Ireland, which was deeply unpopular with conservative liberals, and, for example, split the Cowan family. By this time his father-in-law Charles Cowan was too old to take an active interest, but of Mabel's uncles James broke with the party leader, whereas John supported him and managed Gladstone's third Midlothian campaign, for which he was rewarded with a baronetcy. Those for whom Irish Home Rule was seen not only as an affront to the union of the United Kingdom but as a threat to the supremacy of the Protestant church broke away in 1886 to form the Liberal Union party. Cleghorn, along with many of his neighbouring landowners, was among these and it was in this context that his name appeared in newspaper reports as attending Liberal Unionist demonstrations and meetings in Cupar in 1889, and in Cupar and Anstruther in 1893.[65]

Death of Mabel

From the first year of their marriage, Mabel's health had been an ever-present source of worry to her husband, and on return for his final stretch in India in September 1865 he had had to leave her behind in Scotland. At this point she was in the medical charge of Halliday Douglas, though in 1866 there are references to a trip to the Continent in June, and a period of staying with her sister Charlotte Wilson at Mansfield in October. By December of the same year she was being treated in Edinburgh by one of Cleghorn's old teachers, by now Sir James Young Simpson, which suggests that her problems were at this stage primarily gynaecological, undoubtedly the reason for her visit to Marion Sims in Paris in 1868. A diagnosis is impossible, as there is so little evidence other than the occasional mentions in Cleghorn's letters: for example, in January/February 1875 'a severe pulmonary attack' had confined her to her bedroom for several weeks.[66] One has a picture of the chronic Victorian invalid, though in her case at all times borne up by her strong Christian faith. In the autumn of 1887 a taste of south-coast air at Brighton had been to no avail,[67] and Mabel's earthly sufferings came to an end at Stravithie, on 22 December, aged only 49. The letter that her death elicited six months later from Thomas Farquhar, a 'brother in Christ' and medical colleague who had known the couple in Calcutta, by now retired and living in Aberdeen, is so touching and suggestive that, despite its partial illegibility, it is worth quoting in its entirety, requiring no additional gloss:

> Dear Cleghorn
>
> I plead sad selfishness as my only excuse for not writing you long ago – it was painful to begin. It is an effort yet to speak of the sad loss of your dear sainted wife. What a power she was for good by her bright lovely bearing and example to the [?world] of India & Home before the cloud settled on her shattered nerves– Some inherent weakness in the bodily organization that clothed the purest of all spirits prevented her from being of more use–
>
> I can never forget one or two scenes – one especially in your [?hired] house in Calcutta when on a Sunday evening she sang "Nearer my God". I never heard anything that lifted me out of self or nearer to the Beatified in the [illeg.] than the accents of her voice that night– I thought she had run her course then & was singing almost within the veil– She made religion more real & lovely than anybody except my mother who in so many ways was so like her–
>
> You are lonely dear fellow but I know you are busy & find pleasure in occupation which is a great mercy for you– Her memory will often steal quietly into your soul & beckon you upwards. God bless & keep you I will feel ever so more comfortable after writing this for the reproof I gave myself for not writing a line of sympathy to you for whom I felt so much has [illeg].
>
> Yours ever affectionately
>
> T. Farquhar[68]

Cleghorn's Death & Will

No more is known of the state of Cleghorn's own health in his later days than the little that can be gleaned from his death certificate.[69] This suggests illness for the last four years of his life, which would conform with a reduction in involvement with his various societies, and his making of a will in June 1891. In Robert Moir (1831–1899), son of the Musselburgh poet and physician David McBeath Moir, he found a sympathetic physician. Moir, like his patient, was an Edinburgh medical graduate who had spent most of his career in India – in his case as a Bengal surgeon, followed in 1877 by retirement to St Andrews where he continued to practise as a physician. The causes of Cleghorn's death certified by Moir were 'Prostatic & vesical hypertrophy 4 years [and] Chronic diarrhoea'. The swollen prostate and bladder, common conditions in elderly men,[70] are unlikely to have been the cause of death, unless cancerous, but this cannot be known in the absence of the results of a post mortem. The bowel complaint may well have been a result of bugs picked up many years earlier in India and one can only sympathise with such long-running complaints in the days before antibiotics. Whereas his father had lived to be 80 and his grandfather 85, Cleghorn reached only 75, but the genes for longevity seem to have run out in this third generation. In his own home at Stravithie, at 6.25 in the early evening of 16 May 1895, Cleghorn 'was summoned

14. GRAND OLD FORESTER

… to the activities of immortal youth in Christ, in his seventy-fifth earthly year'.[71] He was not alone, and attending the deathbed was his wife Mabel's younger spinster sister Margaret Menzies Cowan. Interment took place in the kirkyard of Dunino, where, having outlived his three siblings (all buried elsewhere – Hong Kong, Rome and London) he rejoined Mabel, his father, his aunts and his grandparents.

Cleghorn's will is a fascinating document and what it reveals in terms of his wealth and financial acumen has already been discussed. Given the very substantial sum involved (equivalent to at least £4 million in present-day terms, but very considerably more in terms of 'status' value) the document is surprisingly short and simple – virtually everything was left to his late sister Rachel's only son, Alexander Sprot. Sprot was also the sole executor and the whole estate was wound up within two months of Cleghorn's death, an unimaginably short time for dealing with an estate of such a size today. The only cash left outwith the family was three minor bequests to former servants totalling a miserly £70 per year. During his life he had been a generous supporter of (especially religious) charities, and one of his fears of impoverishment in retirement was that whereas his income while employed had enabled him 'to assist schemes of charity … in the retiring allowance we could not continue to do, except on a humble scale'.[72] This turned out not even to be true – he was extremely comfortably off in retirement – which makes it the more surprising that he did not make up for this later neglect post-mortem, when his only responsibility was to an already wealthy nephew. The *only* item excluded from Sprot's vast inheritance was his uncle's treasured and valuable library of botanical and forestry books, to be discussed in the next chapter. Given his commitment to philanthropy during his lifetime

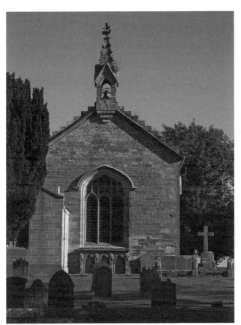 

Fig. 71. Dunino Kirk and the Cleghorn burial plot. (Author's photographs, October 2015).
The west end of Gillespie Graham's church: the four stones (and one hidden one) in red granite against the right-hand corner of the church, and the adjacent cross were erected by Cleghorn in memory of his grandparents, aunts, uncle and parents, and also to his own siblings (left).
The sandstone cross erected by Cleghorn in 1868, inscribed on its base 'The burying place of Hugh Cleghorn, Esq. of Stravithie', and also bearing inscriptions to Cleghorn and his wife Mabel (right).

CHAPTER 14

this must surely show the depth of the passion with which his grandfather impressed upon him the keeping of his estate together and in the family – an extreme example of the principle of blood being thicker than water. An inheritance such as this would commonly have had an attached condition that the heir should change his name, or at least add his benefactor's to his own surname. It seems very surprising that Cleghorn made no such stipulation, so after a mere 90 years the designation 'Cleghorn of Stravithie' died with him.

As this is Cleghorn's story it is not the place to go into subsequent family history, which (because of his childlessness) was in any case indirect, but as Alexander Sprot (1853–1929) was his nephew, heir and executor it is necessary to say a little about him.[73] With increasing Anglicisation of the Scottish upper classes, and given that his mother Rachel was a wealthy widow who ended her days in Mayfair, it is not surprising that the educational system she chose for her only son (Harrow and Trinity College, Cambridge) was very different from her own and her brother's, nor that he should have become a pillar of the establishment: Cavalry officer, Member of Parliament, Baronet and Master of Fox Hounds. In 1874 he was commissioned into a smart cavalry regiment, the 6th Dragoon Guards ('The Carabiniers'), in which he pursued a distinguished military career in Afghanistan and South Africa. He divided his time between London, Stravithie (where he greatly extended the house) and Garnkirk, and served as Conservative MP for East Fife (1918–22) and subsequently for North Lanarkshire (1924–29). In 1879 (possibly in India, where he was for the Afghan War) Sprot married Ethel Florence Thorp (1856–1930), whose father Edward Courtenay Thorp, a Bengal surgeon with a St Andrews MD, was probably known to Cleghorn. Sir Alexander collapsed in the House of Commons and died in February 1929; the following year his wife Ethel suffered the indignity of dying as a result of a car crash: the Rolls Royce in which she was being driven was hit by another vehicle at the Melville crossroads in Fife. The Sprots had nine daughters (several married into the squirearchy but one was more interesting and her religiously minded great uncle might, (with some posthumous ecumenism), have taken pride in the fact that Alix Isabella (1885–1972), as Mother Martha, became a Russian Orthodox nun who spent the last 40 years of her long life in Jerusalem as a member of the Bethany Community of the Resurrection). There were no sons, so the baronetcy awarded to Alexander Sprot in 1918 died with him. The estate passed to his eldest daughter Ethel who was married to Hereward Sadler (also a Carabinier, son of another knighted MP). Their youngest son, who eventually inherited Stravithie, was in 1919 baptised Gerard Hugh Cleghorn Sadler, but in 1931 his father changed his name by deed poll, though to Sprot and not Cleghorn. Hugh's eldest son, the present laird, is Geoffrey Cleghorn Sprot, so the name carries on, if not as a formally hyphenated surname.

Obituaries

Despite the lack of honours in his lifetime Cleghorn did well in terms of obituaries (see Appendix 1). Ones of various lengths appeared in six British, one Indian and an American botanical or forestry journals, in three British medical journals; there were death notices in botanical journals in Lund, Kassel, Karlsruhe and Leipzig, and obituaries in four British newspapers (two national, two local). Of these only two are significant – detailed and extensive accounts by people who knew him well: Andrew Taylor's for the Botanical Society of Edinburgh and William McIntosh's for the Royal Society of Edinburgh.

15
Cleghorn's Legacy

For a committed Christian such as Cleghorn, who believed in the immortality of the soul and resurrection, it would be particularly inappropriate to leave an account of his life with illness and bodily death. So this final chapter looks at the ways in which Cleghorn's memory continues to live and inspire. Such memories are to be found in different ways in two distinct spheres – one tangible in two Scottish institutions; the other less tangibly in India. The first concerns artefacts – books and drawings – that he was keen should be used for the benefit of future generations. The second is less obviously visible, but I would discover that it is still possible to find traces of his memory in the landscapes that he knew, and the influence of his ethos is still to be found in the Forest Department that he helped to initiate.

As explained in the Introduction the idea of writing a life of Cleghorn came about in response to a challenge provided by ignorance about his books and drawings that now form part of the core collections of RBGE. During the course of the project much time was spent in cataloguing these, but in order to tell the story of how they ended up there, and the discovery that there were also many of his books in Edinburgh University library, it is first necessary to backtrack.

The Tangible
The Cleghorn Forest Library

Following the Scottish Arboricultural Society's AGM it was customary to repair to a local hotel for dinner. The event on 7 August 1888 was a special one as a subscription had been raised for the presentation of a 'testimonial' and a portrait to Cleghorn.[1] The reason for the timing is unknown – why did the Society not wait until his seventieth birthday two years later, or did they at this stage fear that he might not make this? The dinner was held in the Waterloo Hotel, the city's oldest purpose-built hotel patronised by notables such as Charles Dickens, which stood opposite the General Post Office in Waterloo Place. The number of guests is unknown, though it seems doubtful that there would have been enough to fill the large ballroom at the back of the building. The dinner was chaired by Isaac Bayley Balfour, then only recently appointed Regius Keeper of RBGE following the early death of Alexander Dickson. Among the guests were several of Cleghorn's long-standing friends – two of the Maclagan brothers, Sir Douglas and Colonel Robert; and Sir William Muir by now Principal of Edinburgh University. Cleghorn's accountant John Dovey is also on the short list of notable guests given in the write-up of the occasion in the society's *Transactions*, as are foresters from some of the big Scottish estates (Robert Baxter from Dalkeith, John McLaren from Hopetoun and James Kay from Bute). Malcolm Dunn, a former Dalkeith forester, was denoted 'croupier' and acted as master of ceremonies. Other known guests are Colonel Peter Dods a retired Bombay soldier, William Erskine an engraver from the firm of McFarlane & Erskine, and James Watt, owner of the Knowefield Nurseries at Carlisle. Brandis could not be there and sent his apologies. The eulogy was given by Muir who reminisced about his introduction to Cleghorn in Cairo in 1861 and the happy times their families had shared at Barrackpore long before the 'cloud of bereavement' that had come with Mabel's death the previous year. Cleghorn replied with a summary of his own life and career, toasts were drunk and the evening ended with the singing of Auld Lang Syne.

CHAPTER 15

The portrait was a charcoal drawing by Theodore Blake Wirgman made on 6 July 1888, a date that suggests it was probably made in London, where Cleghorn certainly was the previous month for the forest examinations (Fig. 66). The drawing may have been made in the India Office itself and shows a patriarchal figure seated at a desk covered in papers; on top of the deeply carved tallboy behind him sits what appears to be a small bronze Buddha. The fee Wirgman charged is unknown but was probably only ten or at most twenty guineas.[2] Although Wirgman did also paint in oils, he is best known as a draughtsman who worked for the popular illustrated periodical *The Graphic* and specialised in portrait drawings of scientists (including Joseph Hooker and Thomas Henry Huxley) and senior Indian administrators, many of whom had been known to Cleghorn in his Simla days. These included Sir Richard Strachey, Sir Henry Yule, Sir Henry Maine, Colonel Craven Dickens and Sir Henry Norman, some of which, though not Cleghorn's, were published as etchings.

Cash donations to the testimonial had reached £200, and Cleghorn was especially pleased that these had come not only from the great and the good in Britain and the Subcontinent, but from some of the junior forest officers he had personally selected for India. Sir William announced Cleghorn's intention of donating the money for the formation of 'the nucleus of a library of suitable books, to be called "The Cleghorn Forest Library", to be placed in the Museum of Science and Art' in Chambers Street, which Cleghorn himself stated was to be 'for the benefit of foresters in general, and young foresters in particular'. However, from an undated document that was found with Cleghorn's will at Stravithie after his death,[3] it can be inferred that there was a delay in handing over the £200, though this had certainly happened prior to his death. This document stated that the money was to be used to acquire additional books and that the library was to be for 'the special benefit of the Professor and Students of Forestry at Edinburgh University' (though also open to the general public). Given its intended users it is very odd that the Museum was chosen for its location, rather than either the University or the Botanic Garden. Perhaps the University would not guarantee its being maintained as a separate collection, and the issue of its not going to the Botanic Garden will be discussed later. First it is worth saying something about the Museum of Science and Art (its name has changed many times: once before and at least twice after), where the Cleghorn Library was to be housed.

The Museum of Science and Art

As the Industrial Museum of Scotland this institution was founded in 1854 with the heavy involvement of Edinburgh University, as its professor of Natural History was to become the Keeper of the university's natural history specimens on their transfer to the new body.[4] But as, in some ways, a culmination of the Scottish Enlightenment principle of 'useful knowledge' the museum was also to exhibit arts and manufactures, one of its chief guiding lights being George Wilson, the Museum's first Director, for whom a new university chair of technology was created. Administratively the Museum came under the South-Kensington-based Department of Science and Art, whose secretary for science was Cleghorn's old friend Lyon Playfair. The Department commissioned a magnificent new building, designed by the engineer Captain Francis Fowke (better known for his Albert Hall in London), with Robert Matheson of HM Office of Works in Scotland as the local executive architect (Fig. 72).[5] The building was to be by far the largest public building of its era in Scotland, with a grand central glazed atrium whose architecture resembled an enormous department store. Contrasting with the light-filled interior the style of the exterior was brooding Italian Renaissance, with a richly detailed facade of polychromatic

15. CLEGHORN'S LEGACY

Fig. 72. Print celebrating the laying of the foundation stone of the Industrial Museum of Scotland by Prince Albert on 23 October 1861. The library, containing the Cleghorn Memorial Library, was in the ground floor of the right-hand (western) wing, by which time it was known as the Museum of Science and Art. Anonymous wood engraving, *Illustrated London News* 21 December 1861.

stone – in striking contrast to the stark minimalism of the exterior of Adam's and Playfair's Old College with which it shared the south side of Chambers Street. The street itself was a new one, created by the destruction of two Georgian squares (Argyle and Brown) and, more seriously, the exquisite villa that William Adam had built for the Elliots of Minto, which Cleghorn's old teacher James Syme had later used as his surgical hospital. The monumental building was built in three stages, the laying of the foundation stone of the first (the eastern part of the great hall, adjacent to the University) in 1861 being the last public act of Prince Albert, in a ceremony presided over by Cleghorn's old friend and patron, the University's Principal, Sir David Brewster. This first section of the building was opened by Prince Alfred in 1866; the western part of the great central hall was completed in 1875, and the west wing in the year of the Cleghorn dinner, 1888. This new wing, which housed the Geological Survey, had public galleries on three levels, including economic botany on its top floor. On the ground floor, facing Chambers Street, was the Museum's library, which coincidentally, but appropriately in view of some of the contents it would shortly receive, lay immediately adjacent to the site of the then still surviving part of 'Society', the Cleghorns' ancestral Edinburgh property.

Wilson, the founding director of the Museum, dogged by ill-health, occupied the post for only four years; he was followed by the economic botanist Thomas Archer who died in 1885.[6] Archer was succeeded by the archaeologist and Persian expert Sir Robert Murdoch Smith who built up the oriental collections, and it was Smith who agreed that the Cleghorn Library should come to his expanding museum. India was already conspicuously represented in the Museum, not least by one of the largest exhibits in the central hall, a life-size cast of the eastern gateway (*torana*) of the great stupa of Sanchi.[7] After undignified squabbles between the University and the Museum, and the blocking of the bridge that linked them, the post of Keeper of Natural History had been separated from the university chair, and since 1872 the museum keepership had been held by Ramsay Traquair, the expert on fossil fish who had explored Pitcorthie with Cleghorn.

With the opening of the 1888 wing, development of the library became a priority. The first librarian, already on the museum staff, was Charles Muston (1848–1909), who therefore became the first curator of the Cleghorn Forest Library. From the accession registers no books seem to have been purchased for this until 1893, perhaps the date when Cleghorn eventually handed over the £200 testimonial money, though the books given by the Government of India to the 1884 Forestry Exhibition had already been incorporated.[8]

CHAPTER 15

A division of the spoils

On Cleghorn's death in 1895 virtually his entire property, as already explained, was left to his nephew Alexander Sprot. The sole exception specified in the will was of 'such books on special subjects as shall be enumerated in Lists thereof which may be found in my private repositories after my decease'.[9] Unfortunately not only could no such lists be found, but it turned out that Cleghorn had said different things to different people as to what was to happen to his forestry books after his death – and publicly so. Although at the launch of the Cleghorn Forest Library in 1888 he had not actually said anything about the ultimate destination of his own books, it would have been a reasonable assumption on the part of the Museum that they were to receive them, otherwise all that they could do was to buy books from the £200 he had eventually given them. But only three years later, in January 1891, when acknowledging his recent election as an Honorary British Fellow of the Botanical Society he had 'intimated his intention to present to the Herbarium at the Royal Botanic Garden his collection of dried specimens of plants, and *to the library any botanical books in his possession that are wanting in the Library at the Royal Botanic Garden*'. This intention had been recorded in print,[10] and from subsequent discussions with his friend, the Regius Keeper, Bayley Balfour believed that Cleghorn had made the necessary provision for such a bequest in his will. Although the will was made only five months later, in June 1891, it proved, much to Balfour's chagrin, to contain no such instruction. Another complication was that by the time of Cleghorn's death a Forestry Department had, at last, been started at Edinburgh University, and its lecturer Frederick Bailey also took an interest in the fate of Cleghorn's books. In other words there was a mess to be resolved by Sprot and Cleghorn's solicitor Charles Grace. Sprot behaved well: he seems genuinely to have wanted to obey his uncle's wishes in the best way possible despite their muddle. He had no need of the money and as a professional soldier had no interest in either the books or specimens. All appears to have been resolved in a civilised manner – RBGE was given the herbarium, though only a handful of books on economic botany, and the bulk of the books were divided between the Museum, which ended up with almost three-quarters of the volumes, and the University Library with the remainder.

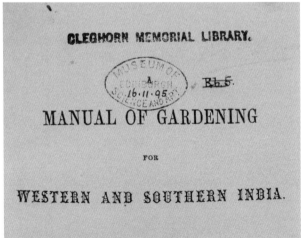

Fig. 73. Library stamps on books from Cleghorn's library. Edinburgh University Library, Cleghorn Bequest (left). The Cleghorn Memorial Library, now at RBGE (right).

15. CLEGHORN'S LEGACY

The Cleghorn Memorial Library & RBGE

The effective foundation of the Forest Library at the Museum therefore came with Sprot's gift of the books from Stravithie after Cleghorn's death. These were accessioned on 6 November 1895, at which point the collection became the Cleghorn Memorial Library. A contemporary newspaper report, which recorded the other notable natural history acquisition of the year as one of the finest known specimens of the extinct great auk, stated that there were 907 volumes, excluding an unspecified number of pamphlets.[11] The total number of volumes (as opposed to titles) recorded in the register, in fact, comes only to about 800, but perhaps the number 907 included the 149 portfolios of 'botanical plates'. There are no photographs of the interior of the museum's library, so nothing is known about how or where the Cleghorn Library was shelved within it, though care appears to have been taken with the binding of at least some of the books in half-calf of an appropriately sylvan shade of dark green.

The high value initially placed on this collection by the Museum is shown by the publication in 1897, by the Science & Art Department, of a catalogue of all of its books on forestry and botany, of which the last section comprised a separate listing of the Cleghorn Memorial Library (the titles also being included in the main listing).[12] In the foreword to the catalogue Murdoch Smith paid tribute to Cleghorn, and the 'very valuable character' of the collection, which appears only recently to have been made 'available for reference'. Within 30 years, however, attitudes and priorities had radically changed both at the Museum and at RBGE.

By the 1920s the Museum was apparently keen to divest itself of botany altogether – both its economic botanical collections and the 4000 botanical and forestry books, which included the Cleghorn Memorial Library. This was not unreasonable as the Museum was legally allowed to transfer collections to other government departments and RBGE with its own library and museum was a logical home. Negotiations began with William Wright Smith, Balfour's successor as Regius Keeper. Remarkably there appears to be no surviving documentation for this transfer of the library,[13] but the botanical collections appear to have come to RBGE shortly before World War II,[14] and receipt of the books was acknowledged by the Garden's librarian James Todd Johnstone on 18 September 1940.[15] In the words of M.V. Mathew, historian of the RBGE library, this represented 'the biggest event in the history of the library', even though at this point it had nowhere adequate to shelve them, and a pitifully inadequate staff: a single librarian with one assistant, Betty Sloane. Even this was short-lived as, having married Harold Fletcher, Betty left in 1942 and Johnstone retired in 1946. It was Fletcher, Wright Smith's successor, who eventually achieved Bayley Balfour's vision of a new building, where the herbarium and library could at last be housed, each in a single, systematic sequence, and in a setting befitting their international status and quality, but it came at a cost.

Fletcher was a modernist who appears to have had little time for the past – it was he who ordered the destruction in the 1960s not only of the RBGE Museum of Economic Botany (which included Balfour's collection and the specimens only recently transferred from the Royal Scottish Museum), but also of the old front range of glasshouses that would today have been lovingly restored. His librarian from 1961 to 1967 was William Hunter Brown who seems to have been similarly brisk, though ironically it was this pair who jointly wrote the authoritative history of RBGE,[16] which, astonishingly, includes not so much as a name check for Cleghorn. Less surprisingly a veil of silence was drawn over the whole shameful episode of the destruction of the Museum, which, hardly surprisingly, is also unmentioned in 'Fletcher & Brown'. Brown embarked on an extensive programme of 'binding and repair of [the book] stock which had been neglected for many years',[17] though the question of their being 'professionally restored' is a moot one, even allowing for the primitive state of book conservation at the time. While antiquarian books were bound in half-calf to look attractive in the glass-fronted book cases when the Queen opened the building in time for the Tenth International Botanical Congress in 1964, those deemed less important (even treasures such as G.B. Ferrari's *Flora* of 1646 from the Marquess of Hastings's collection) were rebound in cheap buckram, their text blocks ruthlessly trimmed in a mistaken aim for neatness. In this process bookplates were usually respected and transferred, but the importance of notes on flyleaves (which were usually removed) or pastedowns was ignored, and marginalia trimmed. Cleghorn as an 'eminent Victorian' was doubtless deemed part of the past that it was necessary to sweep away, and it was not until 1986, that the Kerala-born Manjl V. Mathew appears to have been the first person ever to appreciate Cleghorn's significance as a pre-eminent benefactor to RBGE, even if his wording was bizarrely inaccurate. Among the 4000 books that arrived in 1940, including the Cleghorn Memorial Library:

> were many classics and extremely rare books which the Garden could never have hoped to acquire. The Garden and Library were fortunate in having a friend like Dr Cleghorn who, first through the Botanical Society and later directly [sic], gifted more books than any other single individual during his life time [sic, but in fact 45 years after his death].[18]

All this makes the treatment of the books, and the loss of so many, the more tragic. As part of the present project the Cleghorn books have been diligently sought, but only 565 out of the 800 volumes have been found. Why are so many missing? When the present author started working at RBGE in 1986 a large, bookshelf-lined room in Inverleith House housed what were considered to be duplicate books from the library. Shortly after this the Regius Keeper, Douglas Henderson, obtained permission from the Scottish Office for their deaccession and sale to the long-established antiquarian book dealers Wheldon & Wesley. While by no means all of the Cleghorn 'duplicates' were disposed of (some of the obviously valuable colour-plate books, and runs of journals including the *Botanical Magazine* were kept) a substantial number were undoubtedly among the large number of works sold. The biggest casualties were among the Floras, as copies of the standard works were already in the RBGE library by time the Cleghorn collection arrived in 1940. It should be remembered that the library was already rich in historical works notably from the collections of the Botanical Society and of John Ball, and it would appear that provenance of books and annotations of someone who was not considered a taxonomist were regarded as unimportant. The 149 portfolios of illustrations were also roughly treated – with the exception of only a handful that escaped by accident, Cleghorn's individually stitched species fascicles were split up and the drawings incorporated into the illustrations collection, as were thousands of individual botanical prints. However, many pages of Cleghorn's manuscript notes on individual species must have been discarded and one can only be thankful that the copy illustrations were retained.

The Edinburgh University 'Cleghorn Bequest'

The decision to split the books between the Museum and University Library was certainly strange, not least because at that time their premises were adjacent to each other on Chambers Street. However, there seems to have been no acrimony. Colonel Bailey must have negotiated the share for the University, to which Balfour had free access as Regius Professor of Botany, even if the books were not physically located at RBGE as he might have preferred. So, in the University Calendar for 1896/7, under 'Donations to the Library' is noted, from 'Major Sprot of Stravithie, in [the] name of the late Dr Hugh Cleghorn – Collection of Books, mainly dealing with Forestry'.[19] It is impossible to fathom the grounds on which the split was made. Although most of the forestry books went to the Museum so did much else in categories from which other similar works went to the University. For example, although the Museum got most of the fine folio colour-plate botanical books (by Wallich, Royle, Bennett & Brown and Roupell), the University got Blume's magnificent *Rumphia*. The destination of less grand botanical taxonomic works appears almost arbitrary, so that, for example, while the important interleaved copy of the *Prodromus* of Wight & Arnott went to the Museum, the annotated copy of Wight's *Contributions* went to the University. And while economic botany books would have made much more sense at the Museum, the most important of them, Cleghorn's interleaved and richly annotated copy of Drury's *Useful Plants of India* went to the University. Even the illustrations collection is found to have been split. Whereas it is recorded that the Museum received 149 portfolios of octavo and quarto botanical plates, although there is no record of it, some clearly also went to the University.

The University's share of the books, amounting to around 300 volumes, was placed in the attics above Playfair's magnificent library in Old College, and can be identified using old shelf lists and the printed University Library catalogue,[20] in which most (but by no means all) are denoted 'Cleghorn Bequest' (even if this, for reasons explained, is a slight misnomer). At this point, in 1896, the University passed to the RBGE library the Cleghorn books that related to 'vegetable products' (i.e., economic botany), though none of these can now be identified, together with some '[botanical] plates'.[21] As is discussed in the *Cleghorn Collection* (Chapter 7) some of the latter were left behind, almost certainly by mistake. These show that of Cleghorn's illustrations collection the University received the botanical drawings made for Robert Wight, the Madras Nature Prints, some of the European botanical prints (perhaps the folio ones), together with some interesting miscellaneous material. But there appear to have been few original botanical drawings other than Wight's, and two bound volumes of drawings of Malabar trees made for Alexander Walker of Bowland, evidently purchased by Cleghorn from an antiquarian bookseller (for 16 shillings), of which the related text is now with the rest of Walker's papers in the National Library of Scotland. It seems likely that Balfour must have recognised the pre-eminent importance of the Wight drawings and claimed them for RBGE, together with the Madras Nature Prints and probably also, for adding to the Garden's illustrations collection, many other European botanical prints.

The library catalogue

A catalogue of Cleghorn's books surviving in Edinburgh University and at RBGE has been compiled; it mirrors Cleghorn's biography and his changing interests through the course of his life. A description of the books, and the stories they tell of the way he used them – as, for example, indicated by his annotations, has proved too extensive for inclusion in this biography, as has the complete list, but will be made available on the RBGE website.

An autumnal visit to Stravithie

I had visited Stravithie and Dunino many times in the course of my work on Cleghorn. It was an area I had got to know and love on bird-watching trips in childhood, and during three years spent in the botany department of St Andrews in the early 1980s. I often cycled along the East Neuk roads, with unparalleled views across a shimmering North Sea to the distant lighthouse of the Inchcape Rock, more evident at night from its warning flashes. But a late autumnal Saturday afternoon in 2015 found me heading there for a final visit: to the Sprots at Mains of Stravithie for another look at the family portraits, and to the Chalmers family at Stravithie. Along

CHAPTER 15

North Street passing St Salvator's College where Cleghorn undertook his pre-medical studies in the late 1830s, past the magnificent ruins of the cathedral with the tombs of the Playfairs, the Adamsons and Robert Chambers, through the Pends, the abbey's medieval gatehouse, and past St Leonard's School in which Cleghorn had shares. Past the harbour and out of the town, passing the Gatty Marine Laboratory beside the East Sands (with its previously unsuspected Cleghorn connection). Up the hill to the fork where the road divides at Brownhills – east to Crail, west to Anstruther – marked by a cast-iron waymarker, a concrete poem of romantic place names, including the two Cleghorn ones of Wakefield and Denino. The right branch climbs past Balmungo, in Cleghorn's day the home of Alexander Kyd Lindesay, more recently of the artist Wilhelmina Barns-Graham to the summit at Priormuir, the start of the Stravithie Estate. A pause here to look

Fig. 74. The Brownhills waymarker.

back over the magnificent panorama of the inner part of St Andrews Bay, past Tentsmuir and the mouth of the Tay to the Grampian Mountains on the horizon – the landscape evoked so hauntingly in Violet Jacob's poem *The Wild Geese*. On down the gentle south-facing slope to the two burns, with a glance right to the vanished North Plantings, reflecting on the arable fields – the result of the vision of the Cleghorns but also the sweat of their tenant farmers.

The weather was wild and stormy as I reached the wooded den at the lowest part of the estate, the light fading as I entered the gate at Stravithie. The russet beech leaves were cascading from the trees and already formed a rich carpet on the drive. The visit was for a final look at the room that had been built to house Cleghorn's books that I had spent so long seeking out and cataloguing. By now I knew that the tall deal bookshelves had been built for him by his father-in-law's joiner, John Ewart of Penicuik. The upper open shelves, with the remnants of scalloped and gilt-stamped leather dust flaps, now held a quite different collection of books, but David Chalmers picked out a few that had remained when the Sprots moved out in 1979 – some odd volumes of the classics (in French and Latin) bound in full-calf for the first Hugh Cleghorn with WAKEFIELD stamped on their front boards. A German novel that must have belonged to poor Isabella Cleghorn, and two volumes of a *Letters to a Young Lady* that had been presented to Cleghorn's aunt Helen Allan in her Leith childhood, and reinscribed with her married name Wyllie. Most interesting was a pristine three-volume set of Southey's *Life of Andrew Bell*; this linked the three Cleghorn generations, as Bell had been a pupil of the first Hugh and the book included letters both from him and his wife Rachel to Bell. In 1844 the book had been presented to Peter Cleghorn by the Bell Trustees who had commissioned it; and it was signed by the younger

Hugh, one of the first pupils of the Madras Academy built by Bell's spectacular bequest to the town of his birth. At this point David's wife Renata opened one of the cupboards that form the foundation of the book cases – each door minimally ornamented with a diamond moulding. My heart missed a beat, as there on the edges of the shelves, beneath the detritus of modern family life, were Cleghorn's own hand-written labels showing exactly where his unbound journals and pamphlets on particular subjects, all now in Edinburgh, had resided between 1868 and 1895. It was as though an electrical spark had discharged across time – the gnomic abbreviations could mean nothing to the present owners but brought back vivid memories of Cleghorn's life and travels – on one shelf were 'Coffee', 'Cinchona' and 'Cotton', attesting to his economic botanical interests. On another 'Tract Society', 'Bible Society', 'German Mission' and 'Moravian Mission' were reminders of his evangelicism, but more particularly of an isolated mission house high up in the Punjab Himalaya at Keylong. The 'Sporting Rev[iew]' suggests that his interest in shooting continued through life, a salutary reminder of other topics omitted from this 'Life', through lack of evidence and the patchiness of extant sources.

The Intangible

Vestiges of Cleghorn and his legacy in India

Another avatar of Cleghorn's legacy is to be found in India: in administrative terms in the Forest Department that, with Brandis, he helped to establish – and biologically in its forests and gardens, especially in the surviving remnants of forest in the Himalayas and the Western Ghats. It was not enough to seek Cleghorn in libraries and archives and I needed to see for myself the places in which he had lived and worked, and I therefore made several journeys following in his footsteps, for each of which I kept a detailed diary. It seems appropriate, therefore, to end this volume with a trio of sample extracts – a day from each of the three major trips that were chosen to reflect Cleghorn's three major Indian periods: as Mysore surgeon (an excursion in the Shimoga district), as Madras Conservator (an excursion to the Orissa sal tract), and his role in the Punjab and with the Government of India (a journey over the Sach Pass, between the valleys of the Chenab and Ravi rivers).

'The Nugger Division'

One of the most satisfying discoveries in the RBGE archives in 1997 was the visual botanical diary of Cleghorn's Shimoga period. When plotting his routes on a map from details on the drawings the desire to retrace them became overwhelming. These travels were finally achieved a decade later, in October 2007, using Shimoga as a base and travelling by local buses. There were so many sights apparently little changed since his day and so many of his plants still growing there, that it is hard to choose a single representative day – visits to the sublime Falls of Gersoppah would have been a possibility, as would the spectacular fort at Nagar, or the Baba Booden Hills and their moving, subterranean Muslim shrine. The rich tropical forests that survive around Agumbe would also have been characteristic and revealing, but instead I have chosen a day that reflects Cleghorn's interests in agriculture, but also the resolution of a mystery over the identification of one of his plant drawings.

Chikmagalur, Karnataka. Sunday 28 October 2007. 5.15 p.m.

The second fantastic day in a row, the last of these mad bus journeys on rutted cart tracks in wobbly buses in the company of old farmers (probably the source of another snuffle to which I've succumbed). I got to the bus stand about 8 o'clock and bought a *Deccan Herald*. It rained

CHAPTER 15

here most of last night and the paper reports that Ooty has been cut off by landslides. Got a bus to Kadur (see map, p. 23) almost straight away, so back on that road yet again, passing a circus that was just setting up on the edge of town – five dusty elephants and three camels made up the livestock. In Kadur very little delay for a bus to Yagati, but was surprised to find myself back on the road to Birur rather than the direct road shown on the map. Not that it mattered, nor the fact that the bus was feeling poorly and achieving only a snail's pace before giving up the ghost halfway between Birur and Yagati. This was a positive advantage, as it meant I could walk through the countryside and see more plants.

The first excitement was in a garden beside the 'circle' in the centre of Birur: a large bush (about 8 feet tall) – at last the mystery Shimoga composite (Drawing No. 280), in full flower, which I was able later to identify as *Pluchea ovalis* (*Cleghorn Collection* no. 69). The scenery from Birur is very (and surprisingly) different from the route to Harihar – dead flat plains, with only a single upstanding granite 'droog'; very dry, doubtless in the rain shadow of the Baba Booden Hills. The aridity was emphasised by the presence of agaves and on rocky wasteland between the fields grew low bushes of *Dodonaea viscosa* with its papery fruit; also two succulent euphorbias, *E. antiquorum* and *E. tirucalli*. The second great thrill was the dark maroon trumpets of *Argyreia cuneata* in the hedgerows (composed of some sort of *Acacia* and the ubiquitous hedge plant of these parts, the ugly African *Synadenium grantii* that was not here in Cleghorn's day). The main grain crop was *Eleusine* ('ragi'), with a little *Sorghum*, and less of spiked millet (*Pennisetum glaucum*) and fox-tail millet (*Setaria italica*). [When I got home I found that the latter was drawn for Cleghorn at Ekatty but I had misidentified the drawing as *Pennisetum* – it was therefore the 'native informants' who put me right: Sharvani in Mysore had transliterated the Kannada annotation on the drawing as 'navane', the very name provided 160 years later by my fellow bus passengers]. No betel palms but, as elsewhere, plenty of coconuts. It was here that the bus finally ground to a halt due to faulty brakes, so we all got off and started to wander around. I was only too pleased to be among fields, especially when a low-growing cereal, seen nowhere else, turned out to be one of Cleghorn's Ekatty plants – *Panicum miliaceum* ('save'). I acquired a retinue of curious but affable followers who supplied the local names of some of the crops. At the (slightly damp) edge of a ragi field was the yellow composite *Guizotia abyssinica* ('chello'), of which they said the black seeds were eaten – the 'savi' used only as a cattle feed. A small trifoliolate legume, with small, yellowish-green flowers with a red spot at the centre of the standard, grown in large quantity in fields, proved to be *Macrotyloma uniflorum* var. *uniflorum* (horse gram, '*Dolichos biflorus*'). The local name was 'huralli kalu' and it was said to be used in making sambar.

The driver started to fiddle around under the rear axle but after half an hour I got bored of hanging around and, having checked that the bus would stop and pick me up when it eventually got going (or a replacement sent), I started to walk towards Yagati. Some nice plants on the dry roadsides (on which a Japanese-sponsored attempt at some tree planting had been half-heartedly made) – all of them painted by Cleghorn's Marathi artist: the small and very spiny shrub *Barleria buxifolia*, with pretty pinkish-mauve flowers; the exquisite miniature convolvulus *Evolvulus alsinoides*, with bright blue, forget-me-not-like flowers; the dreary 'everlasting' *Polycarpaea corymbosa*; the borage *Trichodesma indicum* with washed-out pinkish flowers; and the handsome spiky globe-thistle *Echinops echinatus*. More crops, including castor oil (*Ricinus communis*) and red gram, *Cajanus cajan* ('togri'), the flowers of which are extremely attractive when seen at close quarters, the buds and the outside of the standard being dark

15. CLEGHORN'S LEGACY

Fig. 75. *Pluchea ovalis*, Birur, Karnataka. (Author's photograph, October, 2007).

red and shiny. In one field was a strange legume – *Cyamopsis tetragonoloba* with glossy dark green leaves, the three leaflets with serrate margins, and small, widely open, very pale pink flowers in short axillary racemes that develop into fascicles of erect pods. In many of the fields the crops were interplanted, in a variety of combinations: *Guizotia*/*Eleusine* as already mentioned, but other mixtures were rows of single plants of *Lablab* among *Eleusine*; rows of *Ricinus* among *Macrotyloma*; alternating rows of *Cajanus* and *Ricinus*; and (nearer Chikmagalur) cotton interplanted with either *Cajanus* or *Ricinus*.

CHAPTER 15

After about thirty minutes the (same) bus came along and picked me up, only to deposit us, not at Yagati, but a small village called Pura shortly before it. The village was on the banks of a biggish river and had what, from various satellite structures, seemed to be an important temple – these included small, square, four-columned mandapas on some flat rock outcrops, and an extremely elegant swing arch now improbably stranded in a farmyard, on the opposite side of the street from the 'west' end of the temple. Here were some tall plants of *Pennisetum glaucum*, and on a roof I photographed the attractive white-flowered form of *Lablab purpureus*. I was just starting to wonder how to get to Yagati, not knowing where I was, when a handsome young man from the bus came to my rescue and escorted me for what turned out to be only about a final mile – we passed some more of the spiky, paired pods of the asclepiad *Pergularia daemia* in a hedge. He worked in a pharmaceutical company in Bhatkal (on the coast half-way between Mangalore and Goa), but his parents live in Tarikere, and he along with numerous relations (50 were expected) were converging on Yagati for the celebration of the first anniversary of his grandfather's death at the age of 95. He pressed me to go home with him and the whole family were quite exceptionally welcoming to a complete stranger – an aged man in white cotton was clearly local, but there were several sophisticated young cousins from Bangalore. The house was extremely simple and sparsely furnished, but spotlessly clean – on a shelf in the living room was a row of chromolithographs of Gods, providing both decoration and objects of devotion; in the dark kitchen the decoration came from something equally functional – neatly aligned stainless steel canisters on ranks of shelves. A moving atmosphere of unaffected tenderness and celebration from the elders down to the sparky little children. The boy's mother, a distinguished-looking woman, gave me a plate of delicious uttapam and a cup of excellent coffee. I couldn't really savour this unexpected social event as much as I would have liked, owing to worries about getting a bus back and leaving time to get off the bus to photograph the *Argyreia* and the mystery Compositae. So my friend took me for a quick walk around the village, including the ancient temple, which had been unfortunately modernised – bathroom tiles, and paint effects on the stone columns – though with a fine image of Vishnu in the sanctuary. There were buses going back to Kadur by the short route, but this would have meant missing the plants and fortunately a bus for Birur was just about to leave. The bus conductor was asked if he would let me off and wait while I took the photos, but this would have been stretching things just too far, so I got out and waved the bus on, and walked for about half an hour, almost back into Birur, photographing *Dodonea* and *Argyreia* and some interesting crop-combos on the way. Of bird interest were a couple of rollers, a spectacular female koel (black, covered with white checkering), several eagle and some small flocks of laughing thrushes. Handsome sheep with dark astrakhan fleeces, herds of goats, old men patiently leading pairs of bullocks, the bullocks ploughing with amazing efficiency given the simplicity of the ploughs.

An autorickshaw with a cage on the back (used for transporting goods in country areas) came past and gave me a lift into Birur, where I photographed the *Pluchea* (Fig. 75), and in a nearby garden another Cleghorn plant, the climbing chenopod-ally *Basella alba*, that had eluded me so far.

The Orissa sal tract

On the eastern coast of the great Peninsula are many sites connected with the history of the European discovery of the Indian flora, so in the autumn of 2010 I travelled (as usual solo, using local buses) from Calcutta to Madras to look at some of them. Part of this was in order to see sites associated with William Roxburgh and J.G. König, but there was also a Cleghornian dimension – places of significance to his friend Walter Elliot (especially the remains of the great Buddhist stupa of Amaravati) and the irrigation works on the Godavery undertaken by Arthur Cotton, another of his friends, which Cleghorn visited on his excursion to the Northern Circars in January/February 1859. The day chosen here includes a visit to the most southerly of the forests of Sal (*Shorea robusta*).

Raj Palace Lodgings, Russell Kondah (now Bhanjanagar), Orissa. 14 August 2010, 6.40 pm.

I left the horrible Hotel Keshari in Bhubaneshwar at 5.50 this morning with the usual rickshaw driver, Hadibhandu Sahoo (blood group AB, distinguishing mark 'mole on nose' according to his permit, displayed for the benefit of passengers); it took a good twenty minutes to reach the bus stand on the very outermost, western edge of the city. Got straight onto a 'sleeper bus', but any air-conditioning came from the open windows. Against the windscreen was a spectacular floral garland, the flowers so gaudy that they looked plastic – golden yellow trumpets of *Thevetia* (better thus than on their ungainly shrub) strung head to mouth, interspersed with alternating clusters of white *Tabernaemontana* and scarlet *Caesalpinia pulcherrima*. Fortunately I was put in the front seat which was fine both for me and my luggage, but the rest of the bus periodically got immensely crowded – no-one is ever turned away from these vehicles. I guess there aren't too many a day, and it served as a major carrier of college students, farmers etc., from one place to the next. After the usual endless delays and negotiations we started the process of edging out of the bus station and left at 6.45. The journey took 5¼ hours (with two stops of 10 to 20 minutes), the distance being 176 km. I discovered yesterday on the Internet (before a power cut terminated the session) that, rather to my dismay, the altitude of Russell Kondah is a mere 270 m – I'd been hoping that it was at least in the foothills. So the journey was almost entirely through wonderfully lush green paddy fields and scenes of ploughing oxen; occasionally there was a bosky stretch, or a tank jewelled with stars of the dark-red water lily – and a single stork. On the bunds of some of the paddy fields were what I took to be sluice gates of the Keladi sort, often in pairs, like miniature shrines with stepped pyramidal roofs, but I think they must be grave markers, presumably Muslim. (There are many Muslims in Russell Kondah – a large mosque, with a plangent muezzin: from the newspaper I learn that we are in Ramadan). Though rice was the main crop, there were smaller amounts of cashew and sugar cane. Roadside clumps of the beautiful *Martynia annua*, with *Gloxinia*-like pinkish flowers with a dark basal spot, and soft velvety sinuately-lobed leaves (an introduction from Mexico). We reached Nyagarh, half-way in distance at about 8.45 then stopped at Oragaon, a largish town for twenty minutes at 9.45; through a wooded stretch, then back into rice paddies. Many of the people in the fields were wearing large Chinese-style coolie hats – almost a metre in diameter, with a conical point in the centre, of plaited bamboo (like a cane chair seat), the hexagonal holes revealing an insulating layer of some sort of leaf. The last two towns before reaching here at noon were Jagannathaprasad and Bellaguntha (where stems/corms of what appeared to be *Alocasia macrorhizos* were being sold in the market).

Bhanjanagar is clearly NOT a tourist place, and, judging from the open-mouthed stares I encountered, the locals may never have seen one of the species before. From desperation I was bold and challenged a particularly nice-looking, well-dressed and smiley man called Siddarth (a yoga teacher) for help, and within minutes he had me installed here, which is perfectly adequate if far from luxurious, so long as the a/c doesn't conk out as it did earlier. Mosquitoes are another down-side. It cost 500 Rs for a night, and the hotel arranged for a car to take me round the local sights for 1100 Rs.

Fig. 76. The Orissa sal forest, Russell Kondah (now Bhanjanagar). (Author's photographs, August, 2010).

The driver was a maniac who drove at 60 mph whenever he could, and the roads are surprisingly good here. Every five minutes his mobile phone rang, which he never failed to answer, being able to do this and change gear at the same time! I had absolutely no idea where I was being taken, but to my relief found that we were heading north towards Kalingia (distance 35 km) into what, according to the *Lonely Planet Atlas*, are the Khondama

Hills. I have always rather wondered if the Eastern Ghats were an invention – they look so unspectacular on maps, but here they really were and we must have risen about 2000 feet. Driving out of town we passed through a whole suburb devoted to the Forest Department – the District Forest Officer's office, his bungalow, ditto of the Forest Conservator, a Forest Colony, a Nursery, timber yard etc. Cleghorn would be very pleased should he return (or be looking down from a cloud). But the biggest thrill was that within about a mile out of the town we arrived in a sal forest! They were unmistakably plantations (Fig. 76), of various ages, the trees fairly to very skimpy, but with characteristic reticulate bark and I stopped to find some old (dipterocarpous) fruit and to collect a leaf just to be sure. The trees here (at their southern limit) have nothing like the stature of those in the Terai at the foot of the Himalayas, being more like the stunted ones of the Pachmhari plateau, and have short leafy branches all the way down the trunk. There were signs indicating that we were in 'Ghumsur North Division', identifying it as 'Reserved Gallery Forest', the altitude a mere 132 metres. Everywhere the teak is in bloom – when young the trees look improbable, supporting such large leaves and huge frothy panicles of small cream flowers.

The road started to ascend the Kalingia Ghat, the slopes thickly forested, but, as usual, I had no idea what the trees were except for a fruiting *Terminalia*, a 'bignon' with slender twisted pods and some figs – it was all a gloomy dark green, but some were probably deciduous, and the forest seemed very diverse in species. I got out and walked a bit seeing a yellow *Globba*, a fruiting *Arisaema*, *Helicteres isora*, a *Leea*, and single young *Cycas* plant, *Costus speciosus*, *Colocasia affinis*, *Rubus moluccanus* and *Bauhinia vahlii* looping its way through the trees. The fish-tail palm here doesn't look like the 'normal' *Caryota urens*, the leaf rather more dissected. We reached the top of the ghat and, as in the Palni Hills, I was shocked to find plantations of eucalyptus and a pine (possibly *Pinus kesiya*), though such plantations are another legacy of Cleghorn. The mystery tour continued and the driver kept stopping passers-by and asking directions to a "halticuchur" establishment. This seemed suitably Cleghornian, but the reality was a down-at-heel place, symbolised by the jak fruit that were allowed to drop from the trees and rot into disgusting black deliquescent heaps, the seeds germinating in the morass of their own filth. There was a guava plantation, some oil palms, a *Michelia* tree and some weedy pot plants: *Bletilla*, *Coleus*, *Pelargonium* and *Zinnia*. We turned and came back, passing a bicycle laden with sal leaves – perhaps for making leaf bowls. We also glimpsed a large lake – the Duha River has been dammed since Cleghorn was here, which would have put paid to his idea of using it for floating logs. We then went off on a wild-goose chase to a village called Kullar, which was clearly regarded locally as picturesque, but was actually rather squalid, the only compensation being a *rath* parked outside a temple, and some ornithology: a couple of rollers, and the strange sight of some weaver birds' nests on telegraph wires rather than the usual wild date palms (earlier in the day I had seen them attached to several of the leaflets of the fan of a *Borassus* frond). By this time I was feeling like death from a nasty cold I have caught, but struggled up a small hill that overlooks the town to get a view. The path ascended from near a large enclosure with a blank wall that I thought might be some sort of temple complex, but which turned out to be the town gaol. On the hill were rocks full of large, decomposed garnets. The town consists of one large street, which crosses a river, with a few smaller side ones. I have just walked down the main street and saw one old British-period building, a hospital made up of a taller central and two smaller outer pavilions. I then had some vegetable chow mein, eaten off a *Bauhinia vahlii* leaf – with three unripe bananas, the only food all day – and I had to leave half of it. I emailed Vinita Damodaran to tell her where I was: as it is due to Cleghorn's 1859 notebook that she found among Richard Grove's papers that I am here.

The Punjab Hills

In Autumn 2014 I followed Cleghorn's footsteps in the Western Himalaya: the area he surveyed in 1862 and 1863, primarily for its deodar forests – the valleys of the Sutlej, the Beas, the Chenab and the Ravi, the hill stations of Simla, Dharamsala and Dalhousie, and the site of the Moravian Mission at Keylong of which sadly no trace remains. What he had to do, with immense effort and danger, on foot or horseback, is now almost all accessible by road, but the majesty of the landscape is rendered none the less impressive for the comparative ease of travel. The sample day chosen is the crossing of the Sach Pass; this was done in the reverse direction to that in which he travelled, but at a very similar time of year, on which I saw many of the plants and landscape features that he recorded in a notebook that survives among his papers at St Andrews.

Hotel Chamunda View, Chamba (980 m), Himachal Pradesh. 27 August 2014, 7.20 p.m.

A thrilling day, but a draining one – a 12-hour bone shaker – but what incredible sights. The first excitement was to wake up this morning at Killar (see map p. 143) and see a light dusting of snow on the peaks above the Sach Pass – the other mountain-tops covered in grey cloud. We set off at 6, but got only half-way to the bridge a few miles downstream when the driver decided to test the brakes of the new jeep. They weren't working, so the old bone-breaker we used yesterday was sent for and we trans-shipped. This hiatus gave a view of the old Killar Bridge across the Chenab. Inevitably not the one Cleghorn used, a later replacement of it (but of wire and wood not concrete), apparently the last of the old, narrow type which donkeys can cross only if unladen, and when dry (otherwise it is too slippery). A good omen came soon afterwards (2150 m), with the most unexpected appearance of a patch of *Pinus gerardiana* (and growing with it walnut, deodar and *Parrotiopsis*). Here the edible pine nuts are not collected, so that I was able to pick up from the road a fully grown, ripe, seed-shed cone.

We crossed the river by a substantial bridge at 7 o'clock (2000 m), and started the ascent up the left bank of a side valley that leads to the pass, with views of the horrifically narrow old track on the opposite (right) bank, down which Cleghorn would have walked. Through wonderful, rich broad-leaved forest with lots of hazel, ash, walnut, maple, horse-chestnut, ash, willow and *Parrotiopsis*. The hazel (*Corylus jacquemontii*) is in fruit, with amazing succulent, pale green, lacerated cupules (*Cleghorn Collection* no. 158). All day we kept running into three jeeps full of an extended family party of pilgrims from Jammu/Kashmir visiting three temples, their present destination being Lake Manimahesh. When I say full, I mean full – one jeep had 15 people inside and a sheep on the roof rack. The first time we met them one of their vehicles had broken down and they were collecting hazel nuts (which must also have explained the presence of a Nutcracker, one of the most handsome of corvids), and the children were feeding the lush leafy hazel shoots to the sheep on the roof. Another of the jeeps was also carrying a single ewe, and the third two gelded rams, the poor creatures trussed up and in agony, especially on the other side of the pass where it got very hot – extremely cruel, but their days are in any case numbered as their destiny is to be offered to Shiva. Of the party is their own Gaddi pujari in traditional woollen garb and with a wry twinkle in his eye.

At about 2450 m the deciduous trees were thinning out. On the opposite slope was plenty of blue pine (*Pinus wallichiana*), but I was distressed to see another forest fire (small so far), which Prem [my *cicerone*] says are started by the gaddis. On our side (the north/left bank) much less forest though a little deodar (*Cedrus deodara*). On turfy slopes (2600 m) there were large stands of dried-up spikes of the asphodel-like *Eremurus himalaicus* – something I had expected to see long before now – this is one of the two plants (the other being *Meconopsis aculeata*) that Cleghorn at least attempted to bring into cultivation. Prem says that it flowers in May/June, is called 'pyuk' and eaten as a vegetable – (according to Jäschke, the young leaves are 'cooked in the form of spinach ... considered by the [Moravian] Mission a very good vegetable'). From 2700 m *Betula utilis* appeared, though only on the opposite side, and by 3000 m the conifers had virtually stopped and we were in alpine turf with a handsome kingcup (*Caltha*) and the usual colourful herbs – *Anaphalis, Pedicularis, Polygonum affine* (*Cleghorn Collection* no. 156), *Aster, Potentilla, Nepeta* etc. At 9 o'clock (3400 m) in a very bleak stony area we stopped at a roadside hut. A large flock of sheep was on the move guided by two gaddis one of whom was carrying a small hand axe (presumably for chopping firewood). The enterprising and hardy men who ran this primitive roadside café had made their hut (stone walls with a tarpaulin roof) cosy – so long as they had fuel – though it was cold enough for me to put a 'fleece' over my jumper.

Patches of dirty snow were starting to appear, some of them forming bridges over the stream below the road; gradually the landscape got bleaker and barer, and more and more snowy. The U-shaped valley with a median moraine. Near the top was a scene of utter devastation – an abode of the gods for an atheist. Shattered rock, moraine and snow banks by the roadside. The line of the Christmas carol 'snow was lying, snow on snow' came to mind, which here did not seem tautologous, as the types were distinct – the old snow like a geological deposit, sculpted by wind and melted into weird fans and fingers, covered in mottled detritus of dry plant fragments, but on the banks beside the road, the icing sugar of the morning's snow traced delicate patterns on its semi-permanent forebear.

At one point (but only one, and I was wary of getting the driver to stop *too* often) I glimpsed a clump of the inflated blue hoods of *Delphinium brunonianum* and there were several clumps of the exquisite ice-blue *Meconopsis aculeata* (Fig. 77). But for Cleghorn's sake I simply had to ask to stop to dig up a root of *Rheum* (3750 m), and was glad that I had thought to buy a cheap knife with which to perform the task in Killar. A rosette of wonderfully crinkled, rounded leaves with red petioles, so perhaps *R. spiciforme*, though no sign of an inflorescence, the root is like a mandrake but bright orange-yellow when sectioned. With it grew the handsome *Primula involucrata* with nodding bells of very pale pink. A pair of lammergeiers cruised below us, so close that I could see the cream markings on their heads. In a flush at 4000 m was more *Meconopsis*, a lovely yellow *Cremanthodium*, a saxifrage and some *Junci* one of them close to *J. sikkimensis*. At 4600 m a big yellow *Corydalis* (perhaps *C. govaniana*) and an umbel that looked like a *Pleurospermum* (which may, from its white bracts, be what Cleghorn called '*Astrantia*').

Fig. 77. *Meconopsis aculeata*, north side of Sach Pass, Himachal Pradesh. (Author's photograph, August, 2014).

15. CLEGHORN'S LEGACY

At 10 a.m. we reached the Sach Pass, at 4680 metres, or 15,400 feet, the high point of the trip literally and perhaps also metaphorically. Tucked into gloomy rocks are two tiny Shiva shrines at which the driver offered a biscuit provided by Prem, who had bought a packet at the tea hut. Hidden in crevices were quite a few plants, but we paused for such a short time that I couldn't take them in – a tiny *Kobresia* that I didn't recognise, a rush that might have been *J. leucomelas*, an egg-yolky *Corydalis*, a minute gentian and a saxifrage that looked remarkably like *S. cernua*, so rather like being in the 'crater' at the top of Ben Lawers.

The view down over the other side of the pass a vision of a completely different world – SO GREEN – the grassy slopes (littered with scree) came much higher (4100 m), much closer to the Pass, and so apparently did the forest. To the east the peaks were all snowy, but to the south they were grassy to their summits; beside the road a single clump of one of the handsome big black sedges, probably *Carex nivalis*. Below the rock zone (at 3700–3600 m) came rich and vivid pastures in which the predominant flower colour was gold with a little magenta – though I was disturbed on a single brief foray (too much hurry!) to find that what I have assumed from moving vehicles to be a large-flowered *Potentilla*, is actually *Geum elatum* (*Cleghorn Collection* no. 156), and here it was dominant. Lots of other beautiful flowers, including a lapiz-blue *Anchusa*, lots of *Meconopsis aculeata*, tussocks of a handsome fern (*Polystichum*), *Salvia hians* with spectacular two-toned white and blue flowers, a magenta *Pedicularis* and also one with spikes of pale lemon moons (*P. hoffmeisteri*), lots of *Morina* and a spicate *Ligularia*. At 3600 m, while any forest had been cleared on the side of the valley with the road, on the opposite slope was a very dramatic tree line – the trees rounded, like those in a model railway, but black, and skirted by a much paler shrub zone.

The descent was very rapid; at 3350 m we came to a police check-post where we were videoed and I photographed the large white-flowered *Swertia petiolata*. When seen from above the treeline was extremely puzzling, but when we arrived at it on our side of the valley the dark, rounded trees turned out to be *Quercus semecarpifolia* and the paler, fringing shrubs *Rhododendron campanulatum*. Some of the rhodo bushes were growing exactly in the way that Cleghorn had drawn them here, with deflexed stems, though I suspect that this is as much due to soil-creep as to the weight of their winter snow-load as he thought (growing with them was *Carex duthiei*). I have never quite believed the story of Gujars taking buffalo – those supremely lowland beasts of tropical mud wallows – up to the high pasture, but here was one of their camps, and here were the beasts themselves, at over three thousand metres! I suppose it is the hare-and-the-tortoise story again – if they plod slowly they will reach.

By now we were in oak forest and I photographed what looked very like *Impatiens glandulifera*, though only a fraction of the height that it reaches on British river-banks. My eagerness to find birch was granted at 3120 m, where we came to an area of these beautiful white-barked trees (growing with *Rosa macrophylla*, *Sorbus* and still *Rhododendron*). I cut some 'paper' from a young tree that had been felled and was lying beside the road – I had hoped for some much larger pieces, but at least these are similar in size to the herbarium labels used by Cleghorn. We then (3000 m) came into beautiful fir forest (*Abies pindrow* and at last there were some with a very few cones), and shortly after this, at 2980 m – being confined to a very narrow zone – both the birch and the rhododendron abruptly stopped. The oak, however, continued and then *Prunus* became common among the fir. The appearance of the slopes is extremely striking, from the narrowly columnar habit of the firs, presumably adapted for ease of

CHAPTER 15

shedding snow. At 2920 m a little blue pine appeared and at 2800 m spruce (*Picea smithiana*) with its handsomely drooping branches, but oak and fir were still dominant. By 2600 m the oak had dropped out leaving spruce, fir and cherry but by this point large areas were cleared and there was also plenty of *Aesculus*, *Juglans* and *Acer*, similar to the Narkanda–Sungri ridge. Some of the horse chestnuts have an incredible 'shag bark': strap-shaped curling flakes with squared off ends. By 2400 m there was more spruce and the fir had stopped and there was a small quantity of *Parrotiopsis*. Then came the relief of a metalled road (though not for long) and at 2300 m the forest had all been cleared and we were into the largely cultivated zone (apples and maize), though lower down there were still some patches of forest.

At 12.20, at Bairaghar (2200 m), we stopped for lunch for 50 minutes, along with the three pilgrim jeeps and their gaddi with whom we had caught up and who were tucking into tsampa – it was burning hot and it seemed impossible to believe that only 2½ hours earlier we had been in an inhospitable world of rock and snow. Four black-and-white griffon vultures circled overhead and the driver mended our brakes with an old nail. Still 95 km to go to Chamba. At 1550 m we were at the base of the blue pine zone – from a distance the conifer zones are of different colours, the pine green, and looking higher up the spruce/fir black. Then having climbed slightly at 1600 m we suddenly came into a large deodar plantation in which, curiously, was a patch of fruiting Jerusalem cherry (*Solanum pseudocapsicum*).

The thing I was really looking forward to on this stretch of the road was Cleghorn's 'Tisa Nullah'. We were now in the Chamba Chora valley, which Prem says is famous for its maize – lots of incongruous *Yucca* naturalised by the roadside. This (at 1600 m) is the chir pine (*Pinus roxburghii*) zone, which continues to Chamba – off and on – depending on pressures of cultivation, which only doesn't happen when the slopes are too steep.

The Tisa Nulla (at 1390 m) was easily found – and lived up to the dramatic cross-section reproduced in Cleghorn's 1864 *Report*. The main road, carrying all the traffic, crosses the gorge by a high-level bridge, but few of those in vehicles can possibly imagine what lies under them, and it certainly came as a surprise to Prem. A relatively small tributary has worn a narrow fissure, with vertical sides, through soft, cream-coloured rock, which Cleghorn said is 162 feet deep, though I'm sure that is an underestimate. I looked down and saw that there was a bridge far down below, at the water-level, and as we had only two hours to go to Chamba we went down. The low 'bridge', in fact consisted of two rusty girders that I crossed with some trepidation in case they weren't firmly anchored at the far side, but it was well worth the risk for the dramatic view up the gorge.

We met a cavalcade with a 'flag car' carrying the Minister of Forests for Himachal (Thakur Singh Bharmouri) to Pangi; unfortunately his department had proved singularly unhelpful when asked to provide the simplest of logistical help for my visit. We then came to the big Seul Valley, rather like the Pabar Valley, with chir pine, grass and scrub on the steep rocky slopes, otherwise cultivated, and with the weedy tree *Ailanthus* common. The river was devoid of water due to its capture upstream by the Sach Dam, though as we went lower it partly filled up from tributaries and formed braided channels between gravel flats on the wide valley floor. On the other side of the river was a large Hydro town, and then we reached the Badho Forest Checkpost. Down and down we drove, but one very unexpected sight (at 900 m) was a grove of pollarded wild olive (*Olea ferruginea*) – Prem said this is called 'kau' (a name recorded by Lindsay Stewart) and that they have very hard wood and the leaves are fed to livestock – but why bother to cultivate them? At last we hit the westward-flowing Ravi, the fourth of Cleghorn's Punjab rivers – also with huge gravel flats, its water glinting silver in the low evening light. Then, at last, at 5.45, Chamba (950 m) appeared – first its modern sprawl, then the old town on a high terrace above the Ravi. We crossed the river and drove up the steep slope of the terrace, through narrow streets, between vegetable stalls and round the Polo Ground (the Chaugan), where Cleghorn camped, which was full of sheep and gaddis, camped out on their way to Lake Manimahesh. I'm just about to go and investigate these.

8.55 p.m.

Wonderful sights on the Chaugan – so long, that is, as one isn't a sheep or a goat. The campers are not gaddis with their flocks as I had thought from the jeep, but pilgrims, and the beasts are the ones the family parties will sacrifice tomorrow – and what a horrid bloodletting there will be. That shadow-side apart, the scene was a moving and highly picturesque one – a real buzz of subdued activities – low drums, quiet singing, the hum of conversation. Family groups each with its own ground-sheet and luggage, some with a gaudy portable shrine with one or more tridents wrapped in scarlet and tinsel. Some have tents. The hiss of pressure cookers and small rings of flame from primus stoves. In the middle of it all a middle-class local in a sari was exercising her pug. The women colourfully dressed, one group was singing to an accompaniment of a strange, linear tambourine – a bar of iron with what looked like bottle tops attached. Another accompanied by drums and a tiny piccolo. Men were dancing together, one elegantly juggling metal dinner-plates as he danced. A large queue for a meal tent, where enormous cauldrons of rice were being boiled up. Most were clearly going to sleep *en plein air*, but it is very warm – this grotty hotel room is very stuffy and the fan inadequate.

The future

These three entries show that it is still possible to find traces of Cleghorn in modern India. Despite ever increasing pressures of population some of the forests he knew have survived and some have even increased. However, constant vigilance is required, and despite the efforts of the Forest Department constant encroachment is occurring. Appearances can also be deceptive. On returning to Britain from the Himalayan trip where I had been thrilled with the amount of coniferous forest visible from the roads I looked at Google Earth to find that what survives are the merest narrow ribbons on the steep slopes, strictly preserved to prevent erosion. Between them are vast tracts of unforested land, areas subject to conflicts of landuse between forestry and grazing. In apparently relatively unaltered areas like the plains of Karnataka there are also threats to habitats and wildlife from the use of chemical pesticides and herbicides in agriculture and the planting of genetically modified crops. But from a narrower perspective, and the privileged position of my RBGE office, Cleghorn's presence is evoked most strongly in his specimens in the herbarium, and in the library in his books and the exquisite and multi-layered paintings he commissioned from artists in India. A selection of these are reproduced in this volume's companion; and with the two together it is hoped that, at last, something has been done to re-establish his memory.

Post Script

One of the greatest fears of a researcher or author is to discover that somebody else is, at the same time, working on the same subject. In August 2010 I received a letter from S. Subbarayalu, a retired Principal Chief Conservator of Forests in Madras who, many years earlier, had studied forestry at Edinburgh. In this he announced that he was collecting information for a biography of Hugh Cleghorn. This was not entirely ungalling, as one of the reasons that awareness of Cleghorn had been raised in India was from a newspaper article published in Madras on 5 November 2004, of which I had become aware on discovering that Wirgman's portrait of Cleghorn was available on Google Images. At this point – it has subsequently become available from a photograph of the drawing in the collection of the National Portrait Gallery – the only source of the image (other than a visit to Stravithie) was my 1999 book and, sure enough, I found that the newspaper article entitled 'A Life for Forestry' was drawn, entirely unacknowledged, almost exclusively from that source, a copy of which I had given to the author of the article.[1] Imitation is said to be the sincerest form of flattery, so I did not remonstrate with the author or the newspaper's editor, but I did not then foresee the consequences. As already noted I first did some preliminary work on Cleghorn in 1998, when Hugh and Elizabeth Sprot generously lent the Wirgman portrait for an exhibition and gave permission to reproduce it in the catalogue. But my studies began in earnest after completing my work on Robert Wight, with a retracing of Cleghorn's Shimoga journeys, partly as a way of forgetting my semi-centenary in October 2007. There was little I could do, and to have withheld information from Mr Subbarayalu (who *did* most generously acknowledge my help) would have seemed curmudgeonly, especially considering the immense help I have received over the years from numerous correspondents in India, and knowing the difficulty of obtaining information trans-continentally. Other projects relating to the Edinburgh collections intervened: on Sir Stamford Raffles in South-East Asia, on the teaching of Dr John Hope, on the amphi-Pacific voyage of Captain Frederick Beechey, and on the Ceylon collections of Colonel and Mrs Walker, to say nothing of cataloguing the three thousand drawings in the Cleghorn Collection and his books at RBGE and in Edinburgh University Library, on field trips following Cleghorn's footsteps in Rathlin (2008), the Northern Circars (2010), Ipswich (2014), the North-West Himalaya (2014) and the island of Malta (2015).

So Subbarayalu won the race.[2] I received a copy of his book on 11 April 2014, but took the decision not to read it until my own manuscript was complete. I therefore read it for the first time on a train journey between Edinburgh and London on 7 June 2015, a journey on which I was spared the misadventure encountered by Cleghorn's future wife and father-in-law on the same line 160 years earlier. As many of the same sources were used there are inevitably major overlaps between the two works, but the material is very differently organised and each must be judged on its own merits.

Acknowledgements

First and foremost my colleagues at the Royal Botanic Garden Edinburgh: Graham Hardy, Leonie Paterson (library). Robyn Drinkwater, Roger Hyam, Lynsey Wilson (photography). Duncan Reddish, Lee Cooper, Alan Sneath, Martin Pullan (IT). Alan Elliott, Lorna Stoddart, Mark Hughes, Colin Pendry, Markus Ruhsam, Mark Watson (discussions and advice, various). And, above all, to Caroline Muir for infinite care and patience designing the two books.

Staff in other institutions and libraries: AgroParisTech – Nancy Centre (Meriem Fournier, David Gasporotto); British Library (especially John Falconer and the ever-helpful and hard-pressed staff behind the desk in the APAC Reading Room); Linen Hall Library, Belfast; Linnean Society of London (Lynda Brooks, Gina Douglas, Elaine Charwat); National Archives of Scotland; National Art Library, London; National Gallery of Australia, Canberra (Gael Newton, Nick Nicholson); National Library of Scotland; National Library of South Africa, Cape Town (Melanie Geustyn); National Museums of Scotland (Mark Glancy: Cleghorn Memorial Library; Michael Taylor and Stig Walsh: Pitcorthie fish fossils; Alison Morrison-Low: Adamson photographs); Royal College of Physicians Edinburgh (Iain Milne, Estela Dukan); Royal Botanic Gardens Kew: Chris Mills, Lorna Cahill, Kiri Ross-Jones, Michele Losse, Virginia Mills (Library); Caroline Cornish, Mark Nesbitt (Economic Botany and Hanbury Collections); Royal Society of Edinburgh (Vicky Hammond); University of St Andrews (Special Collections: Moira Mackenzie, Rachel Hart, Catriona Foote, Rachel Nordstrom); University of Edinburgh (Centre for Research Collections: Joe Marshall, Grant Buttars, Dr Murray Simpson, Tricia Boyd, Elizabeth Quarmby-Lawrence, Sally Pagan); University of Jena (Emanuele Tommasi); University of Oxford, Bodleian Library (Colin Harris, Oliver Bridle); University of Toronto (Hannah Medical Collection: Graham Bradshaw); Wellcome Library, London.

L. Shyamal (Bangalore), Wikipedist extraordinaire, for generous sharing of information, and in particular for finding and accessing references to Cleghorn in the British Newspaper Archive.

For extended hospitality in London on library visits over many years: Susan Henderson and the Askew family, Richard Blurton and the late Martin Williams, Julian Polhill. For friendship, and discussion over many matters botanical, Indological and otherwise: Anna Dallapiccola, Vinita Damodaran, Richard Grove, Pradip Krishen, David Mabberley, Minakshi Menon.

In Fife: the late Aylwin Clark; the Sprot family (the late Hugh and Elizabeth Sprot, Nether Stravithie, Geoffrey and Belinda Sprot (Mains of Stravithie) and Mr & Mrs Edward Sprot (Wiltshire) for showing me the family collections, and David Chalmers, owner of Stravithie House, for allowing me to see Cleghorn's habitat and take photographs.

The following have all helped in various ways, with apologies to anyone whom I have overlooked: David Alexander; Bill Anderson; Henry Baggott (Bonhams); John Bray (Moravians at Keylong); Ian Brooker (identification of *Eucalyptus*); Janet Carolan (Archivist of Dollar Academy: W.G. McIvor); Kevin Chang (Brandis's PhD dissertation); Jamie Compton (discussions on McIvor's Yorkshire stay); Andrew Cook (for putting me onto the history of Fife railways); Paul Cox (National Portrait Gallery, commissioning costs of portrait drawings); Deepali Dewan (Alexander Hunter); J. Margaret Dickson (the Gage family of Rathlin); Florike Egmond (Clusius and early Indian botanical illustration); Frances Fowle (National Galleries of

ACKNOWLEDGEMENTS

Scotland for information on John James Cowan); Margaret Frood (genealogy of the Thomas family); Allan Grant (antiquarian bookseller who sold some of the Stravithie library); Andrew Grout (geology); Angela Howe (British Golf Museum, St Andrews); Sophie Jay (Dharamsala); Professor Matthew Kaufman and Elizabeth Singh (Royal Medical Society); Roger Kelly (Penicuik: Cowan family); Angela Kenny (Royal Commission for the Exhibition of 1851); Andrew Kerr (wills and money); Lionel Knight (Shimoga and Mysore); Margarita Hernandez Laille (J.H. Balfour and evolution); Annie Lyden (Scottish National Portrait Gallery: Alexander Hunter photographs); the Rev. Professor A. Donald MacLeod (Toronto: Cowan family); Bruce Maslin (identification of *Acacia*); Kirsty Noltie (medical matters); Philip Oswald (Latin translation); Christopher Penn (Indian photography: A.T.W. Penn and Alexander Hunter); Heleen Plaisier (St Andrews information on Cleghorn algae); Anantanarayanan Raman (discussions on many aspects of Madras scientific history, especially over Senjee Pulney Andy); Hugh Ashley Rayner (Samuel Bourne and other photographers); Margaret Sargent and Mary Beresford (A.T. Jaffrey); Roger Taylor (Indian photography: Alexander Hunter); Adrian Thomas (J.L. Stewart); Bruce Wannell; Stephen Wildman (Amy Yule); the late Martin Williams (genealogy) and last, but by no means least, Professor Charles Withers for his generous Foreword and for amending the title.

In India: the late Admiral Oscar Stanley Dawson (Bangalore: Alphonso Bertie's tomb, Agram Cantonment Cemetery); P. Jayabalan (DFO Salem, photographing Louis Blenkinsop's memorial tablet); V. Naryanaswami (Madras, for shared interests in the iconography of Madras); Bob Stewart and Tanya Balcar (Anamalai Hills); Subbarayalu Naidu (Madras, for interesting discussions on Cleghorn, and supplies of Nilgiri tea), Rahaab Allana and Jennifer Choudhry (Alkazi Collection of Photography).

On the three Cleghornian excursions. To Shimoga: Dr Jagadish (the British cemetery in Shimoga); Mr Jois (Keladi); Mr Nirvanappa (Shimoga DC's office); Prof. A. Ramegowda and Dr Rajaram Hegde (Kuvempu University, Shimoga), Mathews Philip (SICHREM, Bangalore), Ms Sharvani (Mysore, for transcribing/translating the Kannada plant names on the Shimoga drawings); to the Punjab Hills: Christina Noble, Rahul Noble Singh, Prem Singh, Kunal Condillac and Ashok the driver; to Malta: Daniel Borg, Jospeh Buhagiar.

References

Manuscripts

Bodleian Library, Oxford
BAAS Deposit.

British Library, London (BL)
APAC – many series of papers, individually cited as e.g. BL APAC F/4/2711 no 194652.
Madras Almanacs (printed).

Edinburgh City Archives –
Royal High School Matriculation Books and Subscriptions to School Library (SL) references provided by Peter Clapham.

Evangelisches Missionwerk, Basel
Missionary records, references supplied by Anke Schürer-Ries.

Fife County Archives, Markinch
Summary of Cleghorn references supplied by Andrew Dowsey.

Linnean Society of London
Election certificates and other mss.

National Archives of Scotland (NAS) –
Cupar Sherriff Court Records (Cleghorn's will and inventory).
Justice of the Peace records of Quarter Sessions for Cupar.

National Library of Scotland (NLS)
Walker of Bowland papers.

National Museums of Scotland (NMS)
Adamson photographic album.
Papers referring to Cleghorn Memorial Library, references supplied by Mark Glancy.

Natural History Museum, London (NHM)
Carruthers Correspondence.

Royal Botanic Garden Edinburgh (RBGE)
John Hutton Balfour incoming correspondence, cited RBGE BC vol. no./letter no.
Botanical Society of Edinburgh (BSE) Minute Books and miscellaneous papers.
Herbarium Donations Book 1873–1913.

Royal Botanic Gardens Kew (RBGK)
Directors Correspondence (KDC).
Bentham Correspondence (KBC).
Henslow Correspondence (KHC).

University of Edinburgh (EU) Library CRC
General & Medical matriculation records.
Medical Examinations.
Senatus Minutes.

University of St Andrews, Special Collections (StA)
Photograph Albums 6, 8.
Cleghorn Papers (MS Deposit 53), cited as CP/Box/Envelope or volume/item or folio Box/Envelope/Item).
Court Minutes; Senatus Minutes.

University of Glasgow (GU)
Matriculation and University Library Registers, references provided by Kiara King.

University of Vermont (UV)
Papers of G.P. Marsh (MP), copies of letters supplied by Prudence Doherty.

Web Sources

Wikipedia

British Newspaper Archive (references in local British newspapers supplied by L. Shyamal).

Scotlandspeople (used for checking genealogical data of Scots, also the census records for Stravithie).

National Library of Scotland – Ordnance Survey maps.

Journals and Periodicals
(multiple parts consulted)

Transactions of the Botanical Society [of Edinburgh] (consulted at RBGE).

Transactions of the Highland & Agricultural Society of Scotland (consulted at RBGE).

Illustrated Indian Journal of Arts (consulted at National Art Library and British Library).

Indian Journal of Arts, Science and Manufactures (consulted at National Art Library and British Library).

Madras Almanacs – published in Madras 1843-61 by Edmund Marsden (title in the form *The Madras Almanac & Compendium of Intelligence for 1843*) and from 1862 by the Asylum Press (*The Asylum Press Almanac & Compendium of Intelligence for the year 1862*) (consulted at BL, on open shelf in APAC).

Bengal Almanacs – published in Calcutta by Samuel Smith & Co. (title in the form *The Bengal & Agra Directory & Annual Reporter for 1849*, consulted on open shelf in APAC, BL).

University of Edinburgh Calendars 1868–95 (consulted in CRC Edinburgh University Library).

British Association – Annual Reports (published the year following the date of the meeting covered) (consulted at RBGE).

REFERENCES

Books

Abhishankar, K. (ed) (1975). *Karnataka State Gazetteer: Shimoga District.* Bangalore: Government Press.

Abhishankar, K. (ed) (1981). *Karnataka State Gazetteer: Chikmagalur District.* Bangalore: Government Press.

Aitchison, J.E.T. (1869). Lahul, its flora and vegetable products &c from communications received by the Rev. Heinrich Jaeschke, of the Moravian Mission. *Journal of the Linnean Society.* Botany 10: 69–101.

Anderson, J.M. (1905). *The Matriculation Roll of the University of St Andrews, 1747–1897.* Edinburgh: W. Blackwood & Sons.

Andy, S.P. (1869). On branched palms in southern India. *Transactions of the Linnean Society of London* 26: 661–2; figs 1, 2, t. 51.

Anon. (1851). *The Illustrated Exhibitor … of the Principal Objects in the Great Exhibition of the Industry of All Nations.* London: John Cassell.

Anon. (1856). Art. VIII. 1. Official and Descriptive Catalogue of the Madras Exhibition of 1855. 2. Madras Exhibition of 1855. Catalogue Raisonné … 3. Jury Reports of the Madras Exhibition of 155. *Calcutta Review* 26: 265–84.

Anon. (1857). *Madras Exhibition of 1857. Reports from the Local Committees* [for Chingelput, Coimbatore, Hyderabad, Madras, Malabar, Masulipatam, Pondicherry and Travancore] on the Consignments to the Exhibition. Madras: Scottish Press.

Anon. (1876). Obituary [of W.G. McIvor]. *Gardeners' Chronicle* 1876 (ii): 150.

Anon. (1884). Dietrich Brandis, the Founder of Forestry in India. *Indian Forester* 10(8): 343–57.

Anon. (1884). *Official Catalogue [of the Edinburgh International Forestry Exhibition].* Edinburgh: University Press (printed by T. & A. Constable).

Anon. (1889). Presentation to Hugh Cleghorn of Stravithie, M.D., LL.D., F.R.S.E. *Transactions of the Scottish Arboricultural Society* 12: 198–205.

Anon. (1892). *Royal Scottish Arboricultural Society, Excursion to Fifeshire and Perthshire.* [Pamphlet, copy at RBGE].

Anon. (1897). *List of Books, &c., relating to Botany and Forestry, including the Cleghorn Memorial Library in the Library of the [Edinburgh] Museum [of Science and Art].* Edinburgh: H.M.S.O.

Anon. (1918–23). *Catalogue of the Printed Books in the Library of the University of Edinburgh.* 3 vols. Edinburgh: T. & A. Constable.

Anon. (1935). *Souvenir Booklet of the Madras Medical College Centenary.* Madras: [printed by Solden & Co.].

Anon. (1991). *Glass House: the Jewel of Lalbagh.* Bangalore: Mysore Horticultural Society.

Argyll, Duke of (?1861). Opening Address [session 1860/1]. *Proceedings of the Royal Society of Edinburgh* 4(53): 350–77.

Axelby, R., Nair, S.P. & Cook, A. (2010). *Science and the Changing Environment in India 1780–1920: a guide to resources in the India Office Records.* London: British Library.

Baden-Powell, B.H. & Gamble, J.S. (1874). *Report of the Proceedings of the Forest Conference held at Allahabad in January 1874.* No place of publication.

Balfour, E.G. (1873). *Cyclopædia of India.* Second edition. 5 vols. Madras: Lawrence & Adelphi Presses.

Balfour, E.G. (ed.) (1856). *Madras Exhibition of Raw Products, Arts, and Manufactures of Southern India, 1855. Reports by the Juries.* Madras: for the General Committee of the Madras Exhibition by Pharoah & Co.

Balfour, E.G. (ed.) (1858). *Madras Exhibition of Raw Products, Arts, and Manufactures of Southern India, 1857. Reports by the Juries.* Madras: for the General Committee of the Madras Exhibition by L.C. Graves, Scottish Press.

Balfour, J.H. (1855, 1856). Account of the origin and of some of the contents of the Museum of Economic Botany attached to the Royal Botanic Garden of Edinburgh. *Proceedings of the Botanical Society [of Edinburgh] for 1855*: 56–9, 85–108, 121–4; *for 1856*: 8–9, 18–20.

Balfour, J.H. (1859). *Class Book of Botany: being an introduction to the study of the vegetable kingdom. Part I. Structural & Morphological Botany.* Edinburgh: Adam & Charles Black.

Balfour, J.H. (1860). *A Manual of Botany: being an introduction to the study of the structure, physiology, and classification of plants.* New edition. Edinburgh: Adam & Charles Black.

Balfour, J.H. (1875). *A Manual of Botany: being an introduction to the study of the structure, physiology, and classification of plants.* Fifth edition. Edinburgh: Adam & Charles Black.

Barton, G.A. (2002). *Empire Forestry and the Origins of Environmentalism.* Cambridge: Cambridge University Press.

Beddome, R.H. (1869–73). *The Flora Sylvatica of South India.* 2 Vols. Madras: Gantz Bros.

Beddome, R.H. (1869–74). *The Forester's Manual of Botany for Southern India.* [Madras: no publisher given].

Bhasin, R. (2011). *Simla: the Summer Capital of British India.* New Delhi: Rupa Publications.

Bidie, G. (1869). On the effects of forest destruction in Coorg. *Proceedings of the Royal Geographical Society of London* 13: 74–83.

REFERENCES

Birdwood, G. (1901). Correspondence: Forestry in India. *Journal of the Society of Arts* 49: 757–60 [about disputed history of forest conservancy].

Bourdillon, E.D. (ed). (1866). East India (Chinchona Plant) Blue Book 1866. *Correspondence relating to the Introduction of the Chinchona Plant into India, and to Proceedings connected with its Cultivation from April 1863 to April 1866*. London: House of Commons.

Bowe, P. (2012). Lal Bagh – the botanical garden of Bangalore and its Kew trained gardeners. *Garden History* 40: 228–38.

Bray, J. (2012). *A History of the Moravian Church in India*. Online, updated version originally published in *The Himalayan Mission*, Leh: Moravian Church (1985). Consulted on *ladakhstudies.org/resources/Resources/Bray.MoravianHistory.pdf*, 19 July 2014.

Brandis, D. (1874). *The Forest Flora of North-West and Central India*. London: W.H. Allen & Co.

Brown, J. (1897). Mr Syme in *Horae Subsecivae*, new edition, series 1: 360–72. London: Adam & Charles Black.

Brown, R.N. (1866). *A Hand Book of the Trees, Shrubs and Herbaceous Plants growing in the Madras Agri-Horticultural Society's Gardens and Neighbourhood of Madras arranged according to the Natural System*. Second edition, with a Supplement by J.J. Wood. Madras: J. Higginbotham.

Buchanan, F. (1807). *A Journey from Madras through the Countries of Mysore, Canara and Malabar*. London: W. Bulmer & Co.

Buck, E.J. (1925). *Simla Past and Present*. Second edition. Bombay: The Times Press.

Burkhardt, F. et al. (1997). *The Correspondence of Charles Darwin. Volume 10 1862*. Cambridge: Cambridge University Press.

Butler, P. (2000). *Irish Botanical Illustrators & Flower Painters*. Woodbridge: Antique Collectors' Club.

Cave, R. (2010). *Impressions of Nature: a history of Nature Printing*. London: British Library and New York: Mark Batty Publisher.

Christie, A. Turnbull (1828, 1829). Sketches of the Meteorology, Geology, Agriculture, Botany and Zoology, of the Southern Mahratta Country. *Edinburgh New Philosophical Journal* 5(n.s.): 292–304; 6(n.s.): 98–101; 7(n.s.): 49–65.

Christison, R. (1847) [paper read 8 Jul 1846]. Observations on a New Variety of Gamboge from Mysore. *Pharmaceutical Journal* 6: 60–9; plate (woodcut, copied from a drawing by Cleghorn's artist, though unacknowledged).

Clark, Aylwin (1992). *An Enlightened Scot: Hugh Cleghorn 1752–1837*. Duns: Black Ace Books.

Clark, Aylwin (2004). Cleghorn, Hugh (1752 – 1837) in *Oxford Dictionary of National Biography* 12: 9–10 (eds H.C.G. Matthew & B. Harrison). Oxford: Oxford University Press.

'Committee of Compilation' (1863). *Report of the Punjab Missionary Conference held at Lahore in December and January, 1862–63*. Lodiana: American Presbyterian Mission Press.

Conolly, M.F. (1861). *Memoir of the Life and Writings of William Tennant, LL.D.* Edinburgh: James Blackwood.

Copley, A. (2006). *A Spiritual Bloomsbury*. Lanham: Lexington Books.

Cowan, C. (1878). *Reminiscences*. Edinburgh: for Private Circulation.

Cowan, J.J. (1933). *From 1846 to 1932*. Edinburgh: for Private Circulation, printed by Pillans & Wilson.

Crawford, D.G. (1930). *Roll of the Indian Medical Service 1615–1930*. London: W. Thacker & Co. Biographical details of all EIC surgeons mentioned in the text have been checked in this.

Crawford, R. (2011). *The Beginning and the End of the World: St Andrews, scandal and the birth of photography*. Edinburgh: Birlinn.

Das, P. (2005). Hugh Cleghorn and forest conservancy in India. *Environment and History* 11: 55–82.

Desmond, R. (1992). *The European Discovery of the Indian Flora*. Oxford: Royal Botanic Gardens [Kew] & Oxford University Press. The essential source for the general botanical history of India.

Desmond, R. (1994). *Dictionary of British and Irish Botanists and Horticulturists*. London: Taylor & Francis and Natural History Museum.

Dewan, Deepali (2001). *Crafting Knowledge and Knowledge of Crafts: Art Education, Colonialism and the Madras School of Arts in Nineteenth-century South Asia*. PhD dissertation for the University of Minnesota.

Dewan, J. (2003). *The Photographs of Linnaeus Tripe: A Catalogue Raisonné*. Toronto: Art Gallery of Ontario.

Dickson, J. Margaret. (ed.) (1995). *A History of the Island of Rathlin by Mrs [Catharine] Gage*. Rathlin, 1851. Published for the editor [printed by Impact Printing, Coleraine].

Dobbs, R.S. (1882). *Reminiscences of Life in Mysore, South Africa and Burmah*. Dublin: George Herbert.

Dossal, M. (2012). Nature's treasure, Mumbai's heritage: the Botanical Garden's early years. in H. Rustomfram & S. Nikharge (eds) *Rani Bagh, Mumbai's Heritage Botanical Garden,150 Years*, pp 26-33. Mumbai: BNHS etc.

Drayton, R. (2000). *Nature's Government: Science, Imperial Britain, and the 'Improvement' of the World*. New Haven & London: Yale University Press.

Dunn, S.T. (1916). Notes on the Flora of Madras. *Kew Bulletin* 1916: 61.

Endersby, J. (2003). *Imperial Science: Joseph Hooker and the Practices of Victorian Science*. Chicago & London: University of Chicago Press.

REFERENCES

Fletcher, H.R. & Brown, W.H. (1970). *The Royal Botanic Garden Edinburgh 1670–1970*. Edinburgh: H.M.S.O.

Fowle, F. (2008). *Impressionism and Scotland*. Edinburgh: National Galleries of Scotland.

Francis, W. (1908). *Madras District Gazetteers. The Nilgiris*. Madras: Government Press.

Gadgil, G. & Guha, R. (2014). *This Fissured Land: an Ecological History of India*. 3rd impression. New Delhi: Oxford University Press.

Gamble, J.S. (1915). *Flora of the Presidency of Madras*, vol. 1. London: Allard & Son Ltd.

Gandhi, R. (2013). *Punjab: a History from Aurangzeb to Mountbatten*. New Delhi: Aleph Book Company.

Gifford, J. (1988). *Fife* [Buildings of Scotland series]. London: Penguin Books.

Gillespie, D., Mitchell, A., Tuke, J.B., Fraser, J. & Pagan, G.H. (1873). *Seventh Annual Report of the Fife and Kinross District Board of Lunacy*. Cupar: Fifeshire Journal Office.

Gilmour, D. (2005). *The Ruling Case: Imperial Lives in the Victorian Raj*. London: John Murray.

Gordon, E. (1976). *The Royal Scottish Academy of Painting, Sculpture & Architecture 1826–1976*. Edinburgh: Charles Skilton Ltd.

[Grant, A.] (2008). *Books from the Library of Hugh Cleghorn (1752–1837)*. Edinburgh: Grant & Shaw Ltd.

Gray, J. (1952). *History of the Royal Medical Society, 1737–1937*. Edinburgh: Edinburgh University Press.

Grove, R.H. (1995). *Green Imperialism: Colonial expansion, tropical island Edens and the origins of environmentalism*. Cambridge: Cambridge University Press.

Guglielmo, A. & Buhagiar, J. (eds) (2012). *Giardini Mediterranei tra Sicilia e Malta*. Siracusa: Editore Morrone.

Gunther, A.E. (1977). *The Life of William Carmichael M'Intosh, M.D., F.R.S. of St Andrews 1838-1931: a Pioneer in Marine Biology*. Edinburgh: Scottish Academic Press (for University of St Andrews).

Hajducki, A., Jodeluk, M. & Simpson, A. (2008). *The St Andrews Railway*. Usk: The Oakwood Press.

Hajducki, A., Jodeluk, M. & Simpson, A. (2009). *The Anstruther & St Andrews Railway*. Usk: The Oakwood Press.

Head, R. (1991). *Catalogue of Paintings, Drawings, Engravings & Busts in the Collection of the Royal Asiatic Society*. London: Royal Asiatic Society.

Heniger, J. (1986). *Hendrik Adriaan van Reede tot Drakenstein (1636–1691) and Hortus Malabaricus: a contribution to the history of Dutch colonial botany*. Rotterdam & Boston: A.A. Balkema.

Herbert, E.W. (2011). *Flora's Empire: British Gardens in India*. Philadelphia: University of Pennsylvania Press.

Heyne, B. [1864]. Statistical Fragments on Mysore, (originally published 1814) reprinted in *Selections from the Records of the Mysore Commissioner's Office*. Bangalore: Mysore Government Press.

Hill, A.W. (1926). James Sykes Gamble – 1847–1925 [obituary notice]. *Proceedings of the Royal Society of London*, series B, 99: xxxviii–xliii.

Hoffenberg, P.H. (2001). *Empire on Display: English, Indian and Australian Exhibitions from the Crystal Palace to the Great War*. Berkeley: University of California Press.

Honigsbaum, M. (2001). *The Fever Trail: the hunt for the cure for malaria*. London: Macmillan.

Hooker, J.D. (1869). [Opening] Address in *Report of the Thirty-eighth Meeting of the British Association for the Advancement of Science held at Norwich in August 1868*, pp lviii – lxxv. London: John Murray.

Hooker, W.J. (1855). *Museum of Economic Botany ... of the Royal Gardens of Kew*. London: Longman, Brown, Green, and Longmans.

Iyer, M., Nagendra, H. & Rajan, M.B. (2012). Using satellite imagery and historical maps to study the original contours of the Lalbagh Botanical Garden. *Current Science* 102: 507–9.

Jaffrey, A.T. (?1847). Professional notes upon horticulture, &c., taken during a visit to the Neilgherries. *Indian Journal of Arts, Science, and Manufactures* 1(n.s.), pt. 3: 103–11.

Jaffrey, A.T. (1855). *Hints to the Amateur Gardeners of Southern India*. Madras: Pharoah and Co.

Kalpana, K. & Schiffer, F. (2003). *Madras: the Architectural Heritage*. Chennai: INTACH.

Kaufman, M. H. (2003). *Medical Teaching in Edinburgh during the 18th and 19th Centuries*. Edinburgh: Royal College of Surgeons of Edinburgh.

Killen, J. (ed.) (2013). *Wild Flowers and Plants of Rathlin Island. Painted by Catherine & Barbara Gage, 1840–1850*. Belfast: Linen Hall Library. [Limited edition set of 8 facsimile prints].

King, G. (1876). *Manual of Cinchona Cultivation in India*. Calcutta: Superintendent of Government Printing.

Lechmere, Sir E.A.H. (1887). *Report from the Select Committee on Forestry*. House of Commons Papers: Reports of Committees 1886–7 (246) vol IX.537. London: House of Commons. A summary was printed in *Transactions of the Scottish Arboricultural Society* 12: 104–55. 1888

Low, U. (1936). *Fifty Years with John Company*. London: John Murray.

Lowenthal, D. (2003). *George Perkins Marsh: Prophet of Conservation* (paperback edition). Seattle & London: University of Washington Press.

REFERENCES

Lubbock, Sir J. (1885). *Report from the Select Committee on Forestry*. House of Commons Papers: Reports of Committees 1884–5 (287) vol. VIII.779. London: House of Commons. Note. Cleghorn's evidence (Cleghorn, 1885b), a summary was printed in *Transactions of the Scottish Arboricultural Society* 11: 124–36. 1885.

Lyon, Rev. C.J. (1838). *The History of St Andrews*. Edinburgh: Edinburgh Printing & Publishing Co.

Macdonald, J.A. (1984). *Plant Science & Scientists in St Andrews, up to the middle of the 20th Century*. St Andrews: for the author.

Maclagan, A.D. (1866). Opening address [of 1863/4 session of Botanical Society of Edinburgh, containing obituary of Francis Appavoo, from information supplied by Cleghorn]. *Transactions of the Botanical Society [of Edinburgh]* 8: 10–11.

MacLeod, A.D. (2013). *A Kirk Disrupted: Charles Cowan and the Free Church of Scotland*. Fearn: Christian Focus Publications.

Masani, Z. (2013). *Macaulay: Britain's Liberal Imperialist*. London: Bodley Head.

Mathew, M.V. (1987). *The History of the Royal Botanic Garden Library Edinburgh*. Edinburgh: H.M.S.O.

Matthew, H.C.G. & Harrison, B. (2004). *Oxford Dictionary of National Biography*. 60 vols. Oxford: Oxford University Press. 'ODNB' – the invaluable source used for biographical details of those appearing in cameo roles.

McIntosh, W.C. (1895) [read 1 Jul 1895]. Dr Hugh Francis Clarke Cleghorn. *Proceedings of the Royal Society of Edinburgh* 20: li–lx.

Medley, J.G. (ed.) (1864). *Professional Papers on Indian Engineering, vol. 1, 1863–4*. Roorkee: Thomason College Press.

[Mercer, C.] (1864). *Official Hand-Book of the Punjab Exhibition of 1864*. Lahore: Dependent Press: Henry George.

Michael, J. (1884). *Catalogue of the Indian Exhibit at the International Forestry Exhibition Edinburgh*. Edinburgh: Edinburgh University Press.

Mitchell, A. (1993). *The People of Calton Hill*. Edinburgh: James Thin, the Mercat Press.

Mitchell, R.J. (2014). *The Botanic Garden in St Andrews: the history of 125 years 1889–2014*. St Andrews: for the author.

Muthiah, S. (1999). *Madras Rediscovered*. Chennai, Bangalore & Hyderabad: EastWest Books (Madras) Pvt. Ltd.

Nair, N.C. (1978). Report on the Madras Herbarium in Anon., *The Madras Herbarium 1853–1978 – 125th Anniversary Souvenir*, pp 3–8. Coimbatore: Botanical Survey of India, Southern Circle

Neve, E.F. (1915). *Beyond the Pir Panjal: Life and missionary enterprise in Kashmir*. London: Church Missionary Society.

Nissen, C. (1951). *Die Botanische Buchillustration Geschichte und Bibliographie*. Stuttgart: Hiersemann Verlags-Gessellschaft.

Noltie, H.J. (1999). *Indian Botanical Drawings 1793–1868 from the Royal Botanic Garden Edinburgh*. Edinburgh: Royal Botanic Garden Edinburgh.

Noltie, H.J. (2002). *The Dapuri Drawings: Alexander Gibson and the Bombay Botanic Gardens*. Edinburgh: Royal Botanic Garden Edinburgh.

Noltie, H.J. (2005). *The Botany of Robert Wight*. Ruggell: A.R.G. Gantner Verlag.

Noltie, H.J. (2007). *Robert Wight and the Botanical Drawings of Rungiah and Govindoo*. 3 vols. Edinburgh: Royal Botanic Garden Edinburgh.

Noltie, H.J. (2011). A botanical group in Lahore, 1864. *Archives of Natural History* 38: 267–77.

Noltie, H.J. (2013). *The Botanical Collections of Colonel and Mrs Walker, Ceylon, 1830–1838*. Edinburgh. Royal Botanic Garden Edinburgh.

[Oldham, T.] (1859). Geology in India. *Calcutta Review* 32: 121-61.

Penn, C.F. (2010). *In Pursuit of the Past*, third edition. Worplesdon: C.F. Penn.

Penn, C.F. (2014). *The Herklots Folder of Photographs: the development of Coonoor and the coffee plantations in the Nilgiri[s] ... during the nineteenth century*. Worplesdon: Christopher Penn.

Phillimore, C.B. (ed.) (1871). East India (Forest Conservancy). Session 9 Feb to 21 Aug 1871. Vol 52. 1871. *A Selection of Despatches ... on Forest Conservancy in India ... from ... the 21st day of May 1862 to the Present Time*. London: House of Commons.

Price, F. (1908). *Ootacamund: a history*. 2002 paperback edition. Chennai: Rupa & Co. Dakshin.

Rajan, S. R. (2008) *Modernizing Nature: forestry and imperial eco-development 1800–1950*. New Delhi: Orient Longman.

Raman, A. (2013). First Indian doctor with foreign degree [Pulney Andy]. *Madras Musings* 22(21). Available online at: madrasmusings.com/Vol22No21/first-indian-doctor-with-foreign-degree.html.

Raman, R. & Raman, A. (2013). Surgeon Senjee Pulney Andy's trials in treating smallpox using leaves of *Azadirachta indica* in southern India in the 1860s. *Current Science* 104: 1720–2.

Raven, the Rev. C.E. (1951). William Robertson Smith in *Centenary of the Birth on 8th November 1846 of the Reverend Professor W. Robertson Smith*, pp 3–9. Aberdeen: The University Press.

Rayner, H.A. (ed.) (2014). *Photographic Journeys [of Samuel Bourne] in the Himalayas 1863–1866*. Bath: Pagoda Tree Press.

Ribbentop, B. (1900). *Forestry in British India*. Calcutta: Office of the Superintendent of Government printing.

REFERENCES

Rocco, F. (2003). *The Miraculous Fever-tree: Malaria, medicine and the cure that changed the world.* London: HarperCollins.

Roger, Rev. C. (1849). *History of St Andrews, with a full account of the Recent Improvements in the City.* Ed 2. Edinburgh: Adam & Charles Black.

Royal Society (1867, 1877, 1915). *Catalogue of Scientific Papers* (volumes for 1800–1863, 1864–1873 and 1884–1900). London: H.M.S.O.

Salmond, J.B. (1950). *Veterum Laudes: being a tribute to the achievements of the members of St Salvator's College during five hundred years.* Edinburgh & London: Oliver & Boyd (for University of St Andrews).

Scherzer, K. von (1861). *Narrative of the Circumnavigation of the Globe by the Austrian Frigate Novara, undertaken by order of the Imperial Government in the years 1857, 1858, and 1859.* Vol. 1. London: Saunders, Otley & Co.

Seward, M.R.D. (1976). Yorkshire lichen material of W.G. McIvor in the herbarium of the National Botanic Gardens, Glasnevin, Dublin. *Naturalist* 1976: 125–8.

Sewell, R. (1896). *Sir Walter Elliot of Wolfelee: a sketch of his life, and a few extracts from his note books.* Edinburgh: Privately published.

Shafe, M. (1982). *University Education in Dundee 1881–1981.* Dundee: University of Dundee.

Shaw, J. (1877). *A Synopsis of the Proceedings of the Agri-Horticultural Society of Madras from July 1835 to December 1870.* Madras: Gantz Brothers (copy at Kew).

Shepherd, J.A. (1969). *Simpson and Syme of Edinburgh.* Edinburgh & London: E. & S. Livingtsone Ltd.

Simonds, D. & Prasad, V.P. [2014]. *Flowers of Southern and Western India Painted by Mrs Margaret Read Brown (1816–1868).* London: High Commission of India.

Smart, R.N. (2004). *Biographical Register of the University of St Andrews, 1747–1897.* St Andrews: University of St Andrews.

Smith, C.I. (1854). *Statistical Report on the Mysore.* Edinburgh: Robert Hardie & Co.

Smith, G. (1885). *The Life of William Carey D.D.: Shoemaker and Missionary.* London: John Murray.

Smith, G. (1990). *Disciples of Light: photographs in the Brewster Album.* Malibu: the J. Paul Getty Museum.

Stafleu, F.A. & Mennega, E.A. (1997). *Taxonomic Literature ... Supplement IV: Ce–Cz.* Königstein: Koeltz Scientific Books.

Stebbing, E.P. (1922, 1923). *The Forests of India*, vols 1 and 2. London: John Lane, the Bodley Head Ltd.

?Stent, J. (1848). *A List of Trees, Shrubs, Plants, &c, &c, growing in the Madras Agri-Horticultural Society's Garden – arranged in their Natural Order agreeably to De Candolle.* Madras. (copy at Kew, possibly unique).

Stokes, H[udlestone]. (1864). Report on the Nugur Division of Mysore [written 1838], reprinted in *Selections from the Records of the Mysore Commissioner's Office.* Bangalore: Mysore Government Press.

Stubbs, B.J. (1998). Land improvement or institutionalised destruction? The ringbarking controversy, 1879–1884, and the emergence of a conservation ethic in New South Wales. *Environment and History* 4: 145–67.

Subbarayalu, S. (2014). *Dr. H.F.C. Cleghorn: Founder of Forest Conservancy in India.* Chennai: Notion Press.

Sutton, D. (2011). *Other Landscapes: Colonialism and the predicament of authority in nineteenth-century South India.* New Delhi: Orient Blackswan Private Ltd.

Swinney, G.N. (2013). *Towards an Historical Geography of a 'National' Museum: The Industrial Museum of Scotland, the Edinburgh Museum of Science and Art, and the Royal Scottish Museum, 1854–1939.* University of Edinburgh PhD dissertation.

Taylor, A. (1895). Obituary notice of Hugh Francis Clarke Cleghorn, M.D. LL.D., F.R.S.E., F.L.S. *Transactions & Proceedings of the Botanical Society of Edinburgh* 20: 439–48.

Taylor, R. & Branfoot, C. with Greenough, S. & Daniel M. (2014). *Captain Linnaeus Tripe: photographer of India and Burma, 1852–1860.* Washington & New York: National Gallery of Art and Metropolitan Museum of Art, with Delmonico Books & Prestel.

Terrell, R. (1992). *John Chalmers: Letters from the Indian Mutiny 1857–1859.* London: Michael Russell.

Thurston, E. & Rangachari, K. (1909). *Castes and Tribes of Southern India*, vol 4. Madras: Government Press.

Turner, A. Logan (1933). *History of the University of Edinburgh 1883–1933.* Edinburgh: for the University by Oliver & Boyd.

Veale, L. (2010). *An Historical Geography of the Nilgiri Cinchona Plantations, 1860–1900.* PhD dissertation, University of Nottingham.

Waterston, C.D. (1986). Hugh Miller in *The Enterprising Scot* (ed. Jenni Calder), pp 160–9. Edinburgh: National Museums of Scotland.

Whitehead, P.J.P. & Talwar, P.K. (1976). Francis Day (1829–1889) and his collections of Indian fishes. *Bulletin of the British Museum (Natural History)*, History Series 5: 103.

Wilkinson, J. (1991). *The Coogate Doctors: the history of the Edinburgh Medical Missionary Society 1841 to 1991.* Edinburgh: The Edinburgh Medical Missionary Society.

Wilks, Rev. C. (1835). *Memoirs of the Life, Writings and Correspondence of Sir William Jones by .... Lord Teignmouth, with the Life of Lord Teignmouth ...* London: John W. Parker.

Endnotes

Introduction
1. Fletcher & Brown, 1970.
2. Noltie, 1999.
3. www.st-andrews.ac.uk/media/special-collections/documents/Cleghorn%20Papers/pdf-characteristically anonymous.
4. Grove, 1995.
5. Barton, 2002.
6. Lowenthal, 2003.
7. Das, 2005.
8. Noltie, 2002.
9. Noltie, 2007.

Chapter 1
1. BL APAC N/2/7/403. The church later became the cathedral of the Diocese of Madras.
2. StA CP 2/5/5 HC to HFCC 7 Oct 1831.
3. Richard Clarke (c. 1785–1868) was a Madras Civil Servant who was in charge of the College of Fort St George. The Clarke family of London had been friends of Hugh Cleghorn senior and Richard Clarke retired to London, where, for many years, he acted as Treasurer of the Royal Asiatic Society (Head, 1991, in which incorrect dates of birth and death given); many years later, Cleghorn would stay with him at 17 Kensington Square.
4. Biographical material in this chapter is taken largely from Clark, 1992.
5. Clark, 1992: 1.
6. Clark 1992: 213–5.
7. The incident of the pearls clearly remained in the family consciousness: in the grandson's library is a copy of an 1865 work on the Ceylon pearl fisheries from the Records of the Bombay Government.
8. Clark, 1992: 246.
9. Low (1936) stated her to be a niece of Sir John Malcolm, but this is untrue.
10. The anecdotal details in this section are from the Cleghorn family correspondence in StA CP Box 2.
11. Smart, 2004.
12. Low, 1936: 52.
13. StA CP 2/1/20 HC to Wm. Adam 17 Aug 1827.
14. Clark, 2004: 10.
15. Goodsir had attended John Hope's botany class in Edinburgh in 1770, the same year as did Hugh Cleghorn senior.
16. On *Anne* in 1849, and on the one sponsored by Lady Franklin in 1850.
17. Edinburgh City Archives SL 137/5/1.
18. Edinburgh City Archives SL 137/15/5.
19. Smart, 2004.
20. StA CP 2/2/11 HC to PC [1831].
21. For history of Madras College see Lyon 1838: 206–8; Roger 1849: 137–40.
22. StA CP 2/5/30 Helen Wyllie to Allan Cleghorn 27 May 1834.
23. For classes taken see Anderson, 1905: 86 and Smart, 2004: 173.
24. Gifford, 1988: 377.
25. StA CP 2/5/32.
26. For information on Dollar Academy I am indebted to the school's archivist Janet Carolan.
27. StA CP 9/A/7.
28. StA CP 9/A/6. Coincidentally Cook forms a link between India's first two Conservators of Forests: before being appointed to the St Andrews Chair in 1828 he had been Minister of Laurencekirk in Kincardineshire (from 1795 to 1828). It was in the parish church of Laurencekirk that Alexander Gibson was baptised on 24 October 1800, and Cook almost certainly performed the sacrament.
29. His matriculation records are in EU General & Medical Registers vol. 33, 34, 35A, 35B.
30. Shepherd, 1969.
31. Brown, 1897.
32. For details of his association with the Society see Gray, 1952; for a copy of Cleghorn's unpublished thesis, in the collection of the Society, I am grateful to Professor Matthew Kaufman and Elizabeth Singh.
33. StA CP 9/4/f. 108.
34. Burkhardt *et al.*, 1997: 725.
35. For Edinburgh medical teaching see Kaufman, 2003.
36. For biographical information see Noltie, 2013.
37. Information in this section is taken from the Minute Books of the Botanical Society of Edinburgh at RBGE; the volumes do not have reference or folio numbers but as they are in chronological order, material is easily found by date; information is also found in the printed *Proceedings* and *Transactions*.
38. Cleghorn, 1854b.
39. Crawford, 2011.
40. EU 1840 Medical Examinations vol. 3 no 41.
41. EU 1841 Medical Examinations vol. 2 no 196 – which also gives details of the subjects he studied each year.
42. StA CP 9/1.

Chapter 2
1. BL APAC L/MIL/9/389 ff 321–6.
2. Cleghorn, 1844.
3. The personal and anecdotal details in this chapter are from family letters in StA CP 2/8.
4. Details of Cleghorn's employment/postings for this period are from BL APAC L/MIL/11/73 f 10, L/MIL/79 f 207, Marsden's *Madras Almanac* from 1843 to 1849 and the *Fort St George Gazette* for 1843 (BL APAC V/11/1597 and 1845 V/11/1599).
5. Then also variously spelt as Nuggur and Nagara, now Nagar.
6. Of the 129 castes recorded in the Nugger Division the largest, forming one quarter of the population, were Shaivite Lingayats. Details relating to the Nugger Division in Cleghorn's period are taken from Stokes (1864, but written in 1838) and Smith, 1854.
7. StA CP 2/8/22 PC to HC 1 Dec 1845.
8. Dobbs, 1882.
9. Cleghorn, 1850b: 236.
10. Dobbs, 1882.
11. Ragi (*Eleusine coracana*), which could be stored for many years in subterranean, clay-lined pits.
12. Climate statistics and other contemporary information on Shimoga District taken from Abhishankar, 1975.
13. Cleghorn, 1850b.
14. Encircling 'bound hedges' of thorny shrubs were a traditional method of defence in South India, including, in the 18th century large settlements such as Bangalore and Madras – see Pennant, *View of Hindoostan* 2: 85. 1798 – I am grateful to L. Shyamal for this reference.
15. Dobbs, 1882, though Cleghorn receives no mention in these reminiscences.
16. There is a large literature on *Hortus Malabaricus*; the most authoritative background is in Heniger, 1986.
17. Heyne, [1864].
18. Christie, 1828, 1829. Christie later played a role in the start of the tea industry in the Nilgiris, having noticed a native *Camellia* species growing near Conoor he ordered some tea plants from China, though died before they arrived – Francis, 1908: 178.
19. Cleghorn in Anon, 1889: 202 – reminiscence by Cleghorn in old age, when he mistakenly attributed the advice to Sir Joseph Hooker – correct identification given in McIntosh, 1895: lii.
20. Cleghorn *et al.* 1852: 84.
21. StA CP 2/8/21 PC to HC 16 Oct 1845.

293

ENDNOTES

22. BSE Minutes, 12 Jul 1849.
23. Christison, 1847.
24. Dunn in Gamble, 1915 1: 46 and Dunn, 1916(3): 61.
25. Abhishankar, 1981: 135.
26. Cleghorn, 1850c.
27. Buchanan, 1807 3: 287.
28. See Cleghorn *et al.,* 1852: 83–6; Cleghorn, 1851c; Stebbing, 1922 1: 107.
29. BL APAC P/320/28 f 3852.
30. Cleghorn, 1851b.

Chapter 3

1. For details of the ship see www.collections.rmg.co.uk/collections/objects/149279.html, consulted 29 v 2015, and other web sources on Blackwall Frigates.
2. Information in this section from *Sam Sly's African Journal* [names of passengers given more accurately in *Madras Almanac for 1849*] and the *Cape Town Mail* 15 April 1848, kindly supplied by Melanie Geustyn, National Library of South Africa, Cape Town.
3. These prints and the drawing are in the possession of Mr & Mrs Edward Sprot; a copy of that of the ship as launched in the National Maritime Museum, Greenwich.
4. McIntosh, 1895: lii.
5. StA CP 9/1.
6. Grove, 1995: 453.
7. Information from *Madras Almanacs*.
8. RBGE BC 3 f 186, HC to JHB 13 Apr 1859.
9. StA CP 9/1 f 125r.
10. www.spuddybike.org.uk has useful information on Madras Churches & Chapels, and Priests & Missionaries, consulted 29 v 2015.
11. His son Frank was later a Nilgiri planter – see Penn, 2014.
12. StA CP 9/1 loose item.
13. StA CP 9/1 f 1v.
14. Information in this section from StA CP 2/8/26–30.
15. StA CP 9/1/ f 3v.
16. StA CP 2/8/28 PC to HFC 23 Jan 1849; the quote was doubtless taken from Teignmouth's 'Works of Jones', where it appears on vol. 1 p. 60 of the edition by Wilks, 1835.
17. StA CP 2/8/27 PC to HFC 4 Dec 1848.
18. *Exeter & Plymouth Gazette* 13 Apr 1850.
19. Cleghorn, 1850a.
20. Dickson, 1995.

21. Butler, 2000; Killen, 2013.
22. Cleghorn, 1850a.
23. Reference to forthcoming baptism in RBGE BC 3/154, HC to JHB 6 Jul 1851.
24. BSE Minute Books, RBGE.
25. In the family Apocynaceae, two species, both collected by Wight in Ceylon in 1836.
26. Cleghorn, 1850b, 1851d.
27. Cleghorn, 1850a.
28. Cleghorn, 1851a.
29. Cleghorn, 1850c.
30. Cleghorn, 1851b.
31. Cleghorn, 1851c.
32. Shaw, 1877: 29.
33. The date when it moved to RBGE is uncertain: it appears to have been by 1855, though Fletcher & Brown (1970: 118) gave 1863.
34. This section relies heavily on Crawford, 2011.
35. StA Special Collections List of members of Philosophical Society extracted from UY8525/1 & 2.
36. When asked to contribute to one of Playfair's town improvements in 1856, the building of a new town hall, Peter wrote to Hugh in India describing the project as a 'nonsense', but eventually relented and sent £20 so that he and Hugh might with a clear conscience use the facilities offered by the new building StA CP 2/9/2–4 PC to HC 17 May 1856, 5 Feb 1857.
37. Brewster, quoted by Crawford, 2011: 27.
38. Crawford, 2011: 155.
39. StA CP 2/8/21 PC to HC 16 Oct 1845 – this was exchanged with a 'miniature' of himself that Hugh was sending back to St Andrews, though it is not recorded if this was by a European or a native artist.
40. Adamson Album, National Museums of Scotland T 1942.1.1; I am grateful to Alison Morrison-Low for showing me this and advising on dating; the album also contains portraits of Peter Cleghorn and Alexander Kyd Lindesay.
41. Now at the J. Paul Getty Museum, Malibu, California and reproduced in Smith, 1990: 76.
42. StA Photography Collection Album 8 no 80.
43. Links with St Andrews, its University and Hugh Cleghorn senior, were, as usual in this tightly inter-connected world, close: the Rev. George Hill had been leader of the moderate party of the Church of Scotland, before being succeeded in the leadership by Rev George Cook, Cleghorn's teacher of

Moral Philosophy. Elizabeth, Hill's eldest daughter, had married Cook's brother, Professor John Cook (who was also her cousin, and not to be confused with either his father or his son, both called John, both of whom also held St Andrews chairs over three generations. There are, moreover, connections between the Sprot and Cowan family into whom Cleghorn married in 1861 – Mabel Cowan's step-grandmother Helen Brodie was third cousin of Harriet Hill, wife of Mark Sprot, whose son Alexander married Cleghorn's sister Rachel), so Elizabeth Cook was an aunt of Alexander Sprot.

44. The Reports of the 20th (1850) and 21st (1851) annual BA Meetings, were published in the years following the meetings, for bibliographical details see Cleghorn, 1851d and Cleghorn *et al.*, 1852.
45. StA CP 2/8/2 AC to HC 24 May 1843.
46. Cleghorn, 1850b, 1851d.
47. Bodleian Library, printed pamphlet of reports requested at 1850 meeting – Dep BAAS 32 f 162–5.
48. Anon, 1865: 245.
49. Anecdotal details are RBGE BC 3/154 HC to JHB, 6 Jul 1851, the Report itself is Cleghorn *et al.*, 1852.
50. Now in the National Portrait Gallery, London, a forerunner of that other campaign-canvas, D.O. Hill's Scottish 'Signing of the Deed of Demission'.
51. RBGE BC 3/153 HC to JHB 7 Apr 1851.
52. BA 1851 Report p. xlviii.
53. Later Sir Charles, a geologist, botanist, palaeobotanist and friend of Lyell and, with Richard Owen, a Vice-President for D Section.
54. *The Athenaeum* 19 Jul 1851 p 780.
55. Cleghorn, 1861a: vi – referring to the economic benefits of conservancy, that 'neither the Government nor the community at large were deriving from the Indian forests those advantages which they were calculated to afford' on account of 'wasteful and uncalled-for destruction'.
56. Anon, 1865 (see p. 313).
57. Grove, 1995: 451.
58. l.c. p. 11.
59. No reference is given; Cleghorn's quotation is actually two different ones spliced together with dots. The first, on the 'two calamities' is from the *Personal Narrative*; the second 'Plants exhale fluid … furnished by rains' is untraced. Rajan, 2006: 27 cited it as from an American edition of *Aspects of Nature* but it is certainly not in the British edition of that work. Curiously this

ENDNOTES

second part of Cleghorn's quote was also used verbatim by J.H. Balfour in his *Phytotheology* p. 123 of the same year, but unreferenced other than a general 'Humboldt'; part of it had earlier been used by Robert Dickson in an article 'Contributions to the Natural Theology of the Vegetable Kingdom' (*Church of England Magazine* 11: 374, 1841) showing an interesting link between evangelical thought and climatic concerns. Further work is required to trace the origin and transformation of this interesting quotation.

60. Cleghorn *et al.*, 1852: 79.
61. Information for this section taken mainly from *The Illustrated Exhibitor*. Anon, 1851.
62. Until recently Professor of Chemistry at the EIC military seminary at Addiscombe.
63. Cleghorn's name does not appear in any of the papers of the Royal Commission for the Exhibition of 1851, Angela Kenny pers. comm.; the few details are known from RBGE BC 3/153 HC to JHB 7 Apr 1851.
64. Anon, 1889: 202. McIntosh, 1895: liii stated that the work was for 90 days, preparing the Catalogue of Raw Products.
65. Anon, 1851: xiii.
66. Anon, 1851: xxxvi.
67. Balfour, 1855.
68. Fletcher & Brown, 1970: 142.
69. Cleghorn's donations are listed in Balfour 1855, 1856, with later ones in the BSE unpublished Minutes and published *Transactions*.
70. Hooker, 1855; Cleghorn's donations are listed in the Museum Donation books and are now available online on the Economic Botany Collection database.
71. Noltie 2007 1: 57.
72. KDC 55/64, HC to WJH 19 Aug 1851.
73. Linnean Society of London Fellows' Election Certificates 4 Nov 1851.
74. RBGE BC 3/153 HC to JHB 7 Apr 1851.
75. StA CP 9/4/1–10.
76. In due course Cleghorn contributed one guinea for a memorial to Forbes, sent with similar amounts from Walter Elliot and another Edinburgh-trained medic G.J. Shaw the Madras Assay Master – RBGE BC 4/171 HC to JHB 13 Mar 1855.
77. RBGE BC 3/155, HC to JHB 12 Dec 1851.
78. Information from the accounts of the Madras Fire in the *Madras Athenaeum* 17 Jul 1852: 342 and *Madras Spectator* 19 Jul 1852: 684.

Chapter 4

1. Anon in *Indian Journal of Medical Science* 2: 724. 1855.
2. RBGE BC 4/163 HC to JHB 13 Feb 1855.
3. RBGE BC 4/168 HC to JHB 5 Jul 1854.
4. l.c.
5. RBGE BC 4/179 HC to JHB 26 Jan 1856.
6. StA CP 2/9/20 PC to HC 1 Apr 1858.
7. StA CP 9/4 ff 76–8.
8. *Home & Foreign Record of the Free Church of Scotland* 1 Dec 1856, article on David Paterson.
9. RBGE BC 4/172 HC to JHB 13 Apr 1855.
10. RBGE BC 4/179 HC to JHB 26 Jan 1856.
11. RBGE BC 4/188 HC to JHB 9 May 1859.
12. Copley, 2006: 45.
13. RBGE BC 3/156 HC to JHB 12 Apr 1852.
14. Accounts of the Madras Fire in the *Madras Athenaeum* 17 Jul 1852: 342 and *Madras Spectator* 19 Jul 1852: 684.
15. RBGE BC 4/158 HC to JHB 12 Aug 1852.
16. StA CP 9/4/2(i) WP to HC 19 Jul 1853.
17. RBGE BC 4/158 HC to JHB 12 Aug 1852.
18. RBGE BC 4/159 HC to JHB 10 Oct 1852.
19. KDC 56/56 HC to WJH 5 Jun 1857.
20. RBGE BC 4/165 HC to JHB 9 Dec 1853.
21. Information in this section taken from Anon, 1935 and the Madras Almanacs.
22. There were two brothers of this name at the time, and it is not known whether this was – Charles David, or Augustus Octavius. Crawford, 1930.
23. RBGE BC 4/182 HC to JHB 14 Oct 1856.
24. RBGE BC 3/155 HC to JHB 12 Dec 1851.
25. RBGE BC 4/174 HC to JHB 23 Jul 1855.
26. Cleghorn, 1855b.
27. Ms 'Memorandum on the Vegetable Tallow of Mysore & Canara' dated 10 Dec 1855, StA CP 9/4/f 68.
28. Cleghorn, 1856k.
29. StA CP 2/9/1 PC to HC 30 Jan 1854.
30. Two syllabuses survive for the Botany and Materia Medica lectures – Cleghorn 1852a, 1854c, and there is much detail of the teaching in StA CP 9/4 ff 6–17, 79–81.
31. StA CP 9/4 f 13.
32. RBGE BC 4/175 HC to JHB 14 Aug 1855.
33. RBGE BC 4/176 HC to JHB 13 Oct 1855.
34. RBGE BC 4/166 HC to JHB 25 Jan 1854.
35. Cleghorn, 1861a: 22.
36. Cleghorn, 1861a: 47.
37. With Jaffrey RBGE BC 4/162 HC to JHB 26 Apr 1853, for which Cleghorn had earlier told Balfour he would not grudge paying £20 or £25.
38. Set up in memory of the Madras surgeon Thomas Moore Lane, superintendent of the Eye Infirmary, who had died in 1844.
39. Maclagan, 1863.
40. l.c.
41. RBGE BC 4/159 HC to JHB 10 Oct 1852.
42. StA CP 9/4/ff 74–5.
43. Raman, 2013.
44. StA CP 9/A/2.
45. RBGE BC 4/219 HC to JHB 14 Aug 1864.
46. Raman & Raman, 2013.
47. Andy, 1869.
48. RBGE BC 4/164 HC to JHB 8 Nov 1853.
49. StA CP 9/4/f 66.
50. RBGE BC 4/167 HC to JHB 6 Jun 1854.
51. *Madras Almanac for 1856*: 634. Information for this section taken from Madras Almanacs; Shaw (1877); a printed pamphlet on the Society sent by Cleghorn to Balfour RBGE BC 4/157 8 Jul 1852; and the *Report of the Committee of the Agri-Horticultural Society of Madras for the year 1853*, of which Cleghorn's copy is at RBGE.
52. Anon, 1857: 28.
53. StA CP 9/4/f 24.
54. ?Stent, 1848.
55. Shaw, 1877: 29, report of meeting on 28 Jan 1852.
56. Cleghorn, 1853a.
57. See Noltie, 2007 1: 63–4.
58. RBGE BC 4/161 HC to JHB 9 Mar 1853.
59. Reid & Cleghorn, 1853.

ENDNOTES

60. RBGE BC 4/161 HC to JHB 9 Mar 1853.
61. RBGE BC 4/176 HC to JHB 13 Oct 1855.
62. StA CP 9/4/f 24 – copy letter, not in Cleghorn's hand, seeking Balfour's help in recruiting a new gardener 12 Nov 1852.
63. Details of Jaffrey's early days are given in Cleghorn's letters to Balfour RBGE BC 4/161–3 of 1853, and Jaffrey's own letters to Balfour RBGE BC 7/1–3, 1853–6.
64. Biographical details, pers. comm. from Margaret Sargent and Mary Beresford.
65. Jaffrey, 1855.
66. RBGE BC 4/164 HC to JHB 8 Nov 1853.
67. RBGE BC 4/172 HC to JHB 13 Apr 1855.
68. RBGE BC 4/162 HC to JHB 26 Apr 1853 – his father had earlier amusingly warned him of the pluses and minuses of 'chumming': 'a Chum is something like a wife – delightful when you perfectly understand & agree with each other but horrible when you do not' StA CP 2/8/21 PC to HC 16 Oct 1845.
69. RBGE BC 4/166 HC to JHB 25 Jan 1854.
70. RBGE BC 4/181 HC to JHB 13 Sep 1856.
71. RBGE BC 7/3 ATJ to JHB 13 Sep 1856.
72. Jaffrey, ?1857.
73. KDC 56/58 HC to WJH 16 Aug 1857. According to Alexander Hunter (AH to JHB RBGE BC 7/260 13 Aug 1863) Jaffrey had also taken to the bottle, but after two years on the coffee plantation was by then in Calcutta and 'getting on well'.
74. RBGE BC 4/185 HC to JHB 25 Jan 1859.
75. KDC 56/58 HC to WJH 16 Aug 1857 & RBGE BC 4/183 HC to JHB 14 Dec 1857.
76. www.scotlandspeople.org.
77. The year in which he was re-engaged for another three years, Shaw, 1877.
78. RBGE BC 7/260 Alexander Hunter to J.H. Balfour 13 Aug 1863.
79. StA CP 9/4/f 69.
80. RBGE BC 4/178 HC to JHB 27 Nov 1855.
81. KDC 157/174 HC to JDH 27 Nov 1855.
82. Cleghorn, 1856l; RBGE BC 4/179 HC to JHB 26 Jan 1856.
83. Information in this section taken mainly from a particularly valuable collection of papers relating to the Exhibition in Cleghorn's papers StA CP 9/2; from Balfour (1856) and an anonymous review of this publication (Anon, 1856).
84. William Underwood, the Collector of Sea Customs was a major figure in the 1855 Exhibition; in 1857 the seven major exports from Madras were, in order of value: Dyes, Sugar, Woven Cotton, Hides & Skins, Raw Cotton, Grain and Coffee.
85. Balfour, a surgeon from Montrose, had played a major role in the Madras Local Committee that sent material to the Great Exhibition, and has since been in charge of the Government Museum; he was also a keen investigator of the subject of the effect of trees on climate.
86. RBGE BC 4/176 HC to JHB 13 Oct 1855.
87. RBGE BC 4/177 HC to JHB 27 Oct 1855.
88. These were of copper, the die cut by the leading London designer Benjamin Wyon, the medals doubtless struck locally at the Madras Mint, whose Assay Master G.J. Shaw and his assistant Dr Andrew Scott were involved in the Exhibition.
89. Cleghorn's medals were sent home in 1858 by Mrs Sanderson (probably wife of Cleghorn's colleague Dr James Sanderson), his father was proud to receive them and hoped that they 'may long be preserved as Heirlooms'; (StA CP 2/9/22 PC to HC 1 Jul 1858) their fate is unknown but one of Hunter's came on the market recently. Baldwin's Auction 74 9 May 2012 Lot 2000.
90. Cleghorn, 1855a.
91. Cleghorn later realised this as he annotated one of own of his own copies of the pamphlet, now at EU, with Dalzell's name and reference.
92. Cleghorn, 1858b.
93. Anon, 1856.
94. *Kew Journal of Botany* 7: 314–6, 343–5. 1855.
95. *Edinburgh New Philosophical Journal* 3(n.s.): 365–6. 1856.
96. Cleghorn, 1856h and reprinted in 1861a: 212.
97. The concern in the Nilgiri Sholas was in 1857, after his Conservancy had started.
98. Hoffenberg, 2001.
99. Anon, 1856.
100. l.c.
101. Hoffenberg, 2001: 5.
102. Taylor *et al.*, 2014.
103. Penn, 2014.
104. *Indian Journal of Arts, Sciences and Manufactures* 1 (ser 2, pt 1): 90–1.
105. RBGE BC 4/185 HC to JHB 25 Jan 1859.
106. Information for this section from Anon, 1857 and Balfour, 1858.
107. This copy, although unsigned, probably belonged to Walter Elliot to whom the RBGE copy of the similar volume for the 1855 Exhibition is inscribed.
108. Like the 1855 Timber Report, the 1857 one was also reprinted in Cleghorn 1861a: 245 *et seq.*
109. Geology in India. Art. VII. *Calcutta Review* 32 (Mar 1859): 122–61.
110. John Murray (1809–1898), one of the finest of the early Indian photographers had an Edinburgh MD.
111. Dewan, 2003.
112. Balfour, 1858: 187.
113. StA CP 9/2/ff 90–1 AH to HC 15 Jan 1855.
114. Nissen, 1951.
115. Information from Simonds & Prasad, [2014] and pers. comm.
116. The dates and locations of the exhibitions and press opinions thereon (from the *Art Journal* May 1866, *Athenaeum* 17 May 1866, *Reader* 17 March 1866, *Evening Standard* 11 April & 12 March 1866, *Morning Post* 12 March & 30 April 1866 and *English Mail* 15 April 1866) are cited on a flier soliciting subscriptions for Mrs Brown's publication in the possession of David Simonds.
117. David Simonds and V.P. Prasad, the then Botanical Liaison Officer of the BSI at Kew, also persuaded the Government of India to publish a large-format book of the drawings in 2014.

Chapter 5

1. Cleghorn – annotation on his own copy of *Forests & Gardens* in EU, showing his primary motivation of economic development rather than preservation of trees at all costs.
2. RBGE BC 4/174 HC to JHB 23 Jul 1855; this version of events is corroborated in McIntosh, 1895: liv.
3. Grove, 1995.
4. Anon, 1884: 78.
5. StA CP 2/9/4 PC to HC 5 Feb 1857.
6. StA CP 2/9/9 PC to HC 18 May 1857.
7. For general history see Stebbing, 1922, and more recent works including Grove, 1995, Barton, 2002, Rajan, 2006.
8. Quoted in Sutton, 2009: 119.
9. Another Bengal surgeon, who had acted as Superintendent of the Calcutta Botanic Garden 1846–7.

ENDNOTES

10. Stebbing quoted most, but not all of it, but as throughout his provoking book failed to give its source. It was published at the time in the *Government Selections No IX Papers relating to the Teak Forests of Pegu*. Calcutta, 1855 pp 73–78, of which Cleghorn had a copy.

11. *Bengal and Agra Directory ... for 1854*: 11.

12. Later Sir Arthur Phayre, someone whom Dalhousie deeply trusted.

13. Ribbentrop, 1900: 72.

14. Stebbing, 1922 1: 206.

15. Grove, 1995: 461.

16. Barton, 2002: 57.

17. Many years later this minute would be edited and published by Cleghorn (1868a).

18. Stebbing, 1922 1: 124; in this case 'His Honour in Council', the President of the Supreme Council, was the Hon Sir T.H. Maddock, see *Bengal & Agra Directory ... for 1849*: 9.

19. Stebbing, 1922 1: 207.

20. StA CP 2/9/5 PC to HC 17 Feb 1857.

21. KDC 56/56 HC to WJH 5 Jun 1857.

22. *Botanical Magazine* 83: t. 4998. 1857.

23. Birdwood, 1901: 757.

24. Stebbing, 1922 1: 302.

25. Brandis, 1888: 90.

26. The only way to communicate directly with Indians was to learn their languages. In Cleghorn's case there is tantalisingly scant evidence apart from his learning 'Canarese' in Nugger, and a quotation of something he said in Urdu to a 'Syed' whose face he had to shave to lance an abscess (StA CP 9/5/71v) showing that he was still learning languages in the 1860s; he later referred to 'languages learned with toil' (RBGE BC 4/222 HC to JHB 22 Nov 1864) and must certainly have been fluent in Tamil when in the south.

27. StA CP2/11/7 PC to HC 24 Mar 1862.

28. Cleghorn, 1861a: vi.

29. Cleghorn, 1861a: vii.

30. Sutton, 2009: 116.

31. General details of the Forest Department are to be found in Cleghorn, 1861a, and the Madras Almanacs.

32. Cleghorn, 1861a: 22.

33. Of the seven remarkable sons of Dr David Maclagan; three of John's brothers also went to India: William Dalrymple (1812–1900), for two years from 1847 in the 51st Madras Native Infantry, invalided out and took a maths degree at Cambridge. He was ordained and ended up as Archbishop of York. Dr James McGrigor Maclagan (1830–91), a Bengal surgeon, whom Cleghorn met when he passed through Madras in 1855 on his way back home on sick leave. Colonel (later Sir) Robert Maclagan (1837–1904), a school-friend of Cleghorn who became an Indian engineer; in the Punjab in the early 1860s Cleghorn would catch up with him again. His herbarium, mainly Indian, some Canadian, is at RBGE. Sir (Andrew) Douglas Maclagan (1812–1900) stayed in Edinburgh to become an expert in forensic medicine and Professor of Medical Jurisprudence; he was heavily involved with the Botanical Society and attended the presentation of Cleghorn's portrait.

34. RBGE BC 4/183 HC to JHB 14 Dec 1857.

35. *Madras Almanac for 1858*: 470.

36. Cleghorn, 1875a: 5.

37. Cleghorn, 1879.

38. Cleghorn, 1858c; 1861a: 1–23.

39. Cleghorn & McIvor, 1859.

40. Cleghorn 1861a: 145–9.

41. KDC 56/56 HC to WJH 5 Jun 1857.

42. Cleghorn 1861a: 5. The four size classes, the others being under 3 ft, and 3-4 ½ ft, were those used by Brandis in Burma, see Anon, 1884: 345.

43. Cleghorn, 1856n, o.

44. Cleghorn, 1858b.

45. Cleghorn 1861a: 7 – emphasis added.

46. John Alexander Campbell (1818–1863) of the 7th Madras Light Cavalry was, of course, a Scot, of the Dunstaffnage family, and was Executive Engineer in the PWD Coimbatore (1853–9) – he gave (in Cleghorn 1861a: 172) his reason for planting in straight lines 'of mathematical accuracy, so that one cannot be cut without detection' – ultimately the plantations were a failure, perhaps being planted on former grassland with poorer soil than that which supported shola forest – see Sutton 2009: 127.

47. Cleghorn, 1861a: 17, and given more fully by Sutton 2009: 121.

48. Cleghorn, 1861a: 79.

49. Scherzer, 1861, chapter 9.

50. StA CP 2/9/19, 20, 26.

51. Cleghorn 1861a: 32–53.

52. Account taken from the *Madras Athenaeum* 13 Nov 1858: 543; 20 Nov: 555; 30 Nov: 571; 7 Dec: 583; 18 Dec: 503.

53. StA CP 2/10/3 PC to HC 31 Jan 1859.

54. The large building to the left was the Courthouse with a jail beneath; the buildings are still there – the Courthouse is now the Pazhassi Raja Museum, the adjacent Guesthouse the Krishna Menon Museum and Art Gallery. The Collector of Malabar lived on West Hill (which is where H.V. Conolly was murdered in 1855), other British administrative buildings were on East Hill. From the annotations on this drawing this is where Georgina and Marianne Phillips stayed and entertained by their cousin-in-law William Rose Robinson (see *Cleghorn Collection* Chapter 5). Cleghorn owned a drawing by one of the Phillips sisters, and might have stayed in this house on one of his forest tours: the annotations identify the trees as a banyan to the right, and an introduced Australian *Casuarina* to the left – information that could possibly have been given to Georgina by Cleghorn.

55. *Madras Athenaeum* 7 Dec 1858 p. 583.

56. StA CP 2/10/3 PC to HC 31 Jan 1859.

57. BL APAC L/E/3/737 no. 43.

58. Blenkinsop did not last long, he was 'carried off by cholera at Hurroor in the 34th year of his age, October 23rd 1861', as recorded on his memorial in Yercaud church.

59. Cleghorn, 1861a: 48.

60. Cleghorn, 1861a: 45.

61. Cleghorn 1861a: 289–302; 1861b.

62. BL APAC WD 567.

63. Cleghorn, 1861a: 115–23.

64. Formerly in the possession of Richard Grove, probably given to him by Major and Mrs Sprot, to be deposited in St Andrews University Library with the Cleghorn Papers.

65. Cleghorn, 1859/60.

66. Cleghorn 1861a: 60, 116.

67. In his library Cleghorn had a copy of the Government's 1854 Report on the *Progress of the Operations for the Suppression of Human Sacrifice and Female Infanticide in the Hill Tracts of Orissa*.

68. RBGE BC 4/187 Printed Testimonial to Lord Harris.

69. Masani, 2013.

70. RBGE BC 4/190 HC to JHB 14 Dec 1859.

71. RBGE BC 4/186 HC to JHB 13 Apr 1859.

72. Cleghorn mentioned him in a letter to Balfour of 1 Sep 1859, RBGE BC 4/189.

73. StA CP 2/10/18 PC to HC 2 Jan 1860.

74. RBGE BC 4/191 HC to JHB 9 Jan 1860.

75. RBGE BC 4/190 HC to JHB 14 Dec 1859.
76. Cleghorn, 1861a: 59–96. This report covers only Jan–August 1860, though appendixes supplied by others cover some material from the calendar year 1859, and the regional reports are for the full financial year.
77. Balfour, 1873 5: Trevelyan.
78. Cleghorn, 1861a: 93.
79. Cleghorn forwarded Ward's application to the Linnean Society to Joseph Hooker on 1 Sep 1859, KDC 157/177: he is 'an excellent Ornithologist and Entomologist and possesses a large collection of Zoological specimens … well known to Dr Wight, Mr Adam White, Mr Bennett and other Zoologists'.
80. Cleghorn, 1861a: 126–44.
81. 'Kumri' was the normal spelling for Cleghorn at this period, though in Mysore earlier it had been 'coomri'.
82. Cleghorn, 1861a: 68.
83. Cleghorn, 1861a: 96–105.
84. RBGE BC 4/178 HC to JHB 27 Nov 1855.
85. KDC 157/178 HC to JDH 2 Dec 1860.
86. RBGE BC 4/198 HC to JHB 29 Nov 1860.
87. 1871 Parliamentary Papers 2: 2.
88. Grove, 1992: 461.
89. Cleghorn, 1861a: 19.
90. Anke Schürer-Ries, pers. comm.; a photograph of Müller is available on http://bmpix.usc.edu/bmpix/controller/view/impa-m45112.html?x=1396880143666 consulted 7 Apr 2014.
91. Cleghorn, 1861a: 46.
92. Cleghorn, 1861a: 33, 40.
93. www.leithhistory.co.uk/2005/04/13/the-maclagan-window consulted 7 Jun 2015.
94. Cleghorn, 1861a: 18.
95. Terrell, 1992: 177.
96. Drayton, 2000: 45.

Chapter 6

1. An arrangement that lasted until 1883.
2. Price, 1908: 238–62, see also Noltie 1: 68, 2007; Penn, 2010.
3. I am grateful to Janet Carolan, archivist of Dollar Academy, for information on McIvor and Dollar; see also Penn, 2010.
4. Anonymous obituary in *Transactions of the Botanical Society of Edinburgh* 13: 11–12, 1879; Seaward, 1976.
5. *Hepaticae Britannicae* 1847.
6. They are recorded on the family tombstone in Dollar Kirkyard, as is William.
7. Anon, 1876.
8. Sutton, 2009: 120.
9. l.c.
10. Anon, 1876.
11. Robinson (1822–1886) an additional member of the Viceroy's Executive Council, who had acted as Governor of Madras the previous year prior to the arrival of the Duke of Buckingham. He was also part of the Thomas clan, married to Julia, daughter of the Madras Civil Servant James Thomas, and so a nephew by marriage of E.B. Thomas and J.F. Thomas (see *Cleghorn Collection* Fig. 23).
12. Cleghorn & McIvor, 1859; this, and similar reports for the two succeeding years, are also summarised in Cleghorn, 1861a: 359–73.
13. McIvor also invented a machine for lifting heavy loads, on which he published a photographically-illustrated pamphlet (Cleghorn's copy in EU); he also devised the technique of 'mossing' – wrapping around with moss as a way of encouraging bark regeneration in cinchona, and seems also to have used moss for binding the roots of newly propagated plants – as an alternative to potting-up – to save both soil and space.
14. In 1857 the total value of sales was around 3000 Rs, of which 2/3 came from sales to the public - for fruit and timber trees, shrubs and flowers, bulbs, rooted grasses and seeds. The Medical Board paid 30 Rs for foxglove leaves and E.B. Thomas 405 Rs for timber trees.
15. Veitch paid the garden 300 Rs in 1857 for various flowers and shrubs.
16. Another member of the Thomas clan, married to Emma, daughter of James Thomas, so a niece by marriage of J.F. Thomas and E.B. Thomas, and a brother-in-law of William Rose Robinson; Cleghorn had known him since Shimoga days, when Francis made for him a pencil sketch showing the habit of the gamboge tree. Francis also built the church at Coonoor and the original Jackatalla Barracks.
17. It is shown in a photograph by A.T.W. Penn reproduced in Penn, 2010: 68.
18. Herbert, 2011: 125–6.
19. *Indian Journal of Arts, Sciences and Manufactures* 1 (n.s., pt. 3): 107, ?1857.
20. The sympathetic and supportive Edward Brown Thomas.
21. Cleghorn, 1861a: 364.
22. Cleghorn & McIvor, 1859: 3.
23. Edited version in Cleghorn, 1861a: 330–4, full report BL APAC P/249/57 ff 4852–62.
24. See Anon, 1991 and Bowe, 2012. Desmond, 1992 was, as usual, more reliable and avoided making unproven connections.
25. Composite image recreated in Iyer *et al.*, 2012.
26. Buchanan, 1807 1: 46–7.
27. BL WD4461.
28. BL WD3769.
29. BL P256.
30. BL APAC P/242/21, P/242/40.
31. S. Narayanaswamy in Anon, 1991: 13–4.
32. Iyer *et al.*, 2012.
33. Letters reproduced in Anon, 1991: 17–8.
34. RBGE BC 4/182 HC to JHB 14 Oct 1856.
35. Cleghorn, 1861a: 333.
36. KDC 56/55 HC to WJH 24 Mar 1857.
37. See letters of New to William & Joseph Hooker, Kew DC 56/298, 157/702–3, 1858.
38. Cleghorn, 1861a: 336–43.
39. Cleghorn, 1861a: 344–58; 1861g.
40. Cleghorn, 1858a.
41. KDC 57/32 24 Mar 1865.
42. Obituaries: *Journal of the Kew Guild* 1895: 29–30; *Gardeners' Chronicle* 3 Feb 1866: 102.
43. Cleghorn, 1861a: 370.
44. Price, 1908: 248–62.
45. Veale, 2010.
46. Sutton, 2009: 119, 68; Lascelles thought nothing of breaking promises to Todas and Badagas, given to persuade them to sell him land.
47. Herbert, 2011.
48. *Marsden's Madras Almanac 1860*: 662.
49. *Asylum Press Madras Almanac 1863*: 250.
50. *Indian Journal of Arts, Sciences and Manufactures* 2: 151. 1850.
51. *Asylum Press Madras Almanacs* 1871 onwards.
52. The transliteration of 'ashokam', if that is what it is, is strange, but so too is the 'tree of liberty' – the tree has Indian religious associations, but the only known Indian use of the latter term is for the tree famously planted by Citizen Tipu at Seringapatam in 1798, inspired by Jacobin principles. Could this be a reference to that, and if so what does it say about Baldrey's political persuasion?
53. See Kalpana & Schiffer 2003: 210–2 for later history of the People's Park.

ENDNOTES

Chapter 7

1. StA CP 2/9/24 PC to HC 4 Aug 1858.
2. RBGE BC 4/195 HC to JHB 1 Nov 1860.
3. RBGE BC 4/196 HC to JHB 12 Nov 1860.
4. RBGE BC 4/194 HC to JHB 13 Sep 1860.
5. RBGE BC 4/197 HC to JHB 26 Nov 1860.
6. RBGE BSE Minutes, and published *Transactions*.
7. Sewell, 1896.
8. See Burkhardt *et al.*, *The Correspondence of Charles Darwin* vols 5–7.
9. RBGE BC 4/197 HC to JHB 26 Nov 1860.
10. RBGE BC 4/180 HC to JHB 29 Apr 1856.
11. Sewell, 1896: 57. Curiously John Hope had chosen the same quotation from the First Book of Kings 4: 33 for the gold medal he gave to his best students and collectors.
12. Gordon, 1976.
13. Argyll, ?1861.
14. Cleghorn, 1861a: 277.
15. Balfour, 1859: 8. I am grateful to Margarita Hernandez Laille for these references.
16. Balfour, 1860, dedicated, as was the first edition, to R.K. Greville.
17. Balfour, 1860: xii–xiii.
18. Balfour, 1875: xii–xiii.
19. Balfour, 1875: 408.
20. For a stimulating account of Miller's life and views, see Waterston, 1986.
21. KDC 157/182 HC to JDH 18 Feb 1861.
22. MacLeod, 2013; Professor MacLeod has been extremely generous with information about the Cowan family and kindly supplied a copy of Charles Boog-Watson's 1915 labyrinthine genealogical tables of the Cowan and Chalmers families, which show just how deeply interconnected they were with Scottish intellectual/ecclesiastical life, and (indirectly) with India; unfortunately there are several errors in the dates on the family tree in his own book.
23. Cowan, 1878.
24. Cowan, 1933.
25. Cowan, 1878: 213.
26. MacLeod, 2013: 288.
27. Cowan 1878: 351.
28. Cowan, 1878: 114.
29. Scott gave Alexander Cowan the manuscript of *Heart of Midlothian*, and that of *Old Mortality* was later given to Charles's brother John by the son of John Ballantyne, Scott's printer.
30. Cowan, 1878, chapter 15.
31. MacLeod, pers. comm.
32. Cowan, 1933, plate 12.
33. StA 9/9/9 – typescript of W.C. McIntosh's biographical memoir of Cleghorn for RSE, but not used in printed version.
34. Cowan, 1878: 374.
35. Fowle, 2008.
36. KDC 157/183 HC to JDH 2 Sep 1861.
37. Mitchell, 1993.
38. Puff from *Daily News* cited in advertisement – see note 41 (below).
39. Cleghorn, 1861a: 94.
40. KDC 157/181 HC to JDH 4 Feb 1861.
41. Advertisement in W.H. Allen catalogue of 1 Jan 1862 bound into copy of *The Indian Army & Civil Service List for 1862*, copy in BL (APAC).
42. McFarlane had started as a partner of the lithographer Friedrich Schenck who was a protege of the Cowan family; these were paid for by a grant from the Royal Society, (RBGE BC 4/214 HC to JHB 22 Mar 1864) as they were first published in its *Transactions* in Cleghorn's paper on the Anamalais (Cleghorn, 1861b).
43. KDC 157/186 HC to JDH ?27 Jan 1862.
44. There is also a brief notice in a German periodical; for bibliographic details see Appendix 1.
45. *The Athenaeum* 1771: 450, 1861.
46. *The Spectator* 28 Sep 1861: 1067–8.
47. *Edinburgh New Philosophical Journal* 14(n.s.): 286–9, 1861.
48. RBGE BC 4/200 HC to JHB undated but c 20 Sep 1861.
49. *Gardeners' Chronicle* 1861: 969, 1009, 1861.
50. KDC 157/186 HC to JDH ?27 Jan 1862.
51. *Madras Quarterly Journal of Medical Science* 5: 134–43, ?1862.
52. StA CP 2/10/22 PC to HC 1 Oct 1861.
53. A book he had with him during his own travels in the Punjab Himalaya – see KDC 157/188 HC to JDH 27 Jan 1862.
54. Details in Kew Garden Plants Outwards Book 1860–9, ff 91–2 and KDC 157/184 HC to JDH Sep 1861 and 157/185, 187 8 Oct 1861.
55. KDC 157/178 HC to JDH 2 Dec 1860.
56. Details of the voyage given in KDC 157/185, 187 HC to JDH 8 Oct 1861 & RBGE BC 4/201, 202 BC to JHB 28 Sep & 7 Oct 1861.
57. This occurred in 1853 – for more on this garden see A. Wilkinson, *Garden History* 39: 83–98, 2011.
58. Yule in Cleghorn, 1868c: 22. Yule was an exact contemporary of Cleghorn though a year ahead of him at the Edinburgh High School.
59. Cleghorn never met Canning, he had hoped to do so in Calcutta on his way to the Punjab (to promote Hooker's work on *Flora Indica*), but Canning was 'on the wing & under heavy affliction' from the death of his wife. RBGE AC 1: 82, HC to TA 4 Oct 1862.
60. King, 1876: 3.
61. The major sources used here are the 1866 East India (Chinchona [sic] Plant) Blue Book, ed. Bourdillon, 1866, Cleghorn, 1863b and King 1876; general information from Desmond 1992; Honigsbaum 2001; Rocco 2003.
62. Cleghorn, 1863b; see also Cleghorn, 1861a: 90–1.
63. Appointed 13 Sep 1862, see 1863 Cinchona Blue Book p 296.

Chapter 8

1. KDC 157/188 HC to JDH 27 Jan 1862.
2. StA CP 2/11 MC to PC 8 Feb 1862.
3. Reproduced in Cowan, 1933 plate 3.
4. KDC 157/186, 188 HC to JDH ?27 Jan 1862.
5. Cleghorn had also spoken about this to Sir William Denison in Madras. KDC 157/186 27 Jan 1862.
6. These had recently increased from two to '13 plying with 3 more daily expected'.
7. Review of Cleghorn, 1864b in *Friend of India* 17 Aug 1865: 954, 1865.
8. RBGE BC 4/204 HC to JHB 15 Oct 1862.
9. For reports on Giri, Pabur and Tonse valleys see Cleghorn, 1864b: 1–12; 1865a; 1866b.
10. See Cleghorn, 1864b: 29–40; 1865b; 1866a, c.
11. For information on Bourne's travels see Rayner, 2014.
12. Cleghorn, 1862c.
13. Cleghorn, 1865b, 1866a, c.
14. RBGE BC 4/203 HC to JHB 12 Jul 1862.

ENDNOTES

15. See Cleghorn, 1864b: 97–108 for account of this journey.
16. RBGE BC 4/203 HC to JHB 12 Jul 1862.
17. For report of Pangi and the Chenab Valley see Cleghorn, 1864b: 133–57.
18. StA CP 9/5/f 33r.
19. StA CP 9/5/f 3v.
20. Cleghorn, 1862d.
21. StA CP 9/5/f 6r.
22. Details of this and the journeys from here westwards to places now in Pakistan, up to December 1862, are given in a diary/notebook for the trip, StA CP 9/5.
23. Information about the Moravians at Keylong from Maria Heyde's diary (available on: https://opus4.kobv.de/opus4-th-wildau/files/18/MariaHeyde_FS08.pdf), from Bray, 2012, and an article in *Christian Missionary Gleanings* 1 Apr 1862, issue 4 pp 47–8).
24. Aitchison, 1869.
25. First published in German in 1832, and popular with missionaries in India.
26. His account of the Beas valley given in Cleghorn, 1864b: 69–77.
27. StA CP 9/5/f 42r.
28. Cleghorn, 1864b: 82.
29. Reports of these journeys in Cleghorn 1864b: 172–201.
30. RBGE BC 4/204 HC to JHB 15 Oct 1862.
31. Cleghorn, 1864b: 222–3; StA CP 9/5/f 59 r.
32. StA CP 9/5/f 71 r.
33. Report in Cleghorn, 1864b: 207–21.
34. RBGE BC 4/258 HC & MC to JHB 2 Dec 1862.
35. RBGE BC 4/205 HC to JHB 17 Jan 1863 *et seq.*
36. Committee of Compilation, 1863.
37. Cleghorn, 1863d: 108; the quotation is adapted slightly from Matthew 4:23.
38. Cleghorn, 1863d: 260–1.
39. Wilkinson, 1991.
40. ODNB.
41. RBGE BC 4/208 HC to JHB 25 May 1863.
42. RBGE BC 4/207 HC to JHB 18 Apr 1863.
43. He later wrote a book on Punjab Grasses, and was grandfather of the eponymous artist.
44. Neve, 1915.
45. RBGE BC 4/217 HC to JHB 14 Jun 1864.
46. RBGE BC 4/220 HC to JHB 19 Sep 1864.
47. RBGE BC 4/221 HC to JHB 19 Nov 1864.
48. RBGE BC 4/206 HC to JHB 3 Mar 1863.
49. KDC 157/190 HC to JDH 18 Dec 1862.
50. Stebbing, 1922 1: 521–34.
51. RBGE BC 4/210 HC to JHB 11 Aug 1863.
52. PP1871 p. 94.
53. Cleghorn, 1864b: 242–4 – as a result, in 1863 the rukhs were put under the Forest Department, but the Government changed its mind and they were put back under the revenue officers the following year.
54. BL APAC L/E/3/511 no. 2.
55. RBGE BC 4/210 HC to JHB 11 Aug 1863.
56. A rather scrappy notebook for this trip is at StA CP 9/6.
57. Bhasin, 2011.
58. RBGE BC 4/210 HC to JHB 11 Aug 1863.
59. RBGE BC 4/211 HC to JHB 5 Dec 1863.
60. RBGE BC 4/212 HC to JHB 4 Jan 1864.
61. Gandhi, 2013.
62. Information from this section mainly from [Mercer], 1864.
63. Medley, 1864.
64. l.c.
65. Noltie, 2011.
66. pers. comm.
67. Rayner, 2014.
68. StA CP 9/6/ f 4 v.
69. RBGE BC 4/218 HC to JHB 29 Jun 1864.
70. *Edinburgh Evening Courant* 17 Dec 1864. For Cleghorn's own work on tea in South India see Cleghorn, 1862b.
71. Cleghorn, 1866f.
72. Report in *Edinburgh Evening Courant* 17 Dec 1864.
73. BSE Minutes for 9 Apr 1891.
74. KDC 82/244 HC to JDH 18 Mar 1869.
75. One of his obituaries revealed that 'post-mortem examination revealed extensive tubercular deposit in the brain' (*Journal of Botany* 11: 320, 1873, probably by Cleghorn).
76. University of Glasgow Archives, R7/1/2, R8/1/2 – provided by Kiara King.
77. See Endersby, 2008.
78. Cutting from an unknown Lahore newspaper quoted by Whitehead & Talwar, 1976: 103. According to a letter from George Henderson to J.D. Hooker (KDC 155 ff 208–9), it was the putting of Baden-Powell, Stewart's locum while on leave, 'over his head' that was the final straw. In fairness to Hooker, while Stewart was on leave at Kew Hooker must have got on well with him and had suggested to him that he would be a suitable successor to Thomas Anderson in charge of Calcutta Botanic Garden. Confirming Cleghorn's view on his rough manners Stewart, however, had received the suggestion with 'ineffable scorn' saying that he had 'no wish for any such appointment' (Kew: MR225 f 211, Calcutta Botanic Garden 1830-1928).
79. For biographical information on Brandis see ODNB, anonymous article in *Gardeners' Chronicle* 2 Jan 1875: 13, and Anon, 1884 on his Burmese career. His PhD dissertation 'De Alcaloidum Hydrocyanetis Imprimis Duplicibus' of 1848 is available online at https://download.digitale-sammlungen.de/pdf/1414752844bsb10851357.pdf, consulted 20 Nov 2014 – about which I am indebted to Kevin Chang for helpful discussions.
80. E.g., Stebbing, 1922 1: 207.
81. Brandis (1874: xvi) acknowledged his debt to Schouw, Treviranus, Lantzius-Beninga and to Karl Nikolaus Fraas, Professor of Botany in Athens.
82. Quoted in Stebbing, 1922 2: 43.
83. Cleghorn 1861a: 147.
84. One example of the influence of Brown's book in India is recorded. In 1858 Captain John Campbell, possibly in consultation with Cleghorn, used a diagram from Brown in planning the disposition of trees in the Australian plantations at Jackatalla in the Nilgiris; (Cleghorn, 1861a: 175). Curiously Brandis sent his brother Carl to consult Brown at Arniston in 1864, but from where and about what, is unknown – see RBGE BC 4/223 HC to JHB 18 Dec 1864.
85. RBGE BC 4/213 HC to JHB 5 Mar 1864.
86. RBGE BC 4/212 HC to JHB 4 Jan 1864.
87. Smith, 1885.
88. The Himalayan Brotherhood was founded in 1838 by, among others, Henry Torrens, the Lodge was built in 1846 and rebuilt in 1855, though in 1870 sold and the site occupied by the Government Press and later the Amy HQ – see Buck, 1925 for this and other details on Simla in the 1860s. It is noteworthy that, unlike his grandfather, and many of his friends, Cleghorn was not a Freemason, but doubtless his religious scruples forbade this.

ENDNOTES

89. RBGE BC 4/217 HC to JHB 14 Jun 1864.
90. RBGE BC 4/220 HC to JHB 19 Sep 1864.
91. UV MP HC to GPM, Rome 6 Mar 1868.
92. Lowenthal, 2003.
93. RBGE BC 4/221 HC to JHB 19 Nov 1864.
94. Cleghorn, 1864b.
95. StA CP 2/11/6 PC to HC 10 Mar 1862.
96. Terrell, 1992.
97. [Brandis], 1865.
98. The quote, given in Latin by Cleghorn, 1864b: 168, also mentioned firs and pines from the woods of the 'montium emodorum' – it was actually Alexander's admiral Nearchos who supervised the boat-building.
99. Anon 1867: 56.
100. Stebbing, 1922, 1: 405.
101. Stebbing, 1922, 1: 448.
102. RBGE BC 4/221 HC to JHB 19 Nov 1864.
103. Gilmour 2005: 162.
104. 1871 PP: 95.
105. Stebbing, 1922 1: 118.
106. E.g. Barton, 2002: 65.
107. Of which the Government of India sent a copy to the Court on 3 August 1865 – see BL APAC L/E/3/608 no. 11.
108. The first one had been in Lahore in 1872.
109. Baden-Powell & Gamble, 1874: 3–25.
110. Gadgil & Guha, 2014.
111. Quoted in Ribbentrop 1900: 109.
112. RBGE BC 4/ 204 HC to JHB 15 Oct 1862.
113. RBGE BC 4/ 217 HC to JHB 14 Jun 1864.
114. RBGE BC 4/ 227 HC to JHB 14 Apr 1865.

Chapter 9

1. What he described as 'the small pension', achieved after 20 years' service – see RBGE BC 4/207 HC to JHB 18 Apr 1863.
2. RBGE BC 4/224, 225, 228 HC to JHB Mar to May 1865.
3. See Cleghorn's final Madras report of 1867, BL APAC L/E/3/1771 1868 no. 3 ff 75–80.
4. Balfour, 1873 3: 435.
5. Nair, 1978.
6. RBGE BC 4/225 HC to JHB 14 Mar 1865.
7. RBGE BC 4/228 HC to JHB 15 May 1865.
8. RBGE BC 4/257 HC to JHB undated [June 1865].
9. Cleghorn, 1866c.
10. Cleghorn, 1866a.
11. StA CP 10/13/1 – Chesser's initial letter enclosing the first sketches, which are no longer extant, dated 8 Sep 1865.
12. StA CP 10/6/77, bill 8 Dec 1868.
13. Early in his career he had worked on the Revesby estate in Lincolnshire, though long after the time of Sir Joseph Banks.
14. For whom it was barely large enough – in 1857 Peter Cleghorn said of it that 'the family were packed to the door' – StA CP 2/9/10 Aunt Jane Campbell to HC explaining why no room for his herbarium specimens at Wakefield.
15. The estimates and bills are all in StA CP box 10.
16. The Ewart family was exceptional: of John's five sons David became Chief Dominion Architect for Canada and James Cossar a zoologist, FRS, Professor of Natural History at Edinburgh – and a fellow member with Cleghorn on the 1885 BAAS McIntosh fish committee.
17. As noted in *Friend of India* Nov 1865 No 1610: 1306.
18. Cleghorn, 1866f.
19. RBGE BC 4/233 HC to JHB 19 Oct 1865.
20. Information for this section taken from Bourdillon, 1866; also a letter from HC to JHB of 7 Jan 1866 RBGE BC 4/235.
21. Cleghorn, 1866f, Bourdillon, 1866: 291.
22. Letter from Markham to Under Secretary of State for India, Guindy 14 Feb 1866 in Bourdillon, 1866.
23. *Friend of India* 3 Nov 1866.
24. RBGE BC 4/237 HC to JHB 17 Jun 1866.
25. Phillimore, 1871.
26. Stebbing, 1923.
27. J.L. Stewart's on the Punjab, 31 May 1866 – Phillimore, 1871: 240–4. Rupin/Tonse forests 2 July 1866 – Stebbing 2: 300. Oude, 30 January 1867 – BL APAC L/E/3/608 no. 19. Bengal, 5 February 1867 – BL APAC L/E/3/608 no 14. Central Provinces, 4 August 1866 – Phillimore, 1871: 450–4. Central Provinces, 5 April 1867 – BL APAC L/E/3/608 no. 29. Mysore, 11 April 1867 – BL APAC L/E/3/609 1868 no. 14. There must also have been one for Burma.
28. BL APAC L/E/3/608 no. 11.
29. BL APAC L/E/3/608 no. 29; Phillimore, 1871: 575–83.
30. Kurumba – a caste once powerful who in ancient times had fled to the Western Ghats, especially the Niligiris and Wynad, where their chief occupation was as wood cutters – see E. Thurston & K. Rangachari *Castes & Tribes of Southern India* 4: 155–77. 1909. Madras: Government Press.
31. Cleghorn, 1866e.
32. BL APAC L/E/3/609 no. 5; Phillimore, 1871: 614–8. He had already considered this matter in 1860 in Madras – Cleghorn, 1861a: 64.
33. BL APAC P/440/40 ff 6458–60.
34. BL APAC L/PJ/3/1102 no. 158.
35. Annual disputes on Caveri water take place between the governments of Karnataka and Tamil Nadu.
36. Cleghorn, 1868b.
37. BL APAC P/440/40 f 6459.
38. BL APAC P/440/40 f 6460.
39. BL APAC P/440/40 f 6458.
40. Bidie, 1869; Cleghorn, 1869a.
41. RBGE BC 4/239 HC to JHB 1 Sep 1866.
42. RBGE BC 4/261 MC to JHB 24 Oct 1866.
43. Cleghorn, 1876b.
44. He had earlier sent this to Sir William Hooker from the Rohtang Pass, who had grown it at Kew and illustrated it in the *Botanical Magazine* t 5456, 1864 – see KDC 157/195 HC to JDH 28 May 1867.
45. KDC 157/194 HC to JDH 26 Feb 1867, and Kew Inwards Book 1 Jan 1859 to 31 Dec 1837 f 408.
46. Cleghorn & Stewart, 1867.
47. *Friend of India* 14 Feb 1867 No 1676: 182. Lawson, a keen arboriculturist and owner of the family seed company in Edinburgh, until recently Lord Provost of Edinburgh.
48. For the training question see BL APAC L/E/5/70 nos 58, 60, 62, 63 and Phillimore, 1871: 358–9.
49. RBGE BC 4/237 HC to JHB 17 Jun 1866.
50. RBGE BC 4/244 HC to JHB 20 Jan 1867.
51. RBGE BC 4/244 HC to JHB 20 Jan 1867.
52. BL APAC L/E/3/608 no. 23.
53. RBGE BC 4/245 HC to JHB 28 May 1867.
54. BL APAC L/E/3/771 1868 no. 3 ff 75–80.

ENDNOTES

55. *Flora Sylvatica* (1869–73) and *Forester's Manual* (1869–74); Cleghorn reviewed the former in *Gardeners' Chronicle* 2(n.s.): 491, 1874, in which he stated that some of the drawings were by Govindoo's son.
56. BL APAC L/E/3/771 1868 no. 3 f 83.
57. As summarised by Stebbing, 1923 2: 14.
58. See Gadgil & Guha, 2014: 111.
59. See Stebbing, 1923 2: 15.
60. Michael in Anon, 1884.
61. BL APAC L/E/3/771 1868 no. 3 ff 81–4 para 14.
62. *London Gazette* 16 Nov 1869: 6115.

Chapter 10

1. KDC 157/197 HC to JDH 14 Dec 1867.
2. Cleghorn, 1869c.
3. Sources for the Malta visit are three letters from HC to JHB RBGE BC 4/246–8, between 14 Dec 1867 and 31 Jan 1868; and Cleghorn, 1870b.
4. RBGE BC 4/247 HC to JHB 7 Jan 1868.
5. Cleghorn, 1870b.
6. For more on the Mall and Argotti gardens, and the Orto Botanico in Catania, see Guglielmo & Buhagiar, 2012. These authors attribute the introduction of *Oxalis pes-caprae* to the Carmelite friar Carolus Giacinto, the first professor of Natural History, who started the first botanic garden at Il Mall, Floriana in 1805.
7. See RBGE BC 4/249 HC to JHB 26 Feb 1868; and Cleghorn, 1870b.
8. Sources for this section RBGE BC 4/249 HC to JHB 26 Feb 1868, 4/226 HC to JHB 1 Apr 1868; and Cleghorn, 1869d.
9. UV MP HC to GPM, Rome 6 Mar 1868; in the same collection are two other letters from Cleghorn to Marsh, and one from the following year from Mabel to Caroline Marsh.
10. Lowenthal, 2003.
11. UV MP HC to GPM 12 Feb 1869.
12. Probably the Città di Milano in the Via Cerretani, said in a Baedeker guide of 1882 to be 'patronised by English visitors'.
13. RBGE BC HC to JHB 4/250 19 Apr 1868.
14. Cleghorn, 1869c.
15. A brief account of this visit is given in Cleghorn, 1869c.
16. UV MP Mabel Cleghorn to Mrs Marsh 11 Feb 1869.

Chapter 11

1. See RBGE BC 4/251–3, HC to JHB May 1868.
2. See ODNB. It is worth noting the extent and intricacy of the Scottish web: Fergusson was a member of a well-known improving Ayrshire, land-owning family whose great great uncle Sir Adam Fergusson had attended Hope's lectures in 1763; his wife Lady Edith was the younger daughter of Lord Dalhousie, and Fergusson would later be Governor, successively, of South Australia, New Zealand and (from 1880 to 1885) Bombay.
3. Cleghorn, 1869c.
4. RBGE BC 4/251 HC to JHB 2 May 1868.
5. The Glasgow story is told in letters to J.D. Hooker KDC 82/242, 243, 157/198–206, CAD–COL no. 97 and to Bentham, KBC 2/567, between 23 May 1868 and 7 Feb 1869.
6. Cleghorn, 1868c.
7. *Dundee Courier* 13 July 1868.
8. Hooker, 1869.
9. Endersby, 2008: 267.
10. Cleghorn, 1869b.
11. *Norfolk Chronicle* 29 Aug 1868 p 5.
12. BL APAC L/E/5/70 no. 63, report by Brandis of 1872 on training since 1867.
13. BL APAC L/E/5/70 no. 62, report on 1871 candidates, and see Phillimore, 1871: pp 358–9, 406–11.
14. KDC 157/208 HC to JDH 13 Oct 1870: he was to use two newly published textbooks: Daniel Oliver's *First Book of Indian Botany* and Hooker's *Student's Flora of the British Islands*, and asked for Hooker's help in getting a set of Henslow's published teaching diagrams.
15. Phillimore, 1871: 416.
16. In January 1871, after he had returned to Magdalen College Oxford, Cleghorn recommended Gamble to Balfour as a 'good working [i.e. active] member' of the Botanical Society. RBGE BC (A–C) HC to JHB 21 Jan 1871. For biographical information see Hill, 1926.
17. Another link, given Cleghorn's interest in such events: neither of Cleghorn's two copies of the book has a dedicatory inscription from Gamble.
18. Sir George King, quoted in Hill, 1926.
19. Published in seven parts, 1915–25, completed by C.E.C. Fischer 1928–36.
20. In 1879 this hotel was run by Miss Sommers: the rate for bed, breakfast and dinner was 12 shillings, a modest sum for such a central location that must have appealed to Cleghorn's Scottish thrift.
21. Linnean Society Ms SP 215.
22. BL APAC L/PWD/8/129.
23. Taylor, 1895.
24. Colonel George Pearson had been based in Nancy, in charge of the British students for India 1873–84 (see Lubbock, 1885).
25. BL APAC L/PWD/8/129.
26. See KDC 82/249–254.
27. KDC 157/210 HC to JDH 27 Mar 1871.

Chapter 12

1. Receipts, and annual accounts of his factor C.S. Grace, StA CP Box 10.
2. Angela Howe, Director British Golf Museum pers. comm.; receipts for his membership fee of £5 1 shilling, and thereafter an annual £1 are at StA CP 10/7/9 and 10/10/12.
3. Cleghorn, 1885b.
4. Cleghorn, 1878.
5. Cleghorn, 1887b.
6. StA CP 9/9.
7. NAS Cupar Sherriff Court Inventories SC20/50/73 – will 15 Jul 1895 pp 1389–93; inventory 29 Jul 1895 pp 541–50.
8. In 1863 the moveable assets in Patrick Cleghorn's will were valued at £7942/8/9.
9. Mainly from the relevant entries in ODNB.
10. StA CP 9/9/42.
11. For half-a-year on 1868 this amounted to £11/12/1 – StA CP 10/6/29.
12. This was still based on the ancient system of a cash equivalent for the value of the crops raised on an estate, so varied annually – in 1869 Cleghorn paid £90/15/11½, and in 1874 £87/14/3 – StA CP 10/3/10 and 10/12/28.
13. E.g. in 1868 the Heritors voted to pay £7/14/9 – StA CP 10/6/33.
14. RBGE BC 4/244 HC to JHB 30 Jan 1867.
15. Originally considered by Balfour for sending to India as a forester, RBGE BC 4/211, 212, Dec 1863/Jan 1864, he appears to have been poached by Cleghorn for his own use.
16. RBGE BC 4/242 HC to JHB 4 Nov 1866.
17. StA CP 10/4/15, 10/5/43.
18. £12/19/6 StA CP 10/6/30.
19. Anon, 1892.
20. See Clark, 1992: plate 22, cf. Ordnance Survey six-inch-to-the-mile maps of 1854 and 1894 available online through National Library of Scotland, and modern images from Google Earth.

ENDNOTES

21. Available on www.scotlandspeople.gov.uk.
22. Hajducki *et al.*, 2009.
23. *Fife Herald* 18 July 1883.
24. Hajducki *et al.*, 2008, 2009.
25. Hajducki *et al.*, 2009.
26. Anon, 1892.

Chapter 13

1. Information from the Society's website www.rhass.org.uk, and the Society's *Transactions* for Cleghorn's period.
2. *Dundee Courier* 20 Jan 1870.
3. *Transactions 1870–71* 3(ser. 4) Appendix A: 78–9. 1871.
4. *Transactions 1870–71* 3(ser. 4) Appendix D: 66–8. 1871.
5. Cleghorn, 1885b.
6. Information from the Society's *Transactions* and *Proceedings*, and their unpublished Minute Books at RBGE.
7. Cleghorn, 1870b.
8. Cleghorn, 1870e.
9. He also wrote to Hooker on this subject KDC 82/248 HC to JDH 23 Oct 1869.
10. Published in six volumes 1867–72; these include references to 12 of Cleghorn's papers; two further volumes covering 1864–73 were published in 1877 and 1879, with references to 10 further Cleghorn papers.
11. NHM, Carruthers Correspondence DF400–499.
12. Cleghorn, 1883a.
13. RBGE BSE Minutes 3 Mar 1884.
14. RBGE BSE Minutes 3 May 1881.
15. RBGE BSE Minutes 13 Mar 1890.
16. RBGE BSE Minutes Feb 1871.
17. Cleghorn, 1868d.
18. Cleghorn, 1873f.
19. Cleghorn, 1876.
20. Cleghorn, 1883b.
21. Cleghorn, 1889b.
22. Cleghorn, 1889a – in 1861 Cleghorn had read a paper about Traill's garden at Russelcondah, Orissa.
23. Cleghorn, 1886b.
24. Cleghorn, 1870a, c.
25. Martius was President of the Royal Botanical Society of Ratisbon [Regensburg], which had elected Cleghorn an honorary member, the elaborate diploma for which amused Cleghorn's father when it arrived at Stravithie: 'what is the name of your Correspondent who writes to you of your German Diploma – no body here can read it – are we to understand that you are now an L.L.D., or what?' StA CP 2/10/2 PC to HC 17 Jan 1859.
26. Cleghorn, 1886b.
27. Cleghorn, 1889c.
28. RBGE: Minutes Botanical Society Club 1836–1906, 25 Jul 1890.
29. For history see 'The Scottish Arboricultural Society', Anon, *Transactions of the Scottish Arboricultural Society* 11: 114 (+ fold-out summary history), 1885; and 'The Royal Scottish Arboricultural Society, 1854–1904', Anon, *Transactions* 18: 5–19. 1905.
30. SAS Laws, amended to 5th November 1862, copy bound into incomplete RBGE copy of Vol. 1.
31. An invention of James Kay, forester to the Marquess of Bute.
32. Though not in its own *Transactions*, the pamphlets were printed by the Society's printers McFarlane & Erskine: RBGE has a bound set and loose copies of many.
33. Summary in *Gardeners' Chronicle* 6 Aug 1887: 170–1; fuller account in pamphlet, copy at RBGE.
34. Summary in *Gardeners' Chronicle* 20 Aug 1892: 220, pamphlet, lacking wrappers, in bound set of Excursion Accounts at RBGE.
35. Cleghorn, 1875a.
36. Cleghorn 1875a: 2.
37. Cleghorn, 1875c.
38. Cleghorn, 1875b.
39. *Transactions* 11: foldout on history of the Society following p. 114.
40. See British Association Annual Reports for Liverpool 1870, Edinburgh 1871 and Brighton 1872 meetings.
41. Balfour *et al.*, 1874.
42. Cleghorn, 1885a.
43. Cleghorn, 1887b.

Chapter 14

1. StA Senatus Minutes 18 Jan 1885; this must have pleased him as his grandfather had been awarded the same honour in 1827.
2. Cleghorn, 1875d.
3. Cleghorn, 1879.
4. Cleghorn appears to have met Huxley in 1871, when 'Darwin's bulldog' with his wife, spent a summer holiday in St Andrews, NHM HC to William Carruthers 12 Aug 1871.
5. See ODNB and Raven, 1951 for an interesting oration given in Aberdeen on the occasion of Robertson Smith's centenary in 1946.
6. Cleghorn, 1885b.
7. For example, since his 1872 Presidential Address (Cleghorn, 1875a).
8. Stubbs, 1998.
9. Information from Official Catalogue, Anon 1884, contemporary newspaper articles, and an account in the SAS *Transactions* Anon 11: 68–113, 1885.
10. An International Fisheries Exhibition had been held in Edinburgh in 1882, in which the Highland Society was heavily involved, and may have been a spur if not the inspiration.
11. Taylor, 1895.
12. *Glasgow Herald* 1 July 1884.
13. The publications from the Government of India, of which there was a separate catalogue, were specially bound in green, with a gold-lettered scarlet label on the front board; at the end of the Exhibition these were presented to the Museum where they were later incorporated into the Cleghorn Memorial Library, and eventually came to RBGE in 1940.
14. Cleghorn's St Andrews friend Professor W.C. McIntosh appears to have provided notes on these – Gunther, 1977: 78.
15. Most probably from Trichinopoly and acquired by his father, the 8th Marquess, while Governor of Madras.
16. This covered the Nilgiris as well as the areas of Bellary and the Northern Circars as far north as Ganjam.
17. Chris O'Brien http://www.edinburghtrams.info/ediburghs_1884_electric tram, consulted 31 Dec 2014.
18. *Leeds Mercury* 2 July 1884.
19. Birdwood in Michael, 1884: 7.
20. See ODNB; in 1869 HC had heard Birdwood read a paper on *Balsamodendron* to the Linnean Society – see KDC 157/206 HC to JDH 7 Feb 1869. There were connections by marriage as one of Cleghorn's first cousins, Thomas McKenzie, was married to Clara Birdwood a cousin of George Birdwood.
21. Dossal, 2012.
22. There was reference to it in *Friend of India* 1865, but this source was close to Cleghorn himself, and after he retired his own strong, if not self-important, belief in its influence seems to have grown and had been prominently expressed in his 'Forests' in 9th edition of the *Encyclopaedia Britannica*, with which Birdwood must have been familiar, and which might have influenced his view.
23. Birdwood, 1901.

ENDNOTES

24. Their Report was included as an appendix to the 1885 Select Committee Report; the criticisms of Scottish forestry included: single-aged stands, with no regeneration due to grazing by cattle and sheep; the large decline in forests between 1812 and 1872, though now being redressed; the huge acreage of wasteland that could profitably be planted.
25. Available through Parliamentary Papers online.
26. Vol. 11: 119–54, 1885; 11: 315–63, 1887; 12: 104–55, 1888.
27. RBGE BC 4/245 HC to JHB 28 May 1867.
28. On 30 June 1868 Cleghorn told JDH that he had 'compounded for life' at a cost of £1 15 shillings Kew JDH Corr. CAD-COL no. 97.
29. Information from the annually printed *Edinburgh University Calendar*.
30. McIntosh, 1895.
31. Purchased by the City, but transferred to the Crown the following year.
32. ODNB.
33. Edinburgh University Court Minutes vol. 9 1886/91 p. 452.
34. Edinburgh University Calendar 1895/6 p. 761.
35. *Dundee Courier* 26 March and 17 April 1869.
36. Shafe, 1982.
37. For a biography of McIntosh see Gunther, 1977.
38. McIntosh, 1895.
39. Macdonald, 1984.
40. Mitchell, 2014.
41. Gunther, 1977: 73
42. StA Court Minutes 6 Apr 1891.
43. StA Court Minutes 16 Oct 1891.
44. StA Court Minutes 16 Apr 1892.
45. Cousin once-removed of the first and last Marquess. The Marquess had no sons, so the title died out but his earldom passed to two of his first cousins in succession; the one under discussion was the son of the younger of these. With these changes the main focus of the family had passed to the Maule estates in Angus – to Panmure near Carnoustie (the house since demolished) and to Brechin.
46. BA 1885 Report p. lxxiii.
47. His father's Penicuik joinery firm had refitted Stravithie for Cleghorn in 1865/6.
48. BA 1887 Report, p lxxxvii.
49. Carruthers Correspondence, NHM.
50. Cleland *et al.*, 1887; McIntosh *et al.*, 1889; the full reports and numerous related papers were published elsewhere.
51. In Anon 1889: 201.
52. McIntosh, 1895.
53. *Dundee Courier* 24 November 1870.
54. Fifth, Sixth and Seventh Annual Reports of the Fife & Kinross District Board of Lunacy. Cupar: Fifeshire Journal Office, 1871, 1872, 1873. Tuke, later knighted, was, like Cleghorn, a contributor to the Ninth Edition of the *Encyclopaedia Britannica*.
55. *Dundee Courier* 11 January 1887.
56. *Fife Herald* 15 Feb 1888.
57. *Scotsman* 17 May 1895.
58. *Dundee Courier* 28 Jan 1896.
59. Fife County records FC/CS/7/3/5/24 (supplied by Andrew Dowsey); Anstruther was a great nephew of the Ceylon botanist Anna Maria Walker.
60. *Dundee Courier* 4 Mar 1873
61. *Fife Herald* 26 Feb 1874.
62. *Fife Herald* 2 May 1872.
63. Though the compensation was clearly inadequate this drastic treatment worked and the disease was eradicated in Britain by 1898.
64. NAS JP 28/2/4 1872–93.
65. *Fife Herald* 16 January 1889, *Dundee Courier* 5 and 8 April 1893.
66. UV MP HC to G.P. Marsh 3 Feb 1875.
67. StA CP 9/9/10.
68. BL Mss A177. SPCK missionary papers also include two letters from another of Cleghorn's evangelical friends Col Michael John Rowlandson (c 1805–1894), a staunch member of the CMS who after retiring from the Madras Army taught languages at Addiscombe.
69. Copy made 11 March 2010 (no 4546590 CE), kindly provided by Alex May of ODNB.
70. Dr A.C. Noltie, pers. comm.
71. Taylor, 1895: 446.
72. RBGE BC 4/222 HC to JHB 22 Nov 1864.
73. A Sprot family tree has recently become available on the internet: https://home.comcast.net/~BrodieofCaithness/McCormick-Hill-Cleghorn-Sprot.pdf, consulted 9 Feb 2015.

Chapter 15

1. Anon, 1889.
2. George Richmond was paid 30 guineas for a portrait drawing of Charlotte Brontë commissioned by her publisher in 1850, but Richmond was of an earlier generation, was much better known and produced much more highly finished work than Wirgman's rapid sketches – I am grateful to Paul Cox of the National Portrait Gallery for a discussion on the subject.
3. StA CP 9/8/5/1.
4. Swinney, 2013.
5. Matheson had previously worked at RBGE for Balfour, building a new classroom and converting the old one into the Museum; in 1854 the iconic temperate palm-house; and in 1872 the extension to the herbarium to house the Botanical Society's library).
6. Several books from his library were purchased, presumably after Archer's death, by Cleghorn.
7. With the rest of the monumental cast collection this was dumped in Granton docks in the 1960s when the atrium was refitted with then-fashionable travertine goldfish ponds.
8. I am grateful to Mark Glancy, librarian, National Museums of Scotland, for much of the information in this and the following sections.
9. StA CP 9/8/5/1–4.
10. Emphasis added: *Transactions & Proceedings of the Botanical Society of Edinburgh* 19: 41, 1891.
11. *Edinburgh Evening News*, 13 August 1896.
12. Anon, 1897.
13. There is nothing in the Museum's Disposal Board minutes for 1837–41, M. Glancy pers. comm.
14. Matthew, 1987: 77.
15. Copy of letter in NMS archives.
16. Fletcher & Brown, 1970.
17. Matthew 1987: 81.
18. Mathew, 1987: 77.
19. Edinburgh University Calendar for 1896/7, p. 727.
20. Anon, 1918–23.
21. See Mathew, 1987: 70.

Post Script

1. This included a significant mistake perpetrated in my 1998 book – that Cleghorn's inspiration was Joseph Hooker, as he himself claimed in confused old age (Anon, 1889), whereas it is now known to have been *William* Hooker who so advised him in 1842.
2. Subbarayalu, 2014.

Appendix 1

Bibliography of Cleghorn's Publications

As with similar catalogues of the publications of Alexander Gibson and Robert Wight, EIC surgeons who also worked in the field of economic botany, the making of a complete bibliography is an impossible task, since they published widely, often in obscure works and sometimes anonymously. In the present compilation many reports submitted to, and printed by, the Fort St George (Madras) Government will have been omitted, as, will articles that Cleghorn almost certainly published in newspapers (often anonymously or under pseudonyms) including the government *Gazettes* in Madras and the Punjab, the *Friend of India*, the *Madras Athenaeum*, and also evangelical and missionary publications (especially the *Madras Christian Herald*). A starting point was the list of partial publications compiled by himself for the testimonials published when he applied for the Glasgow botanical chair in 1868. Relevant material, however, continues to come to light, and recently, while looking through the RBGE copy of the 1874 edition of G.P. Marsh's *Man And Nature*, the first page of a manuscript 'Contents' list of what must have been a set of his own bound reprints, no longer extant, was discovered. This is not in Cleghorn's own hand, and does not give full publication details, but includes several important, otherwise unknown items. Cuttings and annotations on books from his library have also revealed previously unknown works or identified him as author. However, as Cleghorn appears to have been somewhat obsessed with the importance of his work and its dissemination (many papers being published in more than one *avatar*, in several different journals and as a 'separate') the most significant of his publications are likely to have been recorded.

The work has been greatly helped by the collections at RBGE and the Universities of Edinburgh and St Andrews, between which substantial parts of Cleghorn's own library and archive survive. Earlier authors who have attempted a bibliography have not had the privilege of access to these collections over a prolonged period, and their work has inevitably been somewhat sparse: the Royal Society Catalogue (1867, 1877, 1915) listed 23 papers, Grove (1992) noted 19 journal articles, Stafleu & Mennega (1997) included eight, primarily taxonomic works, Subbarayalu (2014) 18 articles, and Das (2005) a mere five papers in addition to the two books.

Many of Cleghorn's memoranda and reports for his period of employment with the Government of India were reprinted in the East India (Forest Conservancy) 1871 Parliamentary Papers (*A Selection of Despatches ... on Forest Conservancy ... [from] the 21*st *day of May 1862 ... [to] August 1871*, ed. C.B. Phillimore), and these have not been itemised here.

Items first read at meetings of the Botanical Society of Edinburgh were often not published until several years later in the Society's *Transactions* but in many, probably most, cases the material was published close to the date of reading in contemporary newspapers or journals. These have been sought where possible, but not exhaustively; for which reason, in cases where there is a large gap, the date of reading is given after the publication date (with the likelihood of an earlier date of first publication being discovered at some point).

Cleghorn, H.F.C. (1839). What are the processes by which the reparation of wounds is effected, and how far are they to be regarded as the results of inflammation. Unpublished dissertation for the Royal Medical Society, Edinburgh.

Cleghorn, H.F.C. (1841). On the Reparation of Wounds and Regeneration of Lost Parts. Doctoral Thesis.
Note. It was a university requirement that MD theses were printed, but no copy is extant in the University of Edinburgh library.

Cleghorn, Dr. (1844) [read 12 i 1843]. Extract from a letter on the scenery and vegetation of Madeira. *Transactions of the Botanical Society [of Edinburgh]* 1: 197–201.

Note 1. The dates of all BSE publications are uncertain – they are taken from the title pages of the volumes but parts are likely to have been issued prior to this, though in the absence of surviving wrappers there is no way of knowing – these are likely to have been fairly shortly after the date of the reading of the paper, for which reason this date is also cited.

Note 2. On the single surviving MS 'Contents' page of what appears to have been a bound set of Cleghorn's publications is given 'On Vegetation of Madeira. Augt. 1842', which has not been traced; it is likely to have been a version of this letter.

[Cleghorn, H.F.C.] (1850a). "List of plants found in the island of Rathlin," by Miss C. Gage (read, with other notes on Rathlin by Cleghorn). *Annals & Magazine of Natural History* 5 (ser. 2): 145–6.
Note. Read to BSE 10 Jan 1850.

Cleghorn, Hugh F.C. (1850b). On the hedge plants of India, and the conditions which adapt them for special purposes and particular localities. *Annals & Magazine of Natural History* 6 (ser. 2): 233–250.
Note. Also issued as separate, pp [1–] 2–18. Read to BSE on 13 xii 1849 and printed in *Transactions of the Botanical Society [of Edinburgh]* 4: 83–100, 1853, where it was stated that this (and other articles) had first been published in the *Annals*. This paper was also read, on 1 viii 1850, to the British Association in Edinburgh, and an abstract published in *British Association Report for 1850*, pt 2: 113, 1851.

Cleghorn, H. (1850c). [Culture of American cotton at Cuddoor], in 'Proceedings of Societies – Botanical Society of Edinburgh'. *Botanical Gazette* 2: 273–4.

Note. Extracts of a letter from Captain Onslow, Superintendent of the Nuggur Division, read by Cleghorn to BSE on 11 vii 1850.

Cleghorn, Hugh F.C. (1851a) [read 9 i 1851]. Biographical notice of the Rev. Dr Rottler.

Note 1. Printed cutting in BSE archive, probably from *North British Agriculturist*: summary of a paper read to BSE.

Note 2. Accounts of the BSE's meetings ('proceedings') were printed in Annual Reports for the period 1836–46, and in a volume of *Proceedings* in 1855 and 1856. In the intervening period the Society did not publish 'proceedings', but these appear to have been published in the *North British Agriculturist*: a file of cuttings of these survives from the period May 1846 to January 1855 in the BSE archives at RBGE. Some, however, are missing; fortunately summaries were sometimes also reported in *The Phytologist* and *Botanical Gazette*.

Cleghorn, Dr. (1851b) [read 13 ii 1851]. Notice of several new Indian plants. *Phytologist* 4: 84–5.

Note. Summary of paper read to BSE; printed cutting in BSE archive, probably from *North British Agriculturist*.

Cleghorn, H. (1851c) [read 19 vi 1851]. On the Government teak plantations of Mysore and Malabar. *Phytologist* 4: 227–8.

Note. Summary of paper read to BSE; printed cutting in BSE archive, probably from *North British Agriculturist*.

Cleghorn, H. (1851d). On the grass-cloth (chu ma) of India, in *Report of the Twentieth Meeting of the British Association for the Advancement of Science; held at Edinburgh in July and August 1850*. Notices and Abstracts of Communications, p. 112. London: John Murray.

Note. Paper read to the British Association on 6 viii 1850; previously read to BSE on 11 vii 1850, of which an abstract published in the *Botanical Gazette* 2: 275, x 1850. Cleghorn provided further information to the anonymous author of a bigger article on the topic in *Chambers's Edinburgh Journal* of Oct 1851: 214–6.

Cleghorn, H.F.C. (1851e). Remarks on Calysaccion longifolium, Wight [with figure]. *Pharmaceutical Journal* 10: 597–598.

Note. Also issued as separate, pp [1–]2 (copy at RBGE).

[Cleghorn, H. (1852a)] *Syllabus of Lectures on Botany* [given at Madras Medical College], pp [1–]2–27.

Note. Pamphlet, printed in Madras, annotated by Cleghorn; pp 7–27 relate to lectures on Materia Medica and Therapeutics (copy at RBGE).

C[leghorn], H. (1852b). Notes on the medical and economical resources of the forests of India. *Indian Journal of Arts, Sciences, & Manufactures* 1: 663–6.

Note. Also issued as separate with an article by Robert Christison on 'a new variety of gamboge', based on material supplied by Cleghorn, reprinted from *Pharmaceutical Journal* 6: 60–9, 1846, with a lithographic copy by 'Sinclair' of the woodcut in the Christison article (copy at RBGE).

Cleghorn, Hugh, Royle, [J.] Forbes, Baird Smith, R. & Strachey, R. (1852). Report of the committee appointed by the British Association to consider the probable effects in an œconomical and physical point of view of the destruction of tropical forests, in *Report of the Twenty-First Meeting of the British Association for the Advancement of Science; held at Ipswich in July 1851*, pp 78–102. London: John Murray.

Note. A summary read by Cleghorn to 'D Section' on 7 July 1851 and printed in *The Phytologist* 4: 290–2, 1851. Whole report reprinted in *Journal of the Agricultural & Horticultural Society of India* 8 (correspondence & selections): 118–149, 1854.

C[leghorn], H. (1853a). Hortus Madraspatensis: catalogue of plants, indigenous and naturalized, in the Agri-Horticultural Society's Gardens, Madras. Madras: for the Society by the American Mission Press.

Note. Copy at RBGE with dedication from Cleghorn to George Lawson.

Cleghorn, H.F.C. (1853b). Report of the Committee of the Agri-Horticultural Society of Madras, for the year, 1853, pp [1–]2–8. [Madras].

Note. Cleghorn's own copy at RBGE in a bound volume of the transactions of the Agri-Horticultural Societies of both Madras and Western India.

Reid, F.A. & Cleghorn, H.F.C. (1853c). Valedictory address to Dr. Wight on his departure from India (March 3^d, 1853), in *Report of the Committee of the Agri-Horticultural Society of Madras, for the year, 1853*, pp 9–10. [Madras].

Cleghorn, H. (1854a). Note on the Ægle marmelos. *Indian Annals of Medical Science* 2: 222–4, with plate.

Note. Also issued as separate, pp [1–]2–3, with plate (copy at RBGE).

Cleghorn, H.F.C. [1854b]. Directions for Collecting and Preserving Botanical Specimens. Compiled for the use of the Botanical Class. Madras: Oriental Press.

Note. Undated pamphlet based on a paper on plant-drying in *Third Annual Report and Proceedings of the Botanical Society of Edinburgh* pp 80–89, 1840, and another on the preservation of fruits in salt by R. Christison from the *Second Report* pp 64–6, 1840.

[Cleghorn, H. (1854c)]. *Syllabus of Lectures on Botany* [given at Madras Medical College], pp [1–]2–18. Pamphlet, printed in Madras.

Note. Copy annotated by Cleghorn at RBGE.

[Cleghorn, H. (1854d)]. 'Desiderata List of Raw Products, Madras Exhibition'.

Note. Listed on the single surviving MS 'Contents' page of what appears to have been a bound set of Cleghorn's publications, but no copy has been seen.

Cleghorn, H. (1855a). Note on the varieties of "Chiretta" used in the hospitals of Southern India. *Indian Annals of Medical Science* 2: 270–2, plates 1–3 [1 and 2 'Govindoo, del'; 3 'A. Hunter; all 'Dumphy, Lith'].

Note. Part of a series of papers entitled 'Report on Febrifuges', including papers on Cinchona by Hugh Falconer and Thomas Anderson. Also issued as a separate, pp [1–]2–3, plates 1–3 (copy at RBGE), and printed in *Edinburgh New Philosophical Journal* 3(n.s.): 364–5, 1856.

Cleghorn, H.F.C. [1855b]. *Memorandum on Indian Grasses, and the mode of cultivating and improving pasture lands*, pp [1–]2–3 + plate. Madras: Fort St George Gazette Press.

Note. A copy of this undated pamphlet (with a lithographed plate of *Cynodon dactylon*) was sent by Cleghorn to J.H. Balfour in a letter dated 26 January 1856 (RBGE BC no. 179). It had been submitted to the Military Department on 6 November 1855 (see copy of Memorandum in StA CP 9/4 f. 70).

Cleghorn, H. (1856a). Suggestions relative to the establishment of soldiers' gardens in India. *Indian Journal of Arts, Sciences, & Manufactures.* 2nd series 1: 9–13.
Note. Memo for the Madras Government, probably first published in 1855; also reprinted in *Forests & Gardens* pp 374–8, 1861.

[Cleghorn, H]. (1856b). Report on chemical and pharmaceutical processes and products generally, in *Madras Exhibition of Raw Products, Arts, and Manufactures of Southern India, 1855, Reports of the Juries on the Subjects in the Thirty Classes into which the Exhibitions was Divided*, pp 8–11, [(ed.) ?E. Balfour]. Madras: For the General Committee of the Madras Exhibition, by Messrs. Pharoah & Co.
Note. A summary of this and the following reports (presumably by J.H. Balfour) published in *Edinburgh New Philosophical Journal* 3(n.s.) 158–9. Jan 1856 as 'Report on some of the Products contributed to the Madras Exhibition in 1855'.

Cleghorn, H. (1856c). Report on substances used for food, in *Madras Exhibition of Raw Products, Arts, and Manufactures of Southern India, 1855, Reports of the Juries on the Subjects in the Thirty Classes into which the Exhibitions was Divided*, pp 12–19, [(ed.) ?E. Balfour]. Madras: For the General Committee of the Madras Exhibition, by Messrs. Pharoah & Co.

[Cleghorn, H]. (1856d). Report on gums and resins, in *Madras Exhibition of Raw Products, Arts, and Manufactures of Southern India, 1855, Reports of the Juries on the Subjects in the Thirty Classes into which the Exhibitions was Divided*, pp 20–25, [(ed.) ?E. Balfour]. Madras: For the General Committee of the Madras Exhibition, by Messrs. Pharoah & Co.

[Cleghorn, H]. (1856e). Report on dyes and colours, in *Madras Exhibition of Raw Products, Arts, and Manufactures of Southern India, 1855, Reports of the Juries on the Subjects in the Thirty Classes into which the Exhibitions was Divided*, pp 43–48, [(ed.) ?E. Balfour]. Madras: For the General Committee of the Madras Exhibition, by Messrs. Pharoah & Co.
Note. Includes description of 'process of dyeing cloth practised by Bala Chetty at Madras'.

[Cleghorn, H]. (1856f). Report on tanning materials, in *Madras Exhibition of Raw Products, Arts, and Manufactures of Southern India, 1855, Reports of the Juries on the Subjects in the Thirty Classes into which the Exhibitions was Divided*, pp 49–50, [(ed.) ?E. Balfour]. Madras: For the General Committee of the Madras Exhibition, by Messrs. Pharoah & Co.

[Cleghorn, H]. (1856g). Report on cellular substances, in *Madras Exhibition of Raw Products, Arts, and Manufactures of Southern India, 1855, Reports of the Juries on the Subjects in the Thirty Classes into which the Exhibitions was Divided*, pp 59–60, [(ed.) ?E. Balfour]. Madras: For the General Committee of the Madras Exhibition, by Messrs. Pharoah & Co.

[Cleghorn, H]. (1856h). Report on timber and ornamental woods, in *Madras Exhibition of Raw Products, Arts, and Manufactures of Southern India, 1855, Reports of the Juries on the Subjects in the Thirty Classes into which the Exhibitions was Divided*, pp 60–73, [(ed.) ?E. Balfour]. Madras: For the General Committee of the Madras Exhibition, by Messrs. Pharoah & Co.
Note 1. Includes a 'Classified list of woods, native, or grown in the Madras Presidency'.
Note 2. In an anonymous review of three publications relating to the 1855 Madras Exhibition in the *Calcutta Review* 26: 282, iii 1856, is reference to 'Dr Cleghorn's [Jury] report on timber'; as he appears to have reviewed Hooker & Thomson's *Flora Indica* for the *Calcutta Review* (see below) it is not inconceivable that the review of the Madras Exhibition publications was submitted (if not written) by Cleghorn.

[Cleghorn, H]. (1856i). Report on animal substances, in *Madras Exhibition of Raw Products, Arts, and Manufactures of Southern India, 1855, Reports of the Juries on the Subjects in the Thirty Classes into which the Exhibitions was Divided*, pp 73–75, [(ed.) ?E. Balfour]. Madras: For the General Committee of the Madras Exhibition, by Messrs. Pharoah & Co.

Cleghorn, H. (1856j). Tabular view of the woods used for furniture in Madras, in *Madras Exhibition of Raw Products, Arts, and Manufactures of Southern India, 1855, Reports of the Juries on the Subjects in the Thirty Classes into which the Exhibitions was Divided*, p. 123, [(ed.) ?E. Balfour]. Madras: For the General Committee of the Madras Exhibition, by Messrs. Pharoah & Co.

Cleghorn, H. (1856k). Notulae Botanicae No. I. On the sand-binding plants of the Madras beach. *Journal of the Agricultural & Horticultural Society of India* 9 (part 1, no II): 174–7, with one plate [*Spinifex squarrosus*].
Note 1. This version was also issued as a separate, pp [1–]2–4 + plate.
Note 2. Read to Madras Literary Society on 11 xii 1856 and printed in *Madras Journal of Literature and Science* 1(n.s.): 85–90 + 4 plates, 1857; the latter version also issued as separate, pp 1–6, tt. 2–5; an abstract of it (lacking plates) published in *Hooker's Journal of Botany & Kew Garden Miscellany* 8: 52–4, 1856.

[Cleghorn, H.] (1856l). Art. III. – 1. Flora Indica ... [Review of Hooker & Thomson's *Flora Indica* and other works]. *Calcutta Review* 26: 355–372.
Note. That this is by Cleghorn is proved by a reference to it in a letter to J.H. Balfour of 26 i 1856 and by its appearance on the MS 'Contents' page of what appears to have been a bound set of Cleghorn's publications. The other works reviewed are J.D. Hooker's *Illustrations of Himalayan Plants* and *Himalayan Journals*; Thomas Thomson's *Western Himalaya and Thibet*; Alphonse de Candolle's *Géographie Botanique*; the 1854 Report of the Peradeniya Garden; [Cleghorn's] *Hortus Madraspatensis*, and the Ootacamund Garden Reports of 1853/54 and 1855. It is likely that these other reviews are also by Cleghorn: although reviewing his own works would have required a certain disingenuousness, the wording is not immodest and does not preclude it; the reviews were clearly written by someone with South Indian interests who was in contact with Robert Wight.

Cleghorn, H. (1856m). *General Index of the Plants Described and Figured in Dr. Wight's Work Entitled "Icones Plantarum Indiae Orientalis".* [Errata page], [i–]ii, [1–]2–78. Madras: H. Smith, Fort St George Gazette Press.
Note. Reprinted (with errata incorporated) in 1921 in an edition of 55 copies by Bernard Quaritch (London) (copy in Hunt Botanical Institute).

Cleghorn, Dr. (1856n) [read 10 i 1856]. On the gutta percha plant of India. *Proceedings of the Botanical Society [of Edinburgh] for 1856*: 10–11.
Note. Also printed in *Edinburgh New Philosophical Journal* 3(n.s.): 353–4.

CLEGHORN'S PUBLICATIONS

Cleghorn, Dr. (1856o) [read 8 iii 1855]. Extracts from a letter from Dr Cleghorn, on the discovery by Major Cotton of the gutta percha in Malabar. *Proceedings of the Botanical Society [of Edinburgh] for 1856*: 16.

Elliot, W. & Cleghorn, H. (1856). *List of Articles Required for the Madras Exhibition of 1857.* [Madras].
Note. Pamphlet, copy bound with other papers on the 1857 exhibition at RBGE.

[Cleghorn, Dr.] (1857a). Proposed establishment (or restoration) of an agricultural and horticultural garden at Bangalore, in the Madras Presidency. *Hooker's Journal of Botany & Kew Garden Miscellany* 9: 24–8.
Note. The proposal for the restoration of the Lal Bagh was also printed in *Forests & Gardens* pp 330–334, 1861.

Cleghorn, H. (1857b). [Letter – on the need for a library with scientific reference works in Madras]. *Madras Journal of Literature & Science* 1 (n.s.): 116.

Cleghorn, H. (1857c). Effects of the gale of 20th November 1856 at the Agri-Horticultural Society's Garden, Madras. *Madras Journal of Literature & Science* 1 (n.s.): 123–5.

[Cleghorn, H.] (1857d). On the introduction of the Cinchona or Bark-tree. *Madras Journal of Literature & Science* 1 (n.s.): 208–220.
Note. A preface to a translation of an article by De Vriese on the introduction of cinchona into the Dutch East Indies, 'a general review of the history of this valuable drug' (including a long quote from Weddell, summarising the work of Humboldt, Bonpland, Lambert & Pöppig), and though anonymous must be by Cleghorn.

Cleghorn, H. (1858a). Notice regarding the Moreton Bay Chesnut [sic], introduced into the Government Garden at Bangalore. *Journal of the Agricultural and Horticultural Society of India* 10: 116.

Cleghorn, H. (1858b). *Memorandum Upon the Pauchontee, or Indian Gutta Tree of the Western Coast*, pp [0–]2–10, tt. 1–8. Madras: Fort Saint George Gazette Press.
Note. Report read at BSE meeting 11 xi 1858, of which an abstract printed in *Edinburgh New Philosophical Journal* 9(n.s.): 325, 1859, and in *Transactions of the Botanical Society [of Edinburgh]* 6: 148–50, 1860.

Cleghorn, H. (1858c). [*First Annual Report on Forest Operations in the Madras Presidency*], to the Secretary to Government, Revenue Department, Fort St George; dated Mangalore 1st May 1858, No. 337. [Pp 1–11] + lithographed map titled 'Rough Survey of the Teak Forest of Goond, by Mr S. Muller. Asst. Conservator of Forests'. [Madras: Fort St George Press].
Note. Reprinted (without map) in *Forests & Gardens* pp [1–]2–23, 1861. Summary printed as 'Report on the Conservation of Forests in India' in *Transactions of the Botanical Society [of Edinburgh]* 6: 237–41, 1860; and in *Edinburgh New Philosophical Journal* 10(n.s.): 147–50, 1859.

Cleghorn, H. (1858d). Government Horticultural Garden, Ootacamund [Report for 1856–7]. *Madras Journal of Literature and Science* 2 (n.s.): 297–303.
Note. A summary; a fuller version given in *Forests & Gardens* pp 359–66, 1861.

[Cleghorn, H.] (1858e). Obituaries of the Rev. Bernhard Schmid and Charles Drew. *Madras Journal of Literature and Science* 3 (n.s.): 143–7.

[Cleghorn, H. & McIvor, W.G.] (1859). *Report on the Government Botanical and Horticultural Gardens, Ootacamund for the year 1858*. Pp. [0–iv], [1–]2–29 + chromolithographed garden plan. Madras: Fort Saint George Gazette Press.
Note. The Report is that of the garden is for 1857; an extract given in *Forests & Gardens* pp 366–370, 1861.

[Cleghorn, H.] (1859). Obituary of J. Forbes Royle, M.D., F.R.S. *Madras Journal of Literature and Science* 4 (n.s.): 145–8.

Cleghorn, H. (1859/60). [Letter, about a visit to the garden of the late Surya Prakasa Row at Aulapilly in the Northern Circars]. *Madras Journal of Literature & Science* 5 (n.s.): 204–5.

Cleghorn, H. (1861a). *The Forests and Gardens of South India*. Pp. [i–]vi–xiv, [1–]2–412. London: W.H. Allen & Co.

Cleghorn, H. (1861b). Expedition to the higher ranges of the Anamalai Hills, Coimbatore, in 1858. *Transactions of the Royal Society of Edinburgh* 22(3): 579–88, tt. XXXI–XXXVII.

Note 1. Also issued as separate with same pagination [Neill & Co., Edinburgh, 1861], and reprinted in *Forests & Gardens* pp 289–302, 1861.
Note 2. On the single surviving MS 'Contents' page of what appears to have been a bound set of Cleghorn's publications the date of the report on the 'Anamallai Excursion is given as 1859, and doubtless an untraced version appeared in that year printed in Madras. Reprinted in *The Anamallais*, by C.R.T. Congreve. Madras, 1924. The original drawings for the plates, by Douglas Hamilton, are in BL (APAC WD 567).

Cleghorn, Dr. (1861c). On the timbers suited for railway sleepers in South India. *Edinburgh New Philosophical Journal* 13 (n.s.): 329–30.
Note. In the 1850s and '60s reports of BSE meetings were published in this journal, often before their appearance in the society's own *Transactions*. This paper was read to BSE 10 i 1861, and printed in their *Transactions* 7: 55–6. 1863.

Cleghorn, Dr. (1861d). On the varieties of mango fruit (Mangifera Indica) in Southern India. *Edinburgh New Philosophical Journal* 14(n.s.): 155–6.
Note. Read to BSE 14 iii 1861, printed in their *Transactions* 7: 111–2. 1863.

Cleghorn, Dr. (1861e). On the species of Dioscorea (yams) occurring in South India. *Edinburgh New Philosophical Journal* 14(n.s.): 156–7.
Note. Read to BSE 11 iv 1861, printed in their *Transactions* 7: 152. 1863.

Cleghorn, H. (1861f). Notes of excursions to the Higher Ranges of the Anamalai Hills, South India, in 1858 and 1859. *Edinburgh New Philosophical Journal* 14 (n.s.): 147–50.

Cleghorn, H. (1861g). List of plants growing in the Bangalore Garden, Mysore [drawn up for Cleghorn by William New]. *Edinburgh New Philosophical Journal* 14 (n.s.): 309–21.
Note. Reprinted in *Forests & Gardens* pp 344–58, 1861; read to BSE 11 Jul 1861, printed in their *Transactions* 7: 223–35, 1863.

Cleghorn, Hugh (1861h). Notes upon the Coco-Nut Tree and its uses. *Edinburgh New Philosophical Journal* 14(2) (n.s.): 173–183, tt. 1–3.
Note. Read to BSE 9 v 1861, printed in their *Transactions* 7: 155–65, 1863; also issued as separate, pp [1–]2–11, tt. 1–3.

CLEGHORN'S PUBLICATIONS

Cleghorn, Hugh (1862a). Extracts from letters [to J.H. Balfour] written during a voyage to Alexandria. *Edinburgh New Philosophical Journal* 15 (n.s.): 130–2.
Note. Report of BSE meeting held 14 xi 1861; reprinted in their *Transactions* 7: 250–2, 1863.

Cleghorn, H. (1862b). Cultivation of Tea. *Madras Journal of Literature & Science* 6 (n.s.): 142–8.
Note. Memo to the Madras Government dated 11 August 1860, first published in *The Public Ledger*, 13 Nov 1860. A summary read to BSE 13 xii 1860, printed in their *Transactions* 7: 30–2, 1863.

Cleghorn, H.C. (1862c). Memoranda on Himalayan rhubarb, and daphne fibre. *Madras Quarterly Journal of Medical Science* 5: 465–472.

Cleghorn, H. (1862d). Extracts from letters from Dr. H.W. Bellew, of the Guide Corps, and Dr. H. Cleghorn, Conservator of Forests, to the Secretary to Government Punjab, regarding the produce of Salep, Asafoetida, &c., in *Report on the Trade and Resources of the Countries on the North-Western Boundary of British India*, ed. R.H. Davies, Appendix XXXIV pp. ccclxxiv-vii. Lahore: Government Press.
Note. The other plants discussed are madder, cumin, koot (*Aucklandia veracosta*) and ekulbeer (*Datisca cannabina*). Cleghorn's annotated copy at RBGE.

Cleghorn, Dr. (1863a) [read 13 xii 1860]. Tea culture in Southern India. *Transactions of the Botanical Society [of Edinburgh]* 7: 30–2.

Cleghorn, Dr. (1863b) [read 13 xii 1860]. On the introduction of cinchona trees (Peruvian bark) into Southern India. *Transactions of the Botanical Society [of Edinburgh]* 7: 33–4.

Cleghorn, H. (1863c) [read 14 xi 1861]. Extracts from letters during a voyage to Cairo. *Transactions of the Botanical Society [of Edinburgh]* 7: 250–2.

Cleghorn, H. (1863d). Remarks on medical missionary work (pp 107–9) and on the Moravian missionaries in Lahoul (pp 260–1) in *Report of the Punjab Missionary Conference held at Lahore in December and January, 1862–63*, ed. by 'A Committee of Compilation'. Lodiana: American Presbyterian Mission Press.

Cleghorn, Dr. (1863e) [read 13 v 1863]. Local Museums in the Punjab [i.e., Lahore, Peshawur, Umritsur, Mooltan, Simla]. *Journal of the Agricultural and Horticultural Society of India* 13 (Proceedings): xxxiii–vi.
Note. Brief abstract published in *Transactions of the Botanical Society [of Edinburgh]* 7: 531, 1863.

Cleghorn, Dr. (1863f) [read 12 ii 1863]. Extracts from Indian letters. *Transactions of the Botanical Society [of Edinburgh]* 7: 500–1.
Note. Also printed in *Edinburgh New Philosophical Journal* 17(n.s.) 323–4. iv 1863.

Cleghorn, Dr. (1863g) [read v 1863]. Communication 'relative to the transport of Rose-cuttings to India from Scotland'. *Transactions of the Botanical Society [of Edinburgh]* 7: 567.

Cleghorn, H. (1863h) [read 9 vii 1863]. Notice of cinchona cultivation on the Neilgherry Hills. *Transactions of the Botanical Society [of Edinburgh]* 7: 585–6.

Cleghorn, H.F.C. (1864a). On some economic plants of India. *Edinburgh New Philosophical Journal* 17(n.s.): 154–5.
Note. Read at BSE meeting 10 xii 1863, later printed in their *Transactions* 8: 63–4. 1866.

Cleghorn, H. (1864b). *Report upon the Forests of the Punjab and the Western Himalaya*. Roorkee: Thomason Civil Engineering College.
Note. Cleghorn's own copy, with numerous manuscript corrections and additions, is at RBGE.

Cleghorn, H. (1865a). Excursion to the valleys of the Giri, Pabur, and Tonse Rivers. *Journal of the Agricultural and Horticultural Society of India* 13: 362–72.

Cleghorn, H. (1865b). Notes on the vegetation of the Sutlej Valley. *Journal of the Agricultural and Horticultural Society of India* 13: 372–91.
Note. Also issued as separate, pp [1–] 2–19.

Cleghorn, H.F.C. (1866a) [read 14 i 1864]. Principal plants of the Sutlej Valley with hill, botanical, and English names; together with approximate elevations, and remarks. *Transactions of the Botanical Society [of Edinburgh]* 8: 77–84.

Cleghorn, H. (1866b) [read 8 vi 1865]. Notes of an excursion from Simla to the valleys of the Giri, Pabur, and Tonse Rivers, tributaries of the Jumna. *Transactions of the Botanical Society [of Edinburgh]* 8: 306–8.

Cleghorn, H. (1866c) [read 13 vii 1865]. Supplementary notes upon the vegetation of the Sutlej Valley. *Transactions of the Botanical Society [of Edinburgh]* 8: 309–15.

Cleghorn, H. (1866d) [read 10 v 1866]. Notes on the Travancore Government Garden at Peermade. *Transactions of the Botanical Society [of Edinburgh]* 8: 441–7.
Note 1. Reprinted as 'Memorandum on the Travancore Government Garden at Peermade' in *Journal of the Agricultural and Horticultural Society of India* 14: 184–90, 1867.
Note 2. Peermade a hill station at 915 m, 85 km east of Kottayam, once a retreat of the Raja of Travancore.

Cleghorn, H. (1866e). *Memorandum upon the Supply of Wood Fuel to the Punjab and Delhi Railways*. [?Roorkee].
Note. On 30 vii 1866 Cleghorn sent J.H. Balfour a copy of this 11 page government report (still at RBGE) 'for private information'. Despite this stricture Balfour read the paper to BSE on 13 xii 1866, a summary printed in their *Transactions* 9: 56–60, 1868.

Cleghorn, H. (1866f). On the deodar forests of the Himalaya. *Report of the Thirty-fifth Meeting of the British Association for the Advancement of Science held at Birmingham in September 1865*. Notes & Abstracts pp 79–80. London: John Murray.

Cleghorn, H. (1867a). Memorandum on the timber procurable by the Indus, Swat and Kabul Rivers. *Journal of the Agricultural and Horticultural Society of India* 14: 73–87.

Cleghorn, Dr. (1867b). Memorandum on the naturalization of Australian plants in certain parts of India. *Journal of the Agricultural and Horticultural Society of India* 14: 87–88.

Cleghorn, H. (1867c). Note on the poisonous properties of certain species of Andromeda. *Journal of the Agricultural and Horticultural Society of India* 14: 260–263, with one plate [an unsigned lithograph of *A.* [= *Lyonia*] *ovalifolia*].
Note. Reprinted, without the plate, in *Transactions of the Botanical Society [of Edinburgh]* 10: 410–2. Also issued as a separate, pp [1–]2–4 + plate (copy at St Andrews University).

Cleghorn, H. (1867d). Notes upon the pines of the North-West Himalaya. *Journal of the Agricultural and Horticultural Society of India* 14: 263–72, tt. 1–6. [including Cleghorn & Stewart, 1867].

Cleghorn, H. (1867e). Notice regarding insects injurious to timber trees. *Journal of the Agricultural and Horticultural Society of India* 14: 294–5.

Cleghorn, H. (1867f). Memorandum on the manufacture of tar [from Himalayan conifers]. *Journal of the Agricultural and Horticultural Society of India* 14 (Correspondence & Selections): 6–7.

Cleghorn, H. & Stewart, J.L. (1867). Analytical key of the conifers of the N.W. Himalaya. *Journal of the Agricultural and Horticultural Society of India* 14: 265–7.
Note. This and Cleghorn (1867d) issued, together with the *Andromeda* paper, as a separate, pp [5–]6–14 + tt. 1–6 (copy at St Andrews University).

C[leghorn], H. (ed.) (1868a). Tree Planting in the Punjab in Select Papers on the Agriculture, Horticulture &c of the Punjab pp [1–]2–53. Lahore: Lahore Chronicle Press.
Note. In 1863 Cleghorn edited, and added a few footnotes, to this selection of papers, which date from the early 1850s, but not published as a collection until 1868. The forestry papers include Dalhousie's 1851 Minute on Arboriculture in the Punjab.

Cleghorn, H.F.C. (1868b). Report on increased floods due to forest clearing for coffee cultivation in Coorg, Wynaad and Western Mysore. *Madras Quarterly Journal of Medical Science* 12: 246–9.

Cleghorn, H. (1868c) (ed). *Testimonials in favour of Hugh Cleghorn, M.D. ... candidate for the Professorship of Botany in the University of Glasgow*. Edinburgh: Thomas Constable.
Note. The references are by Robert Wight, George Bentham, Robert Christison, Douglas Maclagan, J.H. Balfour, George Dickie, J. Forbes Watson, G.J. Allman, C.C. Babington, Rev. W.P. Dickson, J.D. Hooker, Richard Owen, W. Baird, Alex Wood, Sir William Denison, Sir Walter Elliot, Henry Yule, Sir Bartle Frere, Sir Robert Montgomery, W.C. Maclean, Clements Markham, Thomas Moore, M.T. Masters, M.P. Edgeworth, R. Kippist, E.A. Parkes, Daniel Oliver, Thomas Anderson, Daniel Hanbury, Lyon Playfair. It also includes a bibliography of some of Cleghorn's works, and extracts of Government of India orders relating to his forest work (copy at RBGE).

Cleghorn, H. (1868d) [read 9 vii 1868]. Biographical notice of the late Dr Walker-Arnott, Regius Professor of Botany in the University of Glasgow. *Transactions of the Botanical Society [of Edinburgh]* 9: 414–26.

Cleghorn, H. (1869a). [Contribution to discussion of George Bidie's paper 'On the effect of forest destruction in Coorg']. *Proceedings of the Royal Geographical Society* 13: 81–82.

Cleghorn, H. (1869b). On the distribution of the principal timber trees of India, and the progress of forest conservancy. *Report of the Thirty-eighth Meeting of the British Association for the Advancement of Science held at Norwich in August 1868*. Notes & Abstracts pp 91–4. London: John Murray.
Note. Also published in *Transactions of the Scottish Arboricultural Society* 5: 91–3. 1869.

Cleghorn, H. (1869c). On the management of European forests. A report by Dr H. Cleghorn ... to Lord Clinton, Under Secretary of State for India [about visit to Sicily, Malta, Italy, Switzerland and France]. *Transactions of the Scottish Arboricultural Society* 5: 94–7.
Note. Also printed in the 1871 Parliamentary Papers, pp 348–51.

[Cleghorn, H.] (1869d). Notes of a brief visit to Italy in the Spring of 1868.
Note. Printed cutting of a paper read to BSE 10 vi 1869 in BSE archives, but place of publication for such reports in this period unknown.

[Cleghorn, H.] (1869e). Notices of Books. Punjab Plants ... By J. Lindsay Stewart. *Gardeners' Chronicle* 20 Nov 1869: 1211.
Note. Anonymous book review, but attribution is given on Cleghorn's copy in EUL.

Cleghorn, H. (1870a) [read 14 i 1869]. Obituary notice of Professor C.F.P. von Martius, of Munich, and Adalbert Schnizlein, Erlangen. *Transactions of the Botanical Society [of Edinburgh]* 10: 30–31.

Cleghorn, H. (1870b) [read 11 iii 1869]. Notes on the botany and agriculture of Malta and Sicily. *Transactions of the Botanical Society [of Edinburgh]* 10: 106–139.
Note. Also issued as separate, pp [1–]4–36; and as appendix (pp 27–36) to 'Papers relative to the plantation of trees in Malta'. Malta: Government Printing Office, 1870.

Cleghorn, H. (1870c) [read 10 vi 1869]. Obituary notices of the late Dr William Seller and of Professor Bertoloni of Bologna. *Transactions of the Botanical Society [of Edinburgh]* 10: 202–206.

Cleghorn, H. (1870d) [read 8 vii 1869]. On the parasites which affect the Government timber plantations in South India. *Transactions of the Botanical Society [of Edinburgh]* 10: 245.

Cleghorn, H. (1870e) [read 11 November 1869]. Address delivered at the opening of the Botanical Society of Edinburgh, session 1869–70. *Transactions of the Botanical Society [of Edinburgh]* 10: 261–284.
Note. Also issued as separate, pp [1–]4–26.

Cleghorn, H. (1873) [read 11 July 1872]. Obituary notice of Dr Robert Wight, F.R.S. (with portrait). *Transactions of the Botanical Society [of Edinburgh]* 11: 363–388.
Note. Also issued as a separate: Frontispiece + pp [0–]4–29 (copy at RBGE).

[Cleghorn, H.F.C.] (1874a). Notices of Books: The Forest Flora of North-West and Central India by J. Lindsay Stewart, and completed by Dietrich Brandis. *Gardeners' Chronicle* 2(n.s.): 426–7.
Note. The review is anonymous, but the reference in it to his review of Beddome's similar work for South India (see below), the historical review of Indian literature included, and the references to minor forest products all strongly suggest Cleghorn's authorship.

[Cleghorn, H.F.C.] (1874b). Notices of Books: The Flora Sylvatica of Southern India by Lieutenant-Colonel Beddome. *Gardeners' Chronicle* 2(n.s.): 491.
Note. The review is anonymous, but Cleghorn's copy of the work at EUL has a cutting of it pasted into the first volume, and also a manuscript copy of a draft of the review (shorter than the printed version but containing verbatim passages).

Balfour, [J.H.], Cleghorn, [H.F.C.], Hutchison, R., Buchan, A. & Sadler, J. (1874 [read ix 1873]). Preliminary note from the committee ... on the influence of forests on the rainfall. *Report of the Forty-third Meeting of the British Association for the Advancement of Science held at Bradford in September 1873*, pp 488–90. London: John Murray.

Cleghorn, H. (1875a) [read 6 xi 1872]. Address delivered at the nineteenth annual meeting [of the Scottish Arboricultural Society]. *Transactions of the Scottish Arboricultural Society* 7: 1–9.
Note. Also issued as a separate, pp [0–] 2–9.

Cleghorn, H. (1875b) [read 7 xi 1873]. Address delivered at the twentieth annual meeting [of the Scottish Arboricultural Society]. *Transactions of the Scottish Arboricultural Society* 7: 115–22.
Note. Also issued as a separate, pp [1–] 4–10.

Cleghorn, H. (1875c) [read 4 xi 1874]. Address delivered at the twenty-first annual meeting [of the Scottish Arboricultural Society]. *Transactions of the Scottish Arboricultural Society* 7: 199–210.
Note. Also issued as a separate, pp [1–] 4–14.

Cleghorn, H. (1875d). 'Arboriculture', in *The Encyclopaedia Britannica: a Dictionary of Arts, Sciences, and General Literature*, ed. 9 (ed. T.S. Baynes) 2: 314–24. Edinburgh: Adam & Charles Black.

Cleghorn, H. (1875e). [Review of *The Forest Flora of North West and Central India* ... by the late J. Lindsay Stewart ... and ... Dietrich Brandis]. *Journal of Botany British & Foreign* 13: 23–25.

Cleghorn, Dr (1875f). Obituary notice of Dr J. Lindsay Stewart. *Proceedings of the Royal Society of Edinburgh* 8: 321–2.
Note. Cleghorn almost certainly also wrote the very similar obituary for *Journal of Botany* 11: 319–20, which contains a few additional details.

Cleghorn, H. (1876a) [read 13 November 1873]. Obituary notice of Dr John Lindsay Stewart. *Transactions of the Botanical Society [of Edinburgh]* 12: 31–33.

Cleghorn, H. (1876b) [read 13 Jul 1876]. [Account of an excursion to the Chor Mountain]. *Transactions & Proceedings of the Botanical Society [of Edinburgh]* Appendix: xlviii.

Cleghorn, H. (1878). Brief account of the Royal Forest School at Vallombrosa. *Transactions of the Scottish Arboricultural Society* 8: 182–9.
Note. Also issued as a separate, pp [1–] 2–8.

Cleghorn, H. (1879). 'Forests', in *The Encyclopaedia Britannica: a Dictionary of Arts, Sciences, and General Literature*, ed. 9 (ed. T.S. Baynes) 9: 397–408. Edinburgh: Adam & Charles Black.

Cleghorn, H. (1883a) [read 8 April 1880]. On the spontaneous introduction of Aristotelia (Friesia). *Proceedings of the Botanical Society [of Edinburgh]*. Session XLIV: xi–xii.

Cleghorn, H. (1883b) [read 13 July 1882]. Obituary notice of Deputy Surgeon-General [William] Jameson, C.I.E. *Transactions of the Botanical Society [of Edinburgh]* 14: 288–95.

Cleghorn, H. (1885a) [read 5 Aug 1884]. Address delivered at the thirty-first Annual Meeting [of the Scottish Arboricultural Society]. *Transactions of the Scottish Arboricultural Society* 11: 1–8.

Cleghorn, H. (1885b). [Evidence of Dr Hugh Cleghorn, M.D., F.R.S.E.] in Report from the Select Committee of the House of Commons on Forestry (chaired by Sir John Lubbock). House of Commons Papers; Reports of Committees 1884-5 (287) pp 19-27. London: Henry Hansard & Son.
Note. Summary printed in *Transactions of the Scottish Arboricultural Society* 11: 124–136 (1887).

Cleghorn, H. (1886a) [read 4 Aug 1885]. President's address delivered at the thirty-second Annual Meeting [of the Scottish Arboricultural Society]. *Transactions of the Scottish Arboricultural Society* 11: 115–8.

Cleghorn, H. (1886b) [read 9 xi 1882]. [Introductory remarks as Vice-President, including obituary notices of George Dickie, Richard Parnell and George H.K. Thwaites]. *Transactions of the Botanical Society [of Edinburgh]* 16: 1–9.

Cleghorn, H. (1887a). Cleghorn on gunta baringi [a letter written from Illundur on 9 i 1862], in the pamphlet *Notice of the late Sir W. Elliot ... of a bazaar drug called gunta baringi ... identified for the first time as Premna herbacea Roxb. ...*, by Wm. Ferguson. Colombo: Ceylon Observer Press.

Cleghorn, H. (1887b) [read 3 Aug 1886]. President's address delivered at the thirty-third Annual Meeting [of the Scottish Arboricultural Society]. *Transactions of the Scottish Arboricultural Society* 11: 287–90.

Cleland, [J.], McKendrick, [J.G.], Ewart [J. Cossar], Stirling, [W.], Bower [F.O.], Cleghorn [H.] & McIntosh, [W.C.] (1887 [read ix 1886]). Report ... continuing the researches on food-fishes and invertebrates at the St Andrews Marine Laboratory. *Report of the Fifty-sixth Meeting of the British Association for the Advancement of Science held at Birmingham in September 1886*. Reports on the State of Science pp 268–70. London: John Murray.

Cleghorn, H. (1889a) [read 11 xi 1886]. Obituary notice of William Traill, M.D. *Transactions & Proceedings of the Botanical Society of Edinburgh* 17: 17–9.

Cleghorn, H. (1889b) [read 8 xii 1887]. Obituary notice of Sir Walter Elliot of Wolflee. *Transactions & Proceedings of the Botanical Society of Edinburgh* 17: 342–5.

Cleghorn, H. (1889c) [read 5 iii 1889]. Dr Boswell of Balmuto [J.T.I. Boswell Syme]. *Transactions & Proceedings of the Botanical Society of Edinburgh* 17: 516–9.

McIntosh, [W.C.], Allman, [G.J.], Lankester [E. Ray], Burdon-Sanderson, [J.S.], Cleland, [J.], Ewart, [J. Cossar], Stirling, [W.], McKendrick, [J.G.], Cleghorn, [H.] & Traquair, [R.H.] (1889 [read ix 1888]). Report ... continuing the researches on food-fishes at the St Andrews Marine Laboratory. *Report of the Fifty-eighth Meeting of the British Association for the Advancement of Science held at Bath in September 1888*. Reports on the State of Science pp 141–5. London: John Murray.

Reviews of Cleghorn's Publications

1. The British Association report (Cleghorn *et al.*, 1852):

Indian Annals of Medical Science 2: 720–2. 1855 [anonymous]. In the same work (pp 724–7) is a notice/review of four of Cleghorn's papers (on Hedge Plants (1850a); *Calysaccion longifolium* (1851e); Medical Resource of Indian Forests (1852b), and *Hortus Madraspatensis* (1853a), with an encomium of his economic-botanical work.

The Scottish Review 1: 47–53, 1853. [Anonymous review titled 'The Forests of India'].

2. *Forests & Gardens of South India* (Cleghorn, 1861a):

The Gardeners' Chronicle 1861: 969, 1009 [by J.D. Hooker, see KDC 157 no 186].

The Athenaeum 1771: 450. 1861 [by William Robertson].

The Spectator 28 Sep 1861: 1067–8. [Anonymous review titled 'The Tropical Forests'].

The Edinburgh New Philosophical Journal 14 (n.s.): 286–9. Oct 1861. Anonymous – possibly by J.H. Balfour, see p. 134.

Madras Quarterly Journal of Medical Science 5: 134–143. ?1862. Anonymous – see p. 134.

Mittheilungen aus Justus Perthes' Geographischer Anstalt über wichtige neue Erforschungen auf dem Gesammtgebiete der Geographie 8: 116. 1862. Notice with short summary, with a notice of Cleghorn (1861b) [possibly inserted by A.H. Petermann, the editor, who had worked with A.K. Johnston in Edinburgh].

Note. In correspondence Peter Cleghorn mentioned reviews in the *Quarterly* and *Edinburgh Review*, but he must have been misinformed, as no reviews appeared in either journal; he also referred to ones in *The Economist* (StA CP 2/11/2 PC to HC 9 Jan 1862) and the *Agricultural Journal*, (StA CP 2/11/3 PC to HC undated) which have not been traced.

3. *Forest Reports of Punjab & Western Himalaya* (Cleghorn, 1864b):

Friend of India 1598: 954. 17 viii 1865. [By Brandis, see p. 170].

Calcutta Review 45 (no 89): 56–105. Anonymous – see p. 170.

Related Material

Contemporary biographical items:

Anon. (1865). Dr. Hugh Cleghorn and forest conservancy. *Friend of India* 1574: 245.

Brandis, D. (1888). Dr Cleghorn's services to Indian forestry. *Transactions of the Scottish Arboricultural Society* 12: 87–93.
Note. Also printed in *Indian Forester* 14: 395–401. 1888, where it is stated to be reprinted from Proceedings of the '*Royal Scottish Agricultural Society*, July 1887'.

Anon. (1889). Presentation to Hugh Cleghorn of Stravithie, M.D., LL.D., F.R.S.E. *Transactions of the Scottish Arboricultural Society* 12: 198–205.

Obituaries & Death Notices

Botanical/Forestry journals:

McIntosh, W.C. (1895) [read 1 vii 1895]. Dr Hugh Francis Clarke Cleghorn. *Proceedings of the Royal Society of Edinburgh* 20: li–lx.

Taylor, Andrew. (1895). Obituary notice of Hugh Francis Clarke Cleghorn, M.D., LL.D., F.R.S.E, F.L.S. *Transactions & Proceedings of the Botanical Society of Edinburgh* 20: 439–48.

Anon. (1895). *Annals of Scottish Natural History* 1895: 262–3.

Anon. (1895). *Garden and Forest* (New York) 8: 260.

Anon. (1895). *The Geographical Journal* 6: 83.

Anon. (1895). *Journal of Botany* 33: 256 [very brief].

Anon. (1895). *Indian Forester* 21: 276–78 [reprinted from *The Scotsman*].

Anon. (1896). *Scottish Geographical Magazine* 12: 29–30.

Medical journals:

Anon. (1895). *Edinburgh Medical Journal* 40: 1138–40.

Anon. (1895). *The Lancet.* 25 May 1895, p. 1347.

Anon. (1895). *British Medical Journal* no 1797: 1304.

Newspapers:

Anon. (1895). *The Times.* 18 May 1895 p. 14.

Anon. (1895). *St Andrews Citizen* 18 May 1895.

Anon. (1895). *Evening Telegraph* 17 May 1895 p 2. [not seen].

Anon. (1895). *The Scotsman* 17 May 1895.

Notices:

Allgemeine Botanische Zeitschrift 1: 200, 1895; *Botanisches Centralblatt* 63: 288, 1895; *Botanische Jahrbüchen* 21 Beiblatt 53: 55, 1896; *Botaniska Notiser* 1896: 43, 1896.

Membership of Societies

OB= Office Bearer
HM = Honorary Member/Fellow

Agri-Horticultural Society of India

Agri-Horticultural Society of Madras (OB & HM)

Royal Society of Edinburgh

Botanical Society of Edinburgh (OB & HM)

Botanical Society of Ratisbon [i.e. Regensburg] (HM)

Highland and Agricultural Society (OB)

Linnean Society of London

Pharmaceutical Society (HM)

Royal Asiatic Society

Royal Geographical Society

Royal Scottish Arboricultural Society (OB & HM)

Society of Antiquaries of Scotland

Scottish Geographical Society (OB)

Appendix 2

Cleghorn Eponymy

Cleghorn was not a taxonomist, and his collections were not widely distributed, so it is unsurprising that few species should commemorate his name – nonetheless there is one genus (described with two species by his friend Robert Wight; four species currently recognised), and three species in other genera though of these two are of unknown taxonomic status.

Cleghornia Wight, Icon Pl. Ind. Orient. 4(2): 5 (viii 1848). (APOCYNACEAE).

The specimens on which this genus and its two species were based were not collected by Cleghorn, but named for him by Robert Wight: 'The genus is dedicated to Dr. Hugh Cleghorn of the Madras Medical Establishment, a zealous cultivator of Botany, but more especially directing his attention to Medical Botany'. The two species (now regarded as synonymous) had been collected by Wight himself on his short visit to Ceylon in 1836.

Cleghornia acuminata Wight, Icon Pl. Ind. Orient. 4(2): 5, t. 1310 (viii 1848).

Cleghornia cymosa Wight, Icon Pl. Ind. Orient. 4(2): 5, t. 1312 (viii 1848).

= **C. acuminata**

Capparis cleghornii Dunn, Kew Bull. 1916: 61 (1916). (CAPPARACEAE).

Loranthus cleghornii Beddome, Madras J. Lit. Sci., ser. 3, 1: 48 (1864). (LORANTHACEAE).

= **Helixanthera cleghornii** (Beddome) Danser

Note. Status unknown, it is treated neither in Gamble's *Flora of Madras* nor in Hooker's *Flora of British India*.

Digyroloma cleghornii Turczaninow, Bull. Soc. Imp. Nat. Moscou 25 II: 330 (1862). (ACANTHACEAE).

Note. Status unknown: it is treated neither in Gamble's *Flora of Madras* nor in Hooker's *Flora of British India*. Bentham & Hooker (*Genera Plantarum* 2: 1071) considered it a dubious genus, but stated that it was a shrub from Mysore. Turczaniow's herbarium is known to have contained Indian material and this may have been based on one of Cleghorn's early specimens distributed by the Botanical Society of Edinburgh. The type specimen should be in Kiev, but enquiries have been unsuccessful.

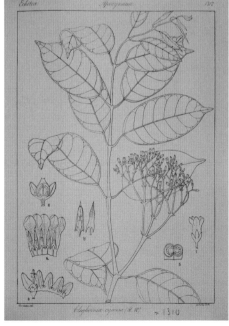

Index

Page numbers in **bold** indicate illustrations

A

Abbottabad (now in Pakistan), 151
Abies pindrow, 151, 188, 218, 281
Acacia longifolia, 109
Acacia pulchella, 107
Acacia robusta, 95, 109
Adam, Sir Frederick (Governor of Madras), 50, 56
Adam, William (Scottish Lord Commissioner), 4, 17
Adamson, Dr John (photographer), 9, 33, 37, 38, 117, 161
Adamson, Robert (photographer), 9, 37, 38
Aden, 176
Agnew, John Vans Agnew (Madras merchant), 114
Agri-Horticultural Society of India (Calcutta), 112
Agri-Horticultural Society of Madras, 55, 61–4, 95
Agri-Horticultural Society of the Punjab (Lahore), 151, 181
Agri-Horticultural Society of Western India, 120, 245
Airy, George Biddell (astronomer), 41, **42**, 44
Aitchison, James Edward Tierney (Indian botanist), 130, 149, 229
Albert, Prince Consort, 41, 46, 127, 265
Alexander, Andrew (Professor of Greek, St Andrews), 8
Alfred, Prince, Duke of Edinburgh, 265
Alison, William Pulteney (Edinburgh medic), 12, 14, 33, 154
Anamalai Forests (teak), 82, 87, 94, 95, 101
Anamalai Hills – C's expedition to (1859), 95–7, **96**
Anderson, John (Indian zoologist), 131
Anderson, the Rev. John (Indian missionary), 31, 54
Anderson, Thomas (Indian botanist), 36, 131, 135, 137, 140, 159, 184, 202
Anderson, Thomas (chemist, Glasgow), 11, 163
Andy, Senjee Pulney (C's Madras medical pupil), 60–1, 71, 72
Anstruther, 5, 258, 259
  Waid Academy, 257

Anstruther & St Andrews Railway – C's involvement with, 223–4
Appavoo, Francis (C's assistant in Madras), 60, 71, 72, 94, 102, 118
Archer, Thomas Croxen (museum director, Edinburgh), 207, 265
Argyll, George Campbell, 8th Duke of (scientist/statesman), 121
  on evolution, 119–20
  Secretary of State for India, 205, 209, 213, 226, 228
Aristotelia racemosa, 228
artists working in Madras in 1857, 77
artists, Indian – drawings of plants in private gardens, 113
artists, Indian – C's employment of, 22
Artocarpus hirsutus (as timber), 100
asafoetida (*Ferula jaeschkiana*), 148
ash, manna (*Fraxinus ornus*), 201
Attock (now in Pakistan), 152
Aucklandia veracosta (koot), 150
Australian trees (*Acacia*, *Eucalyptus*)
  in Bengal, 184
  in Nilgiri Hills, 95, 107, 109, 113
  in Punjab, 151
  plantations in Nilgiris (Jackatalla), 90, 133

B

babool (*Acacia arabica*), 141
Baden-Powell, Baden Henry (Indian civil servant), 172, 235, 300 n 8/78
Bailey, Florence Agnes (née Marshman), 251
Bailey, Frederick (Indian forester, Edinburgh lecturer), 190, 226, 251, 266, 268
Bailey, Frederick Marshman (Himalayan explorer/naturalist), 251
Baines, Edwin E. (Punjab engineer), 160
Baird Smith, Richard (Indian engineer), 41
Baldrey, Robert John (Madras garden designer), 115
Balfour, Andrew Francis (godson), 34, 35
Balfour, Edward Green (Madras surgeon), 46, 73, 100, 133, 175, 296 n 4/85
  Madras Exhibitions, 69, 75
Balfour, Lt. Col. George (Madras soldier), 73
Balfour, Isaac Bayley (later Sir, Regius Keeper, RBGE), 226, 263, 266

Balfour, John Hutton (Regius Keeper, RBGE), **12**, 13, 31, 34, 35, 46, 47, 58, 97, 119, 126, 181, 187, 189, 205, 207, 225, 226, 227, 231, 235, 237, 250, 258
  BAAS committee on forests and climate, 235, 239
  botanical lectures, 58
  botanical lecture diagrams, 58, 59, 65
  botanical textbooks, 50, 59, 121
  C's correspondence with, xi, 30
  review of *Forests & Gardens*, 134
  secretary of Royal Society of Edinburgh, 119
  selecting gardeners for Madras, 63, 65, 66
  views on evolution, 121
  views on Indian forest training, 189
Ballingall, Sir George (military surgeon), 7, 15
Ballochbuie Forest, Deeside, 232
Balmoral – Scottish Arboricultural Society excursion to (1887), 232–3
Bannerman, Helen (author of *Little Black Sambo*), 131
Bartet, Eugène (Nancy professor), 246
Barton, the Rev. John (Indian missionary), 151, **153**
Basel Evangelical Missionary Society, 61, 204
  mission at Mangalore, 99, 102, 205
Bassia latifolia (liquor source), 97
Baynes, Thomas Spencer Baynes (editor of *Encyclopaedia Britannica*), 237
Beadon, Cecil (Indian civil servant), 83, 141, 156
Beddome, Richard Henry (Forest Conservator), 76, 89, 95, 96, 100, 102, 103, 106, 157, 193, 194
Bell Pettigrew, John (anatomist, St Andrews), 230, 237, 252
Bell, Sir Charles (surgeon, Edinburgh), 14
Bell, the Rev. Andrew (benefactor of schools), 2, 7, 270
Bengal Forest Department, 131, 184, 210
Bentham, George (botanist), 50, 205, 207
Bérenger, Adolfo di (Italian forester), 215

INDEX

Bertie, Alphonso (Indian apothecary), 26
Berwick, Thomas (St Andrews gardener), 253
Betula bhojpattra, 148
Betula utilis, 148, 150, 279, 281
Biddulph, Michael Anthony Shrapnel (artist), 188
Bidie, George (museum curator, Madras), 187
Binko, Henry (inventor of electric train), 244
Binny, John (Madras artist), 64
Birdwood, George (later Sir, museum curator), 81, 85, 243, 245, 303 n 14/20
   his version of history of Indian forestry, 245
Black, Allan Adamson (gardener, Bangalore), 113
blackwood (*Dalbergia latifolia*), 89, 95, 101, 243
Blaikley, Alexander (Scottish portrait artist), 1, 220
Blane, Thomas Law (Collector of Canara), 82
Blenkinsop, Louis (Lewis) (Madras forester), 94, 102, 297 n 5/88
Boehmeria nivea, 18, 35, 41
Bonar family (Free Church ministers), 123
Bonar, William, 123, 131
Boppe, Lucien (Nancy professor), 209, 246
Boswell, John Thomas Irvine (né Syme) (Scottish botanist), 229
Botanical Society of Edinburgh, 13, 49, 98, 189
   C's involvement with, 13–4, 17, 25, 34–6, 118, 176, 199, 227–30
Bourne, Samuel (photographer), 145, 161, 162
Boussingault, Jean-Baptiste (French chemist/agriculturist), 43
Brandis, Carl (brother of Dietrich), 300 n 8/84
Brandis, Dietrich (later Sir, Indian forester), 83, 84, 155, 156, **165**, 234, 300 n 8/79
   appointed Inspector General of Forests, 165–6
   appreciation of C, 167
   education (not in forestry), 165
   forest management techniques universally applicable, 166
   later taxonomic work, 164, 167, 188
   missionary connections, 167
   review of C's Punjab Reports, 170
   summoned from Burma, 84, 155, 156
   visits to Britain, 168, 176, 182, 189, 208
Brandis, Professor Christian August (philosopher, Bonn), 127, 129, 130
Brandis, Rachel (née Marshman), 103, 157, 167
Brewster, Sir David (scientist/university Principal), 15, 36, 37–8, 40, 57, 62, 75, 123, 228, 265
Briggs, Robert (chemist, St Andrews), 9
British Association for the Advancement of Science (BAAS), 37, 41
   Aberdeen Meeting 1885 – grant to W.C. McIntosh, 254
   Birmingham Meeting 1865, 162, 181
   Bradford Meeting 1873, 235
   Charles Dickens' view of, 246
   committee under J.H. Balfour on forest and rainfall in Scotland, 235, 239
   Edinburgh Meeting 1850, 39, **40**, 41, 123
   Ipswich Meeting 1851, 41, **42**, 43
   Manchester Meeting 1887 – grant to W.C. McIntosh, 255
   Norwich Meeting 1868, 206–8
   Report on tropical deforestation (C *et al.*)
   C's late views on its significance, 239, 303 n 14/22
   commissioning, 41
   reading & reception, 43–5
British forestry – state in 1880s, 246
Broomhall, Fife (home of Lord Elgin), 158
Brown, James, as author of *The Forester*, 87, 90, 166, 300 n 8/84
Brown, Margaret Read (née Inverarity, botanical artist), 77–8
Brown, Robert N. (Madras gardener), 66, 67, 115
Brown, the Rev. John Croumbie (forester/environmentalist), 205, 249, 250
Brown, William Hunter (RBGE librarian), 267
Brugmansia suaveolens, 107
Buchan, Alexander (Scottish meteorologist), 235
Buchanan (later Hamilton), Francis (Indian botanist), 21, 25, 36, 111
Buchanan, Capt. James (photographer), 75
Buist, Dr George (Bombay journalist), 120
Burckhardt, Heinrich Christian (German forester), 190
Burliar (E.B. Thomas's garden in Nilgiris), 114
Burma
   C's visit to (January 1857), 49, 84
   teak forests of, 56, 82–4
   war with (1852), 55
Bute, John Crichton-Stuart, 3rd Marquess of, 232, 252

C

Calcutta Botanic Garden, 67, 82, 97, 109, 111, 131, 135, 175, 184, 243
Calicut (Malabar coast), 92, **93**
Campbell family, 6
Campbell, Captain John Alexander (engineer, Nilgiris), **86**, 90, 297 n 5/46, 300 n 8/84
Campbell, Jane (née Cleghorn, aunt), 6, 15, 34, 155
Campbell, William Hunter (step-cousin), 6, 13, 14
Candolle, Auguste Pyramus de (Swiss/French botanist), 59, 62, 123
Canning, Charles, Earl, (Governor-General, first Viceroy), 64, 137, 155, 299 n 7/59
Canning, Charlotte, Lady, 106, 114, 191
Cape of Good Hope, 17, 29, 111, 114, 242, 249
Capparis cleghornii, 25, 314
Capparis divaricata, 212
Capparis heyneana, 212
Carey, the Rev. William (missionary), 167
Carruthers, William (botanist, British Museum), 228, 256
Carson, Dr Aglionby Ross (Edinburgh High School), 6
Central Provinces Forest Department, 183
Cesati, Vincenzo de (Neapolitan botanist), 202
Ceylon
   acquisition of by grandfather, 2
   C's brief visit, 137
   quinine plantations, 140, 182
Challenger Expedition, 35
Chalmers family of Stravithie, 269
Chalmers, Jean (née Cowan, sister-in-law), 130
Chalmers, Lt. John (Punjab forester), 103, 169

INDEX

Chalmers, the Rev. Thomas (Scottish theologian), 5, 10, 14, 31, 103, 122, **125**, 130, 154
Chamaecyparis pisifera cv. 'Aurea', 218
Chamba (Himachal Pradesh), 147, 278
Chambers, Robert (Vestiges of Creation), 37
Chenab (Chandrabagha) Valley, 147
Chengulroy, T. (Madras artist), 73
Chesser, John (Edinburgh architect), 177, 217, 301 n 9/13
Chetty, Bala (Madras dyes), 72
Chini (now Kalpa, Himachal Pradesh), 146
Chor Mountain, 188
Christie, Alexander Turnbull (Indian botanist), 22, 36, 293 n 2/18
Christison, Dr David (botanical artist), 229
Christison, Robert (later Sir, toxicologist/physician), 14, 25, 119, 205, 227
Church Missionary Society (CMS), 31, 52, 144, 146, 154
Cinchona calisaya 'Ledgeriana', 140
Cinchona calisaya, 137, 138, 140
Cinchona lancifolia, 140
Cinchona micrantha, 137, 138
Cinchona officinalis, 138, 140
Cinchona pahudiana, 140
Cinchona succirubra (*C. pubescens*), 137, 138
Clark, Aylwin (biographer of Hugh Cleghorn, sr.), ix
Clark, Elizabeth (*née* Browne – medical missionary), 154
Clark, the Rev. Robert (Indian missionary), 52, 151, 152, 154
Clarke, Richard (Madras civil servant), 1, 32, 293 n 1/3
Clavé, Jules (French forestry author), 204
Cleghorn family
  history of, 1
  religious beliefs, 30–1
  portraits of, 39, 220
Cleghorn Forest Library – inauguration, 263
Cleghorn Memorial Library
  at Museum of Science & Art, 264–6
  at RBGE, 267
Cleghorn, Allan Mackenzie (brother), 4, 17, 18, 31, 216
Cleghorn, Hugh (senior – grandfather), **1**, 1–4, 215
  Secretary of Ceylon, 2–3
Cleghorn, Hugh Francis Clarke
artists – employment of in Madras, 59

artists – employment of in Shimoga, 22, 272
attitude to money, 216
benefactor to Edinburgh University for forestry lectureship, 250
benefactor to Edinburgh University library, 251, 268–9
benefactor to St Andrews University, 254
benefactor to RBGE, ix
bibliographer, 228
birth, 1
book collecting and library, 50, **180**, 220
botanical interests, origins of, 12
botanical lectures
  in Madras (1850s), 58
  in St Andrews (1870), 209, 302 n 10/14
botanical library – division between Edinburgh University and Museum, 266
charitable works in Madras
climate and deforestation interests, 44, 72, 73, 91, 133, 186–7, 235, 238–9
Conservator of Forests, Punjab, 151
death of, 260
donations to museums, 47, 49
economic botany interests, 41, 44, 73
editorial work in Madras, 53
Enlightenment and Improvement agenda/motivation, xi, 45, 187
entry into EIC, 17
European tour 1867/8, 195–204
evangelical Christian beliefs, 18, 24, 30–3, 38
evolution, views on, 120–1
'Father of Scientific Forestry' (Birdwood), 246
Forest Reports
  on fuel supply for Punjab railway, 184
  on 'girdling' of trees, 185
  Madras I, 1857–8, 88–91
  Madas II, 1858–9, 92
  Madras III, 1859–60, 9 –101
  Madras final, 1867, 193
  on Madras forest rules, 157, 173
  on regional annual reports, 183–4
  on rukhs in Punjab, 157
Forest Tours
  Madras I (1857), 88–9
  Madras II (1858/9), 92
  Madras III (1860), 99, 101
  Northern Circars/Orissa (1859), 97–8
  Punjab (1862/3), 141–52, **142**, **143**

herbarium collections, 24, 29, 63, 118, 148, 149, 152, 168, **174**, 175, 209, 253
health, 17
honours awarded to, 230, 237, 313
horticultural plants, role in transfer of, 135
illnesses in India, 26, 99, 100, 157, 167
illustrations collection (prints and drawings), 51, 267, 269
  at Edinburgh University, 269
  at RBGE, 268
Indian languages, knowledge of, 19, 94, 297 n 5/26
Indian religions, views on, 58
Joint Commissioner of Forests, with Brandis (1864–5), 141, 157
Justice of the Peace, 258–9
lack of forestry training, 88
lack of recognition by India Office, 237
Madras Forest Conservator 1856–60, 81–102
major publications
  BAAS deforestation report, 43–5, 85, 239, 245
  Encyclopaedia Britannica contributions, 237–9
  Forests & Gardens of South India (1861), 101, 117–8, 132–5
  Punjab Forest Reports (1864), 169–70
manual of Indian botany – intended, 59, 94, 101, 134
marriage to Mabel Cowan, 122, 127
medals, 46, 70, 296 n 4/89
medical posts in Madras (1850s), 54
military surgeon, early appointments, 18
minor excursions
  to Caveri river, 186–7
  to Chor (Chudar) Mountain, 188
minor (non-timber) forest products, interest in, 183, 184, 193, 234
obituaries of, 262, 313
obituarist, 229
Officiating Inspector-General of Forests (1866–7), 183–5
pension, 175, 181, 216, 301 n 9/1
personal qualities as Conservator (Brandis/Stebbing), 85
photographic interests, 39, 75, 162
philanthropic and county activities in Fife, 256–9
politics, 259
portraits of, **1**, **38**, **161**, **165**, **236**, **241**, 263–4
quinine experiment, role in, 110, 137, 138
secretary of Madras Agri-Horticultural Society, 62

316

INDEX

Stravithie – inherits, 155
taxonomist, 188, 212
views on forest rights of Indians, 44, 90, 183, 184, 193
wastage of timber, concerns over, 85, 87, 88, 100, 183, 294 n 3/55
wealth at death, 216
will, 261
Cleghorn, Isabella ('Isa') (sister), 4, 32, 91, 150, 202, 215–6, 270
Cleghorn, Isabella (née Allan) (mother), 4
Cleghorn, John Ross (uncle), 2, 4, 18, 67
Cleghorn, 'Mabel' (née Marjorie Isabella Cowan) (wife), 122, 127–9, **128, 129**, 168, 180, 187, 192
  death of, 260, 263
  her character, 129
  her dowry, 217
  ill-health of, 167, 173, 195, 203, 259–60
  in Nilgiris, 141
  in the Himalayas, 146
  travels with her father, 128–9
  work for husband, 129, 133
Cleghorn, Peter (also Patrick) (father), **1**, 4, 37, **38**, 150, 252
  concern over C's salary, 81
  death of, 155
  his fortune, 5
  opinion on C as author, 135
  religious beliefs, 33
  view of C's role as conservator, 85
  view of the Free Church, 124
Cleghorn, Rachel (née McGill – grandmother), 2, 12
Cleghornia (described by Robert Wight), 35, 314
Cleghornia
  acuminata Wight, 314
  cymosa Wight, 314
coal in India, 185
Cochin (Malabar coast), 93
coffee, cultivation in Mysore, 25
Coimbatore – Botanical Survey of India herbarium, 176
Coldstream, Dr John (Mabel Cleghorn's uncle), 34, 54, 118, 123, 154
Coldstream, William (Punjab civil servant), 118, 154, 159, 188, 229, 243, 300 n 8/43
Colonial Exhibitions, 74
Conolly, Henry Valentine (Collector of Malabar), 26, 71, 82, 91, 297 n 5/54
Constable family (publishers), 127–8
Cook, the Rev. George (St Andrews), 10, 294 n 3/43

coomri (*kumri* – shifting cultivation also 'dhya'), 24, 25, 26, 85, 100, 133, 184
Coonoor (garden, Nilgiris), 110
Cooper's Hill, Surrey – forest training at, 190, **191**, 246, 247
Coorg (and Wynad) – deforestation of, 187
Corylus jacquemontii, 147, 278
Cotton, Arthur (later Sir, Madras engineer), 31, 81, 98, 121
cotton, cultivation in Mysore, 25, 35
Cotton, Frederick (Madras engineer), 71, 81, 82, 89, 105, 114, 121, 245, 248
Cotton, the Rt. Rev. George (Bishop of Calcutta), **153**, 168
Cowan family (in-laws), 32, 122–31
Cowan, Catharine (née Menzies, mother-in-law), 123
Cowan, Charles, of Loganhouse (father-in-law, paper-maker), **123**, 123–6, 130, 180, 238
  and the Disruption, 124
  political career, 126
Cowan, Charles William (brother-in-law), 130
Cowan, James (uncle-in-law), 130, 259
Cowan, John (later Sir John, of Beeslack Bt. – uncle-in-law), 130, 259
Cowan, John James (brother-in-law), 122, 127, 130, 131
Cowan, Margaret Menzies (sister-in-law), 130, 259
Crawford, Robert (Scottish poet), 36
Crinum defixum, 26, **27**
Cross, Robert McKenzie (Kew gardener/collector), 138
Cruickshank, the Rev. James (C's tutor), 5, 7, 8
Cryptomeria japonica, 184
Cubbon, Sir Mark (Commissioner of Mysore), 18, 26, 66, 110
Cullen, General William (Travancore Resident), 71, 72, 90, 93, 114, 182
cumin, black (*Bunium persicum*), 148
Cunliffe, Mrs Brooke (garden owner), 114
Cupar, Fife, 258
Cupressus torulosa, 107, 143, 218
Cynomorium coccineum, **198**, 199, 202
Cytisus scoparius, 229

D

Dalhousie (hill station, Himachal Pradesh), 151, 162, 164
Dalhousie, Christian, Lady (née Broun), 14, 36, 130
Dalhousie, James Andrew Broun-Ramsay, Marquess of (Governor-General), 55, 68, 105, 146

Burmese forests, 83, 146
  minute on Punjab agriculture, 84, 297
Dalhousie, John Ramsay, 13th Earl of (politician), 254, 304 n 14/45
Damodaran, Vinita (historian), ix, 278
Daphne oleioides (jeku, now *D. mucronata*), 150
Dapuri Drawings, 243
Darjeeling – quinine plantations, 66, 140, 184
Darwin, Charles, 11, 14, 51, 118, 120, 121, 206, 228
Datisca cannabina (dye plant), 148
Davis, Ellen (née Sherwood) (missionary), 171
Davis, the Rev. Brocklesby (CMS missionary), 171
Decaisne, Joseph (French botanist), 117, 204
Dehra Dun (forest school), 190, 210, 251
Denison, Sir William (Governor of Madras, Acting Viceroy), 115, 138, **139**, 159, 205
deodar (*Cedrus deodara*), 143, 144, 147, 151, 158, 169, 170, 181, 188, 191, 218, 278, 279
Desmodium elegans (katti, *D. tiliaeifolium*), 150, 243
Dharamsala (hill station), 146, 150, **158**, 159
  church of St John in the Wilderness, **158**
  McLeod's arboretum, 150
Dickens, Craven Hildesley (PWD secretary), 183, **192**, 264
Dickson, Alexander (botanist, later Regius Keeper, RBGE), 11, 187, 206, 207, 228, 263
Dickson, James & Son (Edinburgh nursery), 65, 218
Digyroloma cleghornii
  Turcz., 314
Disruption of the Church of Scotland (1843), 4, 10, 31, 36, 37, 124, **125**
Dobbs, Richard Stewart (Indian administrator), 21, 33, 71
Dodabetta (quinine plantation, Nilgiris), 106, 138
Dollar Institution (later Academy), 8
Dorin, Joseph Alexander (Indian civil servant), 83
Douglas, Dr Andrew Halliday (Edinburgh physician), 34, 181, 187, 259
Douglas fir (*Pseudotsuga menziesii*), 218, 231

INDEX

Dovey, John Edward (C's accountant), 216, 224, 263
Drayton, Richard (historian), 103
Drew, Dr Charles (Madras medic), 51, 67, 71, 90, 175
Drummond, the Rev. David Thomas Kerr (Edinburgh Episcopalian), 34, 171
Drury, Heber (political officer and economic botanist), 93
Dryopteris (*Lastraea*) *cristata*, **208**
Dublin botanical chair – C's interest in, 187
Duncan family, of Parkhill, 77
Duncan, the Rev. Henry (Savings Banks), 257
Duncan, Thomas (mathematician, St Andrews), 9, 151
Dundas, Henry (Viscount Melville), 2, 3
Dunino (Fife), 4, 8, 217, 222, **261**
Dunkeld, Perthshire, **231**
Dunn, Captain (Ooty), 113

E

Edgeworth, Michael Pakenham (Indian botanist), 146, 181
Edinburgh
    Barclay Church, 126
    Caledonian United Services Club, 215
    High School, 5, **6**
    possible site for School of Forestry, 234, 235, 238, 244, 248
    Royal Infirmary, 11, 15
    Royal Terrace, 131
    'Society' (Cleghorn family property), 1, 15, 265
    Waterloo Hotel, 263
Edinburgh Medical Missionary Society (EMMS), 54, 123, 154
Edinburgh Museum of Science & Art (later Royal Scottish Museum), 235, 256, 264, **265**
Edinburgh Sanitary Protection Association, 256
Edinburgh University
    C as undergraduate, 10–5
    C as medical examiner, 250
    Cleghorn Bequest of books, 268–9
    forest lectureship, 250–1
    herbarium – C curates Indian section, 36, 49
Edwardes, Sir Herbert (soldier), 152
Edwardsia tetraptera, 107
Egypt – C's visits, 117, 136, 181
Eisenach forest school, 189, 190
Eleusine coracana (ragi), 21, 150, 272, 293 n 2/11

Elgin, James Bruce, 8th Earl of (Viceroy), 147, 155, 157
    illness and death, **158**
    summons Brandis from Burma, 84, 155, 156
Elgin, Victor Bruce, 9th Earl of (Viceroy), 254
Elliot, Walter (later Sir, Madras civil servant), 46, 51, 63, 64, 70, 73, 75, 76, 81, 84, 115, 117–9, 175, 181, 205, 206, 216, 227, 229, 243
Ellis, Daniel (Scottish botanist), 123
Elmslie, Dr William Jackson (medical missionary), 154–5, 171
Elphinstone, John, 13th Lord (Governor of Madras & Bombay), 119–120
Elwes, Richard Gervase (engineer & photographer), 158, 162
Encyclopaedia Britannica, 9th edition – C's articles, 237–9
Eucalyptus 'perfoliata' (*E. cinerea*), 114
Euphorbia cattimandoo, 46, 70
Escorial (Spanish forest school), 249
Evans, Dr John (medical missionary), 154
evolution – C's views on, 120
Ewart family of Penicuik, 180, 270, 301 n 9/16, 304 n 14/47

F

Falconer, Hugh (Indian surgeon and naturalist), 67, 83, 137, 202, 216
Farquhar, Dr Thomas (Calcutta medic), 154, 171, 216, 260
Farquharson, Dr Robert of Finzean (physician/Liberal politician), 247
Fergusson, Sir James, of Kilkerran (Indian administrator), 205, 302 n 11/2
Ferooz (*Feroze*, steam frigate), 92, 93
Fife & Kinross Board of Lunacy, 256
Fisheries Board of Scotland, 254–5
Fletcher, Harold Roy (Regius Keeper, RBGE), 267
Florence, 203
Fonceca, John Joseph (Indian artist), 73
Fonceca, Simon (Indian artist), 73
Forbes, Edward (natural philosopher), 15, 50, 51, 132, 295 n 3/76
Forest Conservancy in India
    history of, 81, 82
    Conservancy v. 'Conservation', 85
Forest Department and Christianity, 102–3, 251
Forestry in India
    Continental training of officers, 176, 188–9, 204, 208–13, 225, 228, 234

    history of – C's version, 234
    history of – Col. James Michael's version, 245, 248
    history of – Sir George Birdwood's version, 245
    setting up a national department, 155–7
Forests of South India – Map of, **80**
Forsyth, Lt. James (Indian forester), 183
Forsyth, Thomas Douglas (Punjab civil servant), 151, 152
Fothergilla involucrata, 147
Foulis, Susan (Lady) (*née* Low), 4, 7, 34, 66
Fowke, Francis (architect), 264
Fox Talbot, William Henry (photographer), 37
Fraas, Karl Nikolas (botanist in Greece), 165
Francis, Peregrine Madgwick (Madras engineer), 107, 108
Franco-Prussian War – interrupts forestry training (1870), 209
Freemasonry in the Himalayas, 300 n 8/88
Frere, Sir Bartle (Governor of Bombay), 99, 141, 196, 205
Fuchsia 'Princeps', 107
'Fungus melitensis', **198**, 199, 202

G

gaddis (nomadic shepherds), 147, 157, 279, 283
Gage family of Rathlin, 33
Gage, Catherine, 34, 35
Galloway, Anna Maria (C's housekeeper), 220
Galloway, John (nurseryman), 3
Gamble, Harpur (zoologist), 210
Gamble, James Sykes (Indian forester), 209, **210**, 235, 244, 302 n 10/16
gamboge (*Garcinia morella*), 24, 25, 46, 71
Gammie, James Alexander (horticulturist, Darjeeling), 140
Garnkirk (Sprot estate, Lanarkshire), 39
Geddes, Patrick, 237, 252
German forestry methods, 89, 166, 185, 189
Gersoppah (or Jog), Falls of (Karnataka), **16**, 20, 24
Gibson, Alexander (Bombay Forest Conservator), 22, 43, 82, 101, 162, 171, 176, 243, 245, 246, 293 n 1/28
Gillespie, James (St Andrews architect), 177

INDEX

Gillespie, the Rev. Thomas (Professor of Humanities, St Andrews), 8

Gladstone, William Ewart (Prime Minister), 130, 205, 244, 259

Glasgow botanical chair – C's unsuccessful application, 205–6

Gloriosa superba, **24**

Godavery Irrigation Scheme (Arthur Cotton), 98

Goodsir family (Anstruther), 5

Goodsir, John (anatomist), 5, 15, 41

Gordon-Cumming, Constance Frederica (artist), 242, 243

gorse (*Ulex europaeus*) (in Nilgiris), 114

Govan, George (Indian botanist), 13, 37, 137

Govindoo (artist), 59, 63, 64, 70, 71, 73, 89, 114, 118, 192

    C recommends publication of his work, 193

Govindoo junior (artist), 302 n 9/55

Grace, Charles Stuart (C's factor & solicitor), 155, 216, 217, 266

Graham, Robert (botanist, Regius keeper, RBGE), 13, 15, 35, 50

Grahame, W.L. (Indian forester), 189

Grant, Dr Alexander (Calcutta medic), 53, 137

Grant, Sir Patrick (soldier), 84, 196

Great Exhibition (London, 1851), 45, 46, 127

Greville, Robert Kaye (Edinburgh botanist), 13, 14, 31, 34, 35, 41, 43, 55, 58, 131, 227

Grey, George Robinson, Earl de (later Marquess of Ripon, Indian administrator), 121, 186

Griffith, William (Indian botanist), 143

Grisebach, August (German botanist), 166, 167

Grove, Richard (environmental historian), ix, x, 30, 32, 43, 45, 81, 102, 133, 249, 278

Groves, Anthony Norris (missionary), 31, 32, 72

gujars (nomadic buffalo herders), 147, 281

Gundert, Dr Hermann (Basel missionary),102, 204

gutta percha (*Isonandra gutta*), 44, 71

H

Hamilton, Douglas (forester & artist), 95–7, 132, 243 Hampstead Heath, 212

Handyside, Peter David (anatomist, Edinburgh), 11, 14, 33

Harris, George, Lord (Governor of Madras), 53, 63, 64, 67, 69, 74, 75, 91, 92, 98, 119, 132, 196, 216

originator of Madras Forest Conservancy, 81

Harvey, William Henry (Irish botanist), 187

Hasskarl, Justus Karl (Dutch botanist), 140

Havelock, Henry (later Sir, soldier), 157

Hay, Lord William (later 10th Marquess of Tweeddale), 144, 243

Haydon, Benjamin Robert (artist), 43, 294 n 3/50

Hazara, (now in Pakistan), 151

hedge plants in Mysore, 21, 35, 41, 293 n 2/14

Helfer, Johann Wilhelm (German naturalist/traveller), 83

Helixanthera cleghornii (Bedd.) Danser, 314

Henderson, Douglas Mackay (Regius Keeper, RBGE), 268

Henslow, the Rev. John Stevens (Cambridge botanist), 18, 42, 43, 44

Heyde, August Wilhelm (Moravian missionary), 148–9

Heyde, Maria, 149, 157, 300 n 8/23

Heyne, Benjamin (botanist/geologist), 22, 111

Highland and Agricultural Society of Scotland, 225

    forestry education, 226, 248, 251

Hill, David Octavius (artist/photographer), 37, 38, 125

Hill, Alexander (Professor of Divinity, Glasgow), 39

Hill, the Rev. George (Scottish cleric), 39, 294 n 3/43

Hobson-Jobson, 201

Hoffenberg, Peter (historian of exhibitions), 74

Holta (site of tea plantations in Kangra), 146, 161, 162

Home, James (Edinburgh medic), 14

Hooker, Joseph Dalton (later Sir, Director of Kew), 35, 50, 126, 137, 164, 189, 205, 232, 264

    presidential address to BAAS (1868), 206, 227

    review of *Forests & Gardens*, 134

    on religion and science, 206

    work on Indian flora, 141, 213

Hooker, Sir William Jackson (Director of Kew), 22, 49, 50, 84, 112, 176, 205, 228

    his herbarium, 113

    reference for C, 50, 57

Hooper, George Stanley (Madras judge), 64, 105, 115, 118

Hope, John (Regius Keeper, RBGE), 12, 58, 127, 299 n 7/11

Hope, Thomas Charles (Edinburgh chemist), 15

Horsfield, Thomas (EIC naturalist), 50

Hortus Malabaricus (of Hendrik van Rheede), 21, 55, 199, 202

'household of nature' – broader (non-timber) value of forests, 234, 238

Humboldt, Alexander von, 43, 45, 59, 294 n 3/59

Hunneman, John (London bookseller), 51

Hunter, Alexander (Madras surgeon, founder of Art School), 64, 91, 110, 150, 177, 216, 243

    geology, 76

    Great Exhibition, 46

    Madras Exhibitions, 69, 70, 72, 73, 77

    photography, 73, 75

Hutchinson, Charles Waterloo (engineer/photographer), 162, 181

Hutchison, Robert of Carlowrie (laird/forester), 235

Huxley, Thomas Henry (zoologist), 237, 254, 258, 264, 303 n 14/4

Hyder Ali (Sultan of Mysore), 20, 21, 110, 111

I

India Office, London

    C's meetings at, 117, 121, 176, 205

    C's work for, choosing forest officers for Continental training, 208–13

Indian Forest Act of 1865, 172–3, 204, 208

    adoption of in Madras, 194

Indian Mutiny, or Sepoy Revolt, 91, 119, 159, 163

Industrial Museum of Scotland, 264

Instructions for Forest Assistants in Madras, 88

International Forestry Exhibition, Edinburgh (1884), 211, 235, **240**, 241–5

Inzenga, Giuseppa (Sicilian botanist), 201

Ipswich Museum, **42**

Italy, 202–3, 215

J

Jackson, Thomas (Natural Philosopher, St Andrews), 9

Jacquemont, Victor (French botanist and traveller), 117, 146

Jaffrey, Andrew Thomas (gardener), 64–6, 71, 72, 76, 109, 110, 112, 140, 175

Jameson, Robert (Edinburgh naturalist), **12**, 14

INDEX

Jameson, William (Superintendent, Saharunpur Garden), 135, 141, 146, **161**, 163, 229

Janvier, the Rev Dr Levi (missionary), 154

Jäschke, the Rev. Heinrich August (Moravian missionary), 149, 158, 243, 279

Jenkin, Henry Fleeming (engineer and evolutionary critic), 207, 256

Jenner, Charles (Edinburgh storeowner), 227

Jerdon, Thomas Caverhill (Indian zoologist), **161**

Johnston, William & Alexander Keith (Edinburgh cartographers), 132

Johnstone, James Todd (RBGE librarian), 267

Judaeo-Christian legitimation of exploitation of natural resources, 103, 169

Juniperus macropoda (pencil cedar), 148, 188

K

Kabul River, 152

Kalatope Forest (near Dalhousie), 151

Kalhati (garden, Nilgiris), 110

Kashmir – missionary work in, 154

Kelvin, William Thomson, Lord (physicist), 207

Kemp, Kenneth Treasurer (Edinburgh chemist), 11

Kew, Royal Botanic Gardens, 176, 205, 212

    Museum of Economic Botany, **48**, 49, 74

    plants for India, 52, 112, 135

Key, Dr Thomas (Madras medic), 57

Keylong (Lahul, Himachal Pradesh), **145**, 148, 157

Khonds (tribal group of Orissa), 97

Killar (Himachal Pradesh), 147

King, Colonel (Ooty), 113

King, George (later Sir, Director, Calcutta Botanic Garden), 137

Kings Lynn, Norfolk, 208

Kirk, John (African explorer), 176

Kirkpatrick, Dr James (Madras medic), 70, 112

Kodaikanal (Palni Hills), 182

Koders (tribal group, Anamali Hills), 96

Koksur (Himachal Pradesh), 158

Kotgarh (CMS station, Himachal Pradesh), 144, **145**, 149

kumri – see coomri

Kurumba (Koorambers, Koramers, tribal group, Mysore), 19, 184, 301 n 9/30

L

Lahore (now in Pakistan), 141, 151, 152, 162, 164

    missionary/prayer conferences, 152, 159

    Punjab Exhibition (1864), 147, 149, 159, **160**

    Museum, 160

Lake, Colonel Edward (Indian administrator), 146, 152, **153**, 159, 256

Lal Bagh Garden, Bangalore, 66, 95, 99, 110–3, 135, 175

Lankester, Edwin Ray (zoologist), 44, 50, 255

Lantzius-Beninga, Scato (German botanist), 166

larch (*Larix decidua*), **231**, 232, 246

Lascelles, Arthur Rowley William (Nilgiri planter), 113

Lawford, Colonel Edward (Indian engineer), 186

Lawrence Asylums, 117, 149, 154, 159

Lawrence, George St Patrick (soldier), 117, 159

Lawrence, Sir Henry (soldier & administrator), 117, 159

Lawrence, Sir John (Viceroy), 154, 159, 168, 171, **192**

    testimonial to C (1865), 171, 246

    testimonial to C (1867), 192

Lawson, Charles (Edinburgh seed merchant), 189, 226, 301 n 9/182

Lawson, George (botanist), 34, 66

Lees, Dr Leonard Horner (Indian medic), 151

LeHardy, Charles Francis (Superintendent of Nugger), 19, 24

Levinge, Vere Henry (Collector of Madurai), 182

Liberal Union Party, 259

Licopoli, Gaetano (Neapolitan botanist), 202

Lindesay, Alexander Kyd (of Balmungo), 12, 117, 251, 270, 294 n 3/40

Link, Heinrich Friedrich (Berlin botanist), 165

Linnean Society of London, 50, 61, 78, 149, 176, 205, 212

Lobo, Dr Michael Francis (Roman Catholic bishop), 114

Loch, John (EIC director), 17

London – C's visits to, 212

Longden, Capt. (Himalayan forest surveyor), 84

Loranthus cleghornii Bedd., 314

Loranthus longiflorus (*Dendrophthoe falcata*), **79**, 243

Lothian, Schomberg Henry Kerr, Marquess of (Conservative politician), 241, 245

Lowenthal, the Rev. Isidor (Indian missionary), 151, 154

Lubbock, Sir John (scientist/Liberal politician), 206, 246

Lundie, the Rev. Robert Henry (brother-in-law), 123, 127, 257

Lyell, Sir Charles (geologist), 207

Lyons, Dr Robert Spencer Dyer (Irish physician/politician), 247

M

Macaulay, Thomas Babington (Lord, historian/politician), 98, 126

Macdonald, William (Prof. Civil History, St Andrews), 252

Mackenzie, Colin (Indian surveyor), 2, 22, 94, 111

Mackenzie, Dr William (of Culbo), 4, 7, 131

Mackenzie, Margaret (née Allan) (aunt), 4, 7, 131

Maclagan family, 102, 297 n 5/33

Maclagan, Douglas (later Sir, Edinburgh medic), 17, 41, 119, 205, 257, 263

Maclagan, John Thomson (C's assistant, Madras), 64, 76, 87, 94, 99, 102

Maclagan, Margaret Dalziel (née Pearson), 87

Maclagan, Robert (Indian engineer), 151, 152, 159, 169, 253, 263

Macpherson, Captain (Ooty), 114

Macvicar, the Rev. John Gibson (natural historian, St Andrews), 252

Madden, Edward (Indian botanist), 131, 146, 170

Madeira, 17

Madras

    fire at Oakes, Partridge & Co. (1852), 55, 63

    Government Central Museum, 65, 74, 175

    People's Park, 115, **116**

    private gardens, 114–5

    religious establishments, 30

Madras College, St Andrews, **6**, 7

Madras Exhibition of 1855, 49, 65, 67–74, **68**

Madras Exhibition of 1857, 75–7

Madras Forest Act 1882, 194

Madras General Hospital – C's first posting, 17

INDEX

Madras Literary Society, 54
Madras Medical College
  C as Professor of Botany & Materia Medica, 55–61
  history of, 56, **57**
Magnolia campbellii, 184
Magnolia excelsa, 184
Malcolmson, John Grant (geologist), 242
Malsars (tribal group, Anamalai Hills), 87, 96, 99
Malta, 2, 195–9
  Hagar Qim temple, **197**, 199
Marathi artist, C's in Shimoga, 22, 272
Marden, Thomas (Madras headmaster), 52
Markham, Clements (quinine experiment), 98, 138, 181–2, 205
Marsh, Caroline, 203
Marsh, George Perkins, author of *Man & Nature*, 168, 201, 203, 215, 235, 239
  friendship with Henry Yule, 203
Marshman, Joshua (Indian missionary), 103, 157
Martius, Carl Philipp von (German botanist), 229, 303 n 13/25
Matheson, Robert (Edinburgh architect), 47, 264, 304 n 15/5
Mathew, Manjil V. (RBGE librarian), 267, 268
Mayer, John Emilius (Madras medic), 57, 76
McClelland, John (Calcutta naturalist), 83
McCorquodale, William (forester, Scone), 209, 226
McFarlane, William Husband (lithographer), 96, 132, 299 n 7/42
McGregor, John (forester to Duke of Atholl), 226, 230
McIntosh, William Carmichael (marine biologist), 230, 237, 250, 252, 254–6, 262
McIvor, Anne (*née* Edwards), 106
McIvor, James (Nilgiri planter), 106
McIvor, William Graham (gardener, Ooty), 8, 66, 76, 91, 98, 105, 106, 135, 138, **139**, 182, 195
McLeod, Sir Donald Friell (Commissioner of Punjab), 150, 152, **153**, 159, 216
McNab, James (RBGE gardener), 8, 65, 66, 229, 235
  work on Stravithie garden, 191, 217
Meconopsis aculeata, 147, 188, 279, **280**, 301 n 9/44
Meconopsis baileyi, 251

Medical Mission, 33, 98, 152–5
melanoxylon (*Acacia melanoxylon*), 95, 109
Menon, Chatu (Nilambur), 91
Mercer, Charles McWhirter (curator of Punjab Exhibition), **153**, 160, 162
Meriahs (tribal group of Orissa), 98
Merriman family (Dr John and Thomas, apothecaries), 33, 117
Messieux, Samuel (French master, St Andrews), 7, 37
Metz, Johann Friedrich (missionary/plant collector), 99
Michael, James (Indian forester), 82, 106, 245, 248
  role in International Forestry Exhibition (1884), 242, 245
Miller, Hugh (geologist), 76, 121, 242
minor forest products – C's interests in, 184, 193
Mitchell, Jesse (Madras photographer), 75, 76
Moir, Dr Robert (C's physician, St Andrews), 260
monkey puzzle (*Araucaria araucana*), 218, 221
Montgomery, Dr Howard Benjamin (Madras Medical College), 115, 134
Montgomery, Lady (wife of Sir Henry, Madras civil servant), 114
Montgomery, Sir Robert (Lt. Governor of Punjab), 151, 159, 205, 216
Mooroogasen Modeliar, G. (Madras artist), 77
Mooroogasen Modeliar, P. (Madras artist), 59, 64, 71, 73, 76
Mootoosawmy Modeliar, P.S. (assistant, Madras Medical College), 60, 71, 72
Mophlas (Malabar Muslims), 89, 92
Moravian missionaries, 148–9, 152
Moreton Bay chestnut (*Castanospermum australe*), 112
Moreton Bay fig (*Ficus macrophylla*), 201
Morgan, Henry Rhodes (Indian forester), 61, 95, 102, 106
Morham, Robert (Edinburgh architect), 242
Mount Sinai, 176
Muir, William (later Sir, Indian Civil Servant, Principal Edinburgh University), 136, 167, 171, **192**, 216, 256, 263
Müller, Sebastian (forest assistant, North Canara), 89, 95, 99, 101, 102, 204
Multan (now in Pakistan), 141

Munro, Alexander (*Tertius* – anatomist), 15
Munro, Sir Thomas (Governor of Madras), 82, 84
Munro, William (Indian botanist), 111, 181
Murdoch Smith, Sir Robert (museum director, Edinburgh), 265, 267
Murray, J.A. (timber agent, Chenab), 147
Murree (now in Pakistan), 151
Muston, Charles (museum librarian, Edinburgh), 265
Mysore Agri-Horticultural Society, 111, 112
Mysore Commission, 18
Mysore Forest Department, 184

N

Nagar (Bidanur) Fort, Karnataka, **20**
Nancy forest school, 189, **190**, **204**, 208–9, 251
  C's visit to, 204, 248
Nancy professors – visit to British forests (1881), 246
Nanquette, Henri (Nancy forestry professor), 190, 204
Naples, 202
Napoleon's willow, 110
Narcissus tazetta, **196**
Nedivuttum quinine plantation (Nilgiris), 138
Nether Stravithie (formerly Stravithie Mill Farm), 221
nettle cloth, 18, 35, 41
New South Wales (C's influence on conservation), 239
New, William (gardener, Bangalore), 66, 99, 112, 113, 118, 135, 137
Newman, Anna (*née* Cowan, sister-in-law), 130
Nicholson, Alleyne (Prof Civil & Natural History, St Andrews), 252
Nilambur Teak Plantations, 26, 35, 82, 89–92, 99, 110, 133
Nile flood (1861), **136**
Nilgiri Hills – quinine plantations, 106, 138, **139**
North Canara Forests (teak), 82, 87, 89, 94, 99, 101
North, the Hon. Frederick (administrator, Ceylon), 2
Northcote, Sir Stafford (Secretary of State for India), 205, 208
Northern Ireland – C's excursion (1849), 33–4
Novara – Austrian scientific voyage, 91
Nugger (Nuggur, Nagar) Division – see Shimoga

INDEX

O

Oldham, Thomas (Indian geologist), 76, 185
olive (*Olea europaea*), 112
Onslow, William Campbell (Superintendent of Nugger Division), 22, 24, 25, 26, 41
Ootacamund ('Ooty')
  Frederick Price's History of, 105, 113
  Government Garden(s), 95, **104**, 105–10, **108**, 135, 175
  private gardens, 113–4
  St Stephen's Church, 106
Ophelia elegans (*Swertia angustifolia*), 70
Orissa sal tract – author's visit (2010), 274–8
Owen, Richard (zoologist), 43, 205
Oxalis cernua (*O. pes-caprae*), 197, 302 n 10/6

P

Pagell, Eduard (Moravian missionary), 148–9
Palaquium ellipticum, 71, 89
Palgrave, Francis Turner (poet), 207
Paliars (tribal group, Anamalai Hills), 96
Pamplin, William (London bookseller), 50, 51, 55
paper-making
  at Dharamsala, 150
  at Penicuik, 122
papyrus (*Cyperus papyrus*), 201
Paris
  Exposition Universelle (1867), 191
  International Exhibition (1855), 74
  Jardin des Plantes, 117, 204
Parlatore, Filippo (Florentine botanist), 201, 203
Parrotiopsis jacquemontiana (*Fothergilla involucrata*), 148, 278, 282
Pasquale, Giuseppi (Neapolitan botanist), 202
Paterson, David Horn (Madras medical missionary), 54, 58, 134, 154
Patton family (of Kinaldie), 12
paucontheee (*Palaquium ellipticum*), 71, 89
Pearson, George Falconer (Indian forester), 183, 212, 246
Peermade (Travancore Government Garden), 182
Penicuik, 122
Penn, Albert Thomas Watson (photographer), 108
Periploca aphylla, 152
Peshawar (now in Pakistan), 152, 154
Phayre, Arthur (later Sir, Burma), 83, 156, 200, 297 n 5/12
Photographic Society of Madras, 73
photography
  early history in St Andrews, 36–9
  in Madras, 73–6
  in Punjab, 161
Pilkington, Frederick Thomas (Scottish architect), 126, 127
Pinus cembra (Arolla pine), 233
Pinus gerardiana, 188, 278
Pinus roxburghii, 188, 282
Pinus wallichiana (kail), 114, 149, 188, 218, 279
Pitcorthie, Fife (fossil fish), 256
Pius IX, Pope, 203
plantations
  quinine, coffee, tea – effect on climate, 186, 187
  tea and coffee in Nilgiri Hills, 91, 106, 186
Playfair family (St Andrews), 13
Playfair, Hugh Lyon ('The Major', Provost of St Andrews), 13, 15, 36, 37, 117
Playfair, Lyon (later Lord, chemist/politician), 13, 205, 249, 264
Playfair, Robert Lambert (later Sir, soldier/diplomat/naturalist), 13, 118
Pluchea ovalis, 20, **273**
poon spar tree (*Calophyllum angustifolium/polyanthum*), 89
Pottinger, Sir Henry (Governor of Madras), 53, 55, 63
Prinsep, Edward Augustus (Punjab civil servant), 150–2, **153**
Prinsep, Margaret Eleanor (née Hunter, artist), 150, 188
Prison Visiting Committee (Fife), 257
Profeit, Dr Alexander (Queen Victoria's Commissioner), 232
prosopography of EIC surgeons Gibson, Wight & Cleghorn, xi
Punjab
  British acquisition of, 159
  Exhibition (Lahore, 1864), 147, 149, 159, **160**
  Hills – author's visit (2014), 278–83
Purdie, William (botanist, West Indies), 64
Pykara quinine plantation (Nilgiris), 138
Pyper, Willliam (Edinburgh schoolmaster), 6

Q

quinine
  experiment in India, 98, 137–40
  C's role in, 110, 137, 138
  plantations
    Ceylon (Hakgalle), 140, 182
    Darjeeling, 66, 140, 184
    Java, 140
    Nilgiris, 106, 138, **139**, 182
    Travancore, 182
    South America, 137

R

railways
  C's investment in, 222
  in India, **86**, 89, 94, 168, 169, 171, 185
  in Scotland, 117, 222
Raman, Anantanarayanan (historian of science, Madras), 60
Ransome, George (Ipswich Museum), 41
Rao, Goday Surya Prakasa (garden at Ankapilly), 97
Rathlin Island, Co. Antrim, 33, **34**
Ravi Valley (Himachal Pradesh), 146
Rawlinson, Sir Christopher (Madras civil servant), 91, 114
Reid, Francis Archibald (Madras Agri-Horticultural Society), 62, 63, 64, 65
Reid, Robert (Scottish architect), 8, 35, 127
Reserved Forest
  concept of, 157, 183
  origins of, 90
Reuss, Eugène (Nancy forestry professor), 246
Rhododendron arboreum, 107, 109
Rhododendron campanulatum, 146, 281
rhododendrons, Sikkim, 232
rhubarb (*Rheum* spp.), 146, 148, 151, 279
Ribbentrop, Berthold (German forester in India), 83, 191
Robertson Smith, William (theologian), 126, 238
Robertson, Robert Alexander (botany lecturer, St Andrews), 254
Robinson, Sir William Rose (Indian civil servant), 106, 297 n 5/54, 298 n 6/11
Rohde, John (Madras civil servant), 97, 106, 114
Rohtang Pass (Himachal Pradesh), 150, 152, 158
Rome, 202
Roorkee, Thomason College of Civil Engineering, 169
Rottler, Johann Peter Rottler (botanist/missionary), 2, 12, 35
Rottlera tinctoria (dye plant), 97
Roupell, Arabella (botanical artist), 78
Roxburgh, William (Indian botanist), 17, 36
Royal Botanic Garden Edinburgh (RBGE), 13, 15, 105, 176, 242, 248, 250, 252

INDEX

arboretum, 235
   library and herbarium, 13, 35, 36, 164, 210, 227, 229, 253, 254, 266
   Museum of Economic Botany, 46, **47**, 74, 267
Royal College of Surgeons of Edinburgh – C's licentiate, 17
Royal Medical Society (RMS), Edinburgh, 11
Royal Society of Edinburgh, 95, 119, 121, 225
Royle, John Forbes (economic botanist), 41, 45, 50, 62, 112, 134, 137, 146
Rubia cordifolia (dye plant), 95
rukh forests (Punjab), 141, 155, 157, 170, 185, 300 n 8/53
Rungasawmy, T. (Madras artist), 64, 70
Ruskin, John, 152, 201, 258

S

Sach Pass (Himachal Pradesh), 147, 281
Sadler, John (J.H. Balfour's assistant), 133, 227, 231, 232, 235
Saharunpur Botanic Garden, 13, 135, 137, 141, 169
Saiful Maluk Lake (now in Pakistan), 151
sal (*Shorea robusta*), 97, **211, 276**, 277
salinification of soils, 185
sandalwood (*Santalum album*), 25, 89, 98, 99
Sanderson, Dr James (Garrison Surgeon, Madras), 93, 114
Sang, Edward & Son (Kirkcaldy nurserymen), 218
Sankey, Major (later Sir) Richard Hieram (Indian engineer), 186
Saussurea lappa (*Aucklandia veracosta*, koot), 150
scammony (*Convolvulus scammonia*), 181
Schlich, Wilhelm (German forester in India), 190, 191, 212, 247
Schmid, the Rev. Bernhard (missionary, Nilgiris), 105
Schouw, Joakim Fredrik (Danish plant geographer), 59, 166, 167
Scone, Perthshire (estate of Earl of Mansfield), 189, 209, 231
Scots pine (*Pinus sylvestris*), **230**, 231, 232
Scott, Andrew (Madras medic), 65, 70, 73, 75
Scott, James (St Andrews architect), 177
Scott, Sir Walter, 127–8, 130
Scottish (later Royal) Arboricultural Society, 230–5, 241, 263
   C's presidential addresses to, 234

Scottish forestry – nature of in 19th century, 231, 303 n 14/24
Scottish Geographical Society, 225
Scudder, the Rev. John (Indian missionary), 54
Seaton, William John (Indian forester), 189
Select Committee on Foretry (under Lubbock, 1885), 226, 246–9
Sequoiadendron giganteum, 218
Serle, William Ambrose (Madras civil servant), 115
shanny (*Lipophrys pholis*), **255**
Shimoga District
   author's visit (1997), 271–4
   botanical exploration of, 21–2
   C's period in, 19–26
   history and geography, 19–21, **23**
Shimoga Drawings, the, 22, 35
Shortt, John (Madras medic), 61
Sicily, 199–202
Sim, Col. Duncan (Madras engineer), 18
Sim, James Duncan (Madras civil servant), 115
Simla (summer capital of India), 143, 144, 155, 157, 183, **243**
Simpson, James Young (later Sir, gynaecologist), **12**, 15, 181, 259
Sims, Dr James Marion (gynaecologist), 203, 204, 260
Singh, Shamsher, Raja of Bashahr, 146, 169
Singh, Sri, Raja of Chamba, 146, 151, **163**
Sispara creeper, 107
sissoo (*Dalbergia sissoo*), 26
Skirving, Archibald (Scottish artist), 1, 220
Smith, Adam (distant cousin), xi, 4, 122, 242
Smith, Dr George (Madras medic), 57, 69
Smith, James Edward (botanist), 21
Smith, William Wright (later Sir, Regius Keeper RBGE), 267
Smithe, J.D. (timber agent, Chenab), 147, 150
Society of Antiquaries of Scotland, 222, 225 (chemist)
Solly, Edward, 45, 294 n 3/59
Someren, Godlieb James van (Mysore forester), 184
Somerville, William (forest lecturer, Edinburgh), 226, 230, 250–1
Sorghum vulgare, 228
Spanish broom (*Spartium junceum*), 135
Sprot family of Stravithie, ix
   history, 39
Sprot, Alexander (later Sir, first Bt,

Cleghorn's nephew & heir), 39, 177, 222, 251, 261, 266
Sprot, Alexander (of Garnkirk) (brother-in-law), 4, 39
Sprot, Alix Isabella ('Mother Martha'), 262
Sprot, Elizabeth, ix, 221
Sprot, Ethel Florence (Lady Sprot, *née* Thorp) (nephew's wife), 262
Sprot, Geoffrey Cleghorn, 220, 262
Sprot, Gerard Hugh Cleghorn ('Hugh', *né* Sadler), 221, 262
Sprot, Col. Hereward (*né* Sadler), 262
Sprot, Mark, 39
Sprot, Rachel Jane (*née* Cleghorn) (sister), 4, 32, 39, 40, 216
Spruce, Richard (S Amercian botanist), 138, 229
St Andrews, Fife, 5, 230, 258
   Literary & Philosophical Society, 37, 225
   Marine Laboratory (later the Gatty), 230, 252, 254, 255
   Royal & Ancient Golf Club, ix, 12, 215
   St Leonards School, ix, 36, 257
   University
      C's papers in library, ix
      C as Court Assessor, 251, 252
      C as undergraduate, 8–10, **9**
      teaching of natural history, 252
      Botanic Garden, **253**
St Thomé, Madras – home of C, 57
Stebbing, Edward Percy (forestry historian, Edinburgh professor), 83, 84, 85, 156, 170, 251
Stenhouse, William (Indian forester), 189
Stent, J. (Madras gardener), 62, 63
Stewart, John Lindsay (Punjab Conservator), 135, 152, 157, 159, **161**, **163**, 162–4, 170, 188, 209, 213, 228, 229, 300 n 8/78
Stocks, John Ellerton (Indian botanist), 51
Stoliczka, Ferdinand (zoologist), 188
Strachey, Richard (later Sir, engineer), 41, 44, 156, 200, 237, 264
Strathspey – visit of C with Brandis (1865) and J.L. Stewart (1869), 164, 209
Stravithie (Wakefield) Estate, **214**, 217–22
   derivation of names, 219
   purchase/improvement by Hugh Cleghorn (senior), 3
   farms and tenantry, 219–22
   later days with the Sprot family, 262
   visit of Botanical Society Club (1890), 230

visit of Royal Scottish Arboricultural Society (1892), 218, 224, 233, 251
    Stravithie (Wakefield) House garden and policies, 217–8
    in C's time, 219–20
    rebuilding of by C, 177, **178**, **179**
Stravithie Mains, 221
Striharikota Forest (Madras fuel), 88, 92
Subbarayalu, S. (biography of C, 2014), 284
Sullivan, John (Collector of Coimbatore), 82
Sultanpore (now Kullu, Himachal Pradesh), 150
sumac (*Rhus coriaria*), 201
Sutherlandia frutescens, 107
Sutlej – shipwreck (1848), **28**, 29
Sutlej Valley, 144
Switzerland, 204
Syme, Professor James (Edinburgh surgeon), **10**, 10, 11, 14, 123, 154, 265
Syme, Patrick (Edinburgh artist), 105, 229

T

Tanjore, 187
Taylor, Andrew (C's obituarist), 212, 262
teak (*Tectona grandis*), 82, 84, 85, 88, 89, 95, 98, 101, 277
    decline in Shimoga (influence on C), 25, 35, 41
telegraphy – timber demands, 89, 94
Tennant, Professor William (poet & linguist), 5, 8, 19, 127, 134
Tharandt forest school, 189, 190, 215, 248
Thomas family, 298 n 6/11 & 16
Thomas, Edward Brown (Collector of Coimbatore), 63, 91, 106, 114, 133
Thomas, John Fryer (Madras Civil Servant), 63
Thomson, Allen (Edinburgh pathologist), 14
Thomson, Grant (forester at Castle Grant), 209, 226
Thomson, Thomas (Indian botanist), 67, 81, 137, 143, 146, 147 176, 181, 205
Thorp, Edward Courtenay (Bengal surgeon), 262
Thunbergia harrisii, 84
Thwaites, George Henry Kendrick (Peradeniya Garden, Ceylon), 135, 140, 175, 182, 229
Tipu (Sultan of Mysore), 18, 20, 21, 67, 111, 298 n 6/52
Tisa Nullah, 147, 282
Todaro, Agostino (Sicilian botanist), 201
Torenia cordifolia, 107

Tornabene, Francesco (Sicilian botanist), 200
Torquay, Devonshire, 32, 22, 127
traction engines (pollution), 259
Traill, Thomas Stewart (Edinburgh medic), **12**, 14
Traquair, Ramsay Heatley (Edinburgh geologist), 256, 265
Travancore
    C's excursion with Clements Markham, 182, 186
    state visit of Lord Harris (with C), 92–3
Tremenheere, George Borlase (engineer, Burma), 82
Trevelyan, Sir Charles (Madras Governor, later Indian Government), 54, 73, 75, 98, 99, 100, 107, 115, 119, 156, 168
Trevelyan, Sir Walter Calverley, 98
Treviranus, Ludolph Christian (German botanist), 166, 167
Trichinopoly, 18
Tripe, Linnaeus (Madras photographer), 73, 75, 77, 112
Tuke, Dr John Batty (lunacy), 256
Tulloch, Alexander (Madras soldier), 46, 58, 115
Tulloch, Emma (*née* Wahab), 115
Tweeddale, George Hay, 8th Marquess of (Governor of Madras), 105, 144, 225

U

Underwood, William (Madras), 296 n 4/84

V

Valleyfield, Penicuik (Cowan mills and home), 122, 126, 127, 130
Vallombrosa (Italian forest school), 203, 215, 248
Vashist, Himachal Pradesh (hot springs), 150
Veitch of Exeter (Nursery firm), 107, 150, 298 n 6/15
Ventilago madraspatana, **174**
Vesuvius (eruption of), 202
Victoria, Queen, 45, 49, 119, 232, 233, 244
Vilmorin-Andrieux (Paris seed merchants), 204
Voigt, Joachim (Danish botanist), 157

W

Wahab family, 115, 130
Wahab, Kate (*née* Catherine Cowan, sister-in-law), 130
Waid, Andrew (philanthropist, Anstruther), 257
Wakefield, Fife – see Stravithie
Walker-Arnott, George Arnott (Glasgow botanist), 36, 131, 163, 187, 205, 229, 237

Walker, Alexander (of Bowland), 2, 12, 269
Walker, Anna Maria (*née* Patton) (botanist), 12
Wallich, Nathaniel (Superintendent, Calcutta Botanic Garden), 46, 50, 62, 82, 97, 111
Ward, Samuel Neville (entomologist), 100, 298 n 5/79
Waring, Edward (medic, Travancore), 70, 176
Wasse, the Rev. Henry (Anglican Chaplain, Rome), 215
Watson, Captain (Malabar conservator), 82
Watson, John Forbes (economic botanist), 176, 205
wattle (*Acacia dealbata*), 109
Waugh, Col. Gilbert (Bangalore garden), 111
Wester Lea, Edinburgh (Cowan home), 126, 228
Westwood, John (botany teacher, Dollar), 105
Whately, Mary Louisa (missionary, Egypt), 181
Wheldon & Wesley (booksellers), 268
Whistler, James McNeill (artist), 131
Wight, Robert, 17, 25, 26, 35, 36, 43, 46, 50, 62, 63, 73, 109, 131, 176, 181, 229, 269
    Icones Plantarum Indiae Orientalis, 26, 35, 54
    leaves India, 63
Wilkie, David (artist), 1, 14, 220
willow orchards (Lahul), 148
Wilson, Charlotte (*née* Cowan, sister-in-law), 130, 259
Wilson, George (museum director, Edinburgh), 264–5
Wilson, John (Prof. Agriculture, Edinburgh), 226
Wilson, John Hardie (botanical lecturer, St Andrews), 226, 230, 252–4, 256
Wirgman, Theodore Blake (portrait artist), 201, 236, 264
Wood, Sir Charles (Secretary of State for India), 156
Wuddur (tribal group, Anamalai Hills), 87
Wüllerstorf-Urbair, Bernhard von, 91
Wyllie, Helen (*née* Allan) (aunt), 4, 7, 34, 131, 270
Wyville Thomson, Charles (later Sir, *Challenger* expedition), 35

Y

Yule, Amy Frances, 201
Yule, Henry, 137, 141, 199–201, 203, 206, 216, 237, 238, 243, 264, 299 n 7/58
    editor of Marco Polo, 200